T0304703

Energy, Matter, and Change

This textbook serves as an introduction to the field of chemistry, aimed at secondary school students, and it assumes no prior knowledge on the readers' part. As an introductory text, the book emphasizes fundamental skills that are necessary for chemistry and science generally. This includes an emphasis on good writing and a focus on problem solving, with problems incorporated throughout the text. To help prepare students to pursue chemistry further, all information presented is in accord with the International Union of Pure and Applied Chemistry's style and technical guidelines and supported through citations to the primary literature.

Energy, Matter, and Change

An Introduction

William B. Tucker

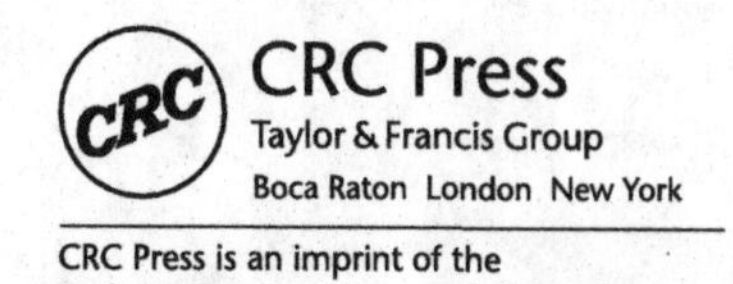

First edition published 2025
by CRC Press
2385 NW Executive Center Drive, Suite 320, Boca Raton FL 33431

and by CRC Press
4 Park Square, Milton Park, Abingdon, Oxon, OX14 4RN

CRC Press is an imprint of Taylor & Francis Group, LLC

Funding Body: Author Funded

ISBN: 9781032782959 (hbk)
ISBN: 9781032781679 (pbk)
ISBN: 9781003487210 (ebk)

DOI: 10.1201/9781003487210

Typeset in Times
by Deanta Global Publishing Services, Chennai, India

This book is dedicated to

Gary Osborn

Beloved Winslow High School (Winslow, ME) chemistry teacher

and inspiration for my love of and interest in chemistry.

Contents

Author's Note/Attribution

All chemical structures were drawn using ChemDraw 22.2 (RRID:SCR_016768). Registered Trademark of Revvity Signals. https://revvitysignals.com/products/research/chemdraw (accessed September 24, 2024).

All computationally derived models were created and calculated using the platform WebMO: Schmidt, J.R.; Polik, W.F. *WebMO*, version 20.0; WebMO LLC: Holland, MI, USA, 20; https://www.webmo.net (accessed April 30, 2023).

Calculations at the B3LYP/6-31G(d) and CCSD(T)/cc-pVTZ level of theory and basis set were performed using Gaussian 09: Gaussian 09 (RRID:SCR_014897), Revision A.02, M. J. Frisch, G. W. Trucks, H. B. Schlegel, G. E. Scuseria, M. A. Robb, J. R. Cheeseman, G. Scalmani, V. Barone, G. A. Petersson, H. Nakatsuji, X. Li, M. Caricato, A. Marenich, J. Bloino, B. G. Janesko, R. Gomperts, B. Mennucci, H. P. Hratchian, J. V. Ortiz, A. F. Izmaylov, J. L. Sonnenberg, D. Williams-Young, F. Ding, F. Lipparini, F. Egidi, J. Goings, B. Peng, A. Petrone, T. Henderson, D. Ranasinghe, V. G. Zakrzewski, J. Gao, N. Rega, G. Zheng, W. Liang, M. Hada, M. Ehara, K. Toyota, R. Fukuda, J. Hasegawa, M. Ishida, T. Nakajima, Y. Honda, O. Kitao, H. Nakai, T. Vreven, K. Throssell, J. A. Montgomery, Jr., J. E. Peralta, F. Ogliaro, M. Bearpark, J. J. Heyd, E. Brothers, K. N. Kudin, V. N. Staroverov, T. Keith, R. Kobayashi, J. Normand, K. Raghavachari, A. Rendell, J. C. Burant, S. S. Iyengar, J. Tomasi, M. Cossi, J. M. Millam, M. Klene, C. Adamo, R. Cammi, J. W. Ochterski, R. L. Martin, K. Morokuma, O. Farkas, J. B. Foresman, and D. J. Fox, Gaussian, Inc., Wallingford CT, 2016.

Electrostatic potential maps were produced using NBO 7: E. D. Glendening, J, K. Badenhoop, A. E. Reed, J. E. Carpenter, J. A. Bohmann, C. M. Morales, P. Karafiloglou, C. R. Landis, and F. Weinhold, Theoretical Chemistry Institute, University of Wisconsin, Madison (2018).

Preface

This book was the result of striving to create an introductory chemistry textbook that truly met students where they are and helped them to build core chemistry skills and foundational knowledge. To that end, I endeavored to accomplish several things to make this book the most useful for the broadest population. First, this book is intended to be accessible to any student who wishes to gain an introduction to the field of chemistry. This means that the content is, by design, self-limited to an introductory coverage. To provide the most definitive information for the reader, all ideas presented are supported with literature citations. These references are also meant to provide an interested reader with an opportunity to pursue these ideas further, though I should caution that a working knowledge of German, French, and English is necessary to read all the references presented. Second, this book is designed to incorporate all problems within the body of the text itself, rather than putting them into a separated section. Readers are encouraged to evaluate their understanding of topics by essaying these problems as they go. Third, it should be noted that there will be discrepancies between what is presented in this book and some other introductory chemistry books (especially around the topic of Lewis structures and periodic trends). The ideas presented here are fully supported by both experiment and theory, and it is hoped that this presentation produces a clear introductory survey of the content that will prepare students for further chemistry study without creating misconceptions that the reader might have to unlearn later.

Acknowledgments

I want to thank the following people for their thoughtful review of this work and their feedback and criticism that has helped to make this work better.

Susan Flink, *Concord Academy*, Editor
Kimberly Kopelman, *Concord Academy*, Editor
Kiley Remiszewski, *Concord Academy*
Jeremy Biggar, *Concord Academy*
Cecily Monahan, *Concord Academy*
Grace Carnahan, *Concord Academy*
Margaret Rapawy, *Concord Academy*
Edie Menard, *Concord Academy*
Mia Bourji, *Concord Academy*
Xander Grossman, *Concord Academy*
Nicole Orangi, *Concord Academy*
Diana Allado, *Concord Academy*
Lila Rhee, *Concord Academy*
Mei Yang, *Concord Academy*
Lira Schwab, *Concord Academy*
Emme Taylor, *Concord Academy*

Author Biography

William Tucker's passion for chemistry was inspired by his high school teacher Gary Osborn. He left Maine to pursue Chemistry at Middlebury College, and after graduating in 2010 he decided to pursue a PhD in Organic Chemistry at the University of Wisconsin-Madison. At the University of Wisconsin-Madison, he worked in the laboratory of Dr Sandro Mecozzi, where he developed semi-fluorinated triphilic surfactants for hydrophobic drug delivery. After earning his PhD in 2015, he took a fellowship at Boston University as Postdoctoral Faculty Fellow. There he co-taught organic chemistry while working in the laboratory of Dr John Caradonna. In the Caradonna laboratory, he worked on developing a surface-immobilized iron-oxidation catalyst for the oxidation of C–H bonds using dioxygen from the air as the terminal oxidant. Throughout all of this work, his passion has always been for teaching and working with students both in and out of the classroom. He has been lucky for the past six years to work at Concord Academy, where his students have, through their questions, pushed him to think deeper and more critically about chemistry. Their curiosity inspires him, and their inquisitiveness inspired his writing.

1 Introduction

Chemistry is – along with astronomy, biology, Earth science, and physics – one of the fundamental natural sciences. Chemistry is the branch of natural science that focuses on the study of matter and change. Matter is anything that has mass and takes up space. Change is any transformation of matter that alters its appearance, state, or composition. The study of matter and change is the traditional definition of chemistry, but it is also not necessarily how an average chemist thinks about their own field. For the author, chemistry is the study of atoms, their properties, and how and why they change. As you read this text, you are encouraged to develop your own working definition of chemistry.

A general presentation of matter and change will be the focus of the second chapter of the book. As you might imagine, given the broad definition, chemistry is a very broad field of study, which has been divided into various specialties: analytical chemistry, physical chemistry, nuclear chemistry, organic chemistry, inorganic chemistry, chemical biology, and materials chemistry. Throughout this book, we will consider fundamental aspects important to all of these subdisciplines of chemistry.

The objective of this book is to provide the reader with an introductory understanding of chemistry. This book could be utilized by individuals on their own, but it is expected that this text will be part of the materials for a high school chemistry course. No previous knowledge of chemistry is assumed, although outside reading is encouraged for any individual who finds a topic of personal interest. Throughout each chapter of this book, you will find text that is meant to introduce and explain each topic, illustrative examples that present approaches to problem solving, and in-text problems to test your understanding as you read.

INCLUSION[1]

If you were to research the individuals cited throughout this text, there would be a stark trend among those individuals: they are overwhelmingly white, male, cisgender, heterosexual, wealthy, and European or American. The field of chemistry, as we know it today, was established by and for men in Europe and America during the 19th century. This means that the academic pursuit of chemistry is the byproduct of exclusion: it is an area of study that was developed by a self-limited, homogeneous group. The effects of chemistry's history of exclusion can be seen in the underrepresentation of Black, Latine, and nonmale chemists receiving degrees, obtaining grants, and working both in academia and in research.

In the 19th and early 20th centuries, only people who were wealthy, white, and male were afforded the privilege of pursuing science. Changes to this structure came during the 20th century. First, governments began to fund the study of science and chemistry. Public funding meant that chemistry became accessible to those who were not from affluent backgrounds. Subsequently, activists in the Civil Rights, Women's Liberation, and LGBTQIA Liberation movements created space for folks who were not cisgender, white, heterosexual, and male to partake in the study of chemistry. Given the changes brought about by the 20th century, chemistry has become more diverse, but it is still harder for some people than it is for others to progress through the system. What is important as you read this book is to consider the folks – nonwhite, nonmale, non-cisgender, and non-heterosexual – who are not cited or recognized in chemical history. These people have always been a part of chemistry and they have done significant research in the field of chemistry. As chemistry slowly becomes more inclusive, it is incumbent upon us to both recognize those folks omitted by chemistry's past and to help expand and make space for those folks who can and should be part of chemistry's present and future,[2] because to not do so is actively harmful.[3]

DOI: 10.1201/9781003487210-1

SUSTAINABILITY AND GREEN CHEMISTRY

An important aspect of modern chemistry to introduce here is sustainability and green chemistry. Chemistry involves the material transformation of matter with two aims: the fundamental understanding of matter and change and the production of materials for modern society. These transformations require raw materials (especially petroleum) and energy, and they produce waste. The production, consumption, and disposal of chemicals are dependent upon the Earth's land, air, and water. Sustainability is ensuring that production, consumption, and disposal are not causing short- or long-term harm to the Earth, the biosphere, or human society. As such, making chemistry sustainable is not a single change in a single process, but it requires a systemic analysis and approach to chemistry and an understanding of how it interfaces with the planet and with society.[4]

Sustainability is a systemic process, but green chemistry is work that contributes to the larger goal by making individual reactions or processes more sustainable. Principles to help make a process sustainable include the following:[5]

1. Preventing waste
2. Reducing hazardous chemicals and using more benign chemicals
3. Increasing energy efficiency by minimizing the amount of heat or pressure used
4. Using renewable raw materials
5. Minimizing environmental impact by designing materials to be degradable

Chemistry is a materially intensive science, which historically has not accounted for its own impact on the Earth. Chemistry's negative impact has been substantial: contamination of water and soil lead from leaded gasoline and paint, contamination of indoor spaces with asbestos, the depletion of the ozone layer with chlorofluorocarbon (CFC) aerosols, the proliferation of plastics and the resulting pollution from plastic waste, and the contamination of water and soil with perfluorinated chemicals (so-called "forever chemicals"). Chemistry has created these environmental challenges, but it is also the solution to these and other problems.

You, as the reader, are also encouraged to consider how to improve the sustainability of chemistry. Advancements in green chemistry and improvements in sustainability will require not only applying existing knowledge to current practices but also new innovations. As learners of chemistry, innovations and new insights will come from those of you who are not yet bound by knowing what is or is not possible. You are encouraged to ask questions and to pursue the questions you have as their answers may lead to new discoveries that will help make chemistry more sustainable.

CLARITY

Clarity, by which we mean unambiguousness, is key in most areas of life; however, it is of paramount importance in science. Whether discussing feet or meters, 0.1 g or 0.01 g, or weight or mass, it is vital that the speaker (or writer) can convey the correct information to the audience.[6] In this chapter, we will consider how to ensure that our writing has clarity. Rules for good analytical writing are potentially a review for the reader but are presented as the first chapter to ensure a common and strong foundation that is necessary both for success in working through this book and in the course this book might accompany.

ANALYTICAL WRITING

Claim, Evidence, and Reasoning

Good analytical writing necessitates several things. First, analytical, or argumentative, writing should always include three pieces: claim, evidence, and reasoning (CER). A claim is any inference that you draw based on your analysis of the information or data at hand. The evidence is the information from the problem or data from empirical observation or an experiment that you are basing that claim on (you need to clearly highlight this evidence to the audience because they may not see or notice the same evidence). Finally, the audience needs to see your reasoning, which is the logical connection that you have made to move from the evidence to the claim you are stating. This is an approach to make sure that each problem presents (1) a statement of what the reader thinks is true (claim), (2) the visual, empirical, and/or quantitative data that supports their statement (evidence), and (3) a logical argument that explains how the data supports the statement (reasoning). This is an effective tool for not only chemistry but science and learning generally.

Consider the following example: when an egg is placed into vinegar, bubbles form on the surface of the shell. Eventually, the eggshell completely disappears, and the egg is left inside a clear, flexible membrane. When asked to explain what happened, the following provides an answer using CER.

> The eggshell underwent a chemical change by reacting with the vinegar (claim). After being placed in vinegar, the appearance of bubbles and disappearance of the shell (evidence) are indicators that a chemical change took place and that matter with a different chemical composition is produced (reasoning).

Note, the example of an eggshell reacting with vinegar utilizes ideas and concepts (chemical change) that we will investigate in the next chapter. At this point, the goal is to get a sense of this framework, not to understand what chemical change is.

Problem 1.1. For each of the following, identify the claim, evidence, and reasoning in each statement.

a. The reaction produced heat (was exothermic) because the temperature of the water increased. The increase in temperature was the result of energy being released by the reaction, and releasing energy is exothermic.
b. Susan is happy. Susan has a beaming smile, and whenever she is smiling like that, it means she is happy.
c. The weather has been unnaturally warm and snowless for the past several winters. A long-term change in weather patterns is evidence of a change in climate. Together we can say that there is local evidence of global warming.

Problem 1.2. For each of the following, identify what piece (claim, evidence, or reasoning) or pieces are missing from each statement.

a. Anna is allergic to mangoes.
b. As the reaction progressed, the temperature decreased from 25.0 °C to 18.2 °C.
c. Will is unhappy because he is frowning.

Prescriptive Grammar

Good analytical writing also follows the rules of prescriptive grammar. Prescriptive grammar is not how people necessarily talk, but it is a convention that is established so that a reader does not become confused by the writing itself. To ensure that a reader can understand, in prescriptive English grammar, there is always (1) proper punctuation, (2) subject–verb agreement, (3) helping verbs used when needed, (4) agreement between pronouns and their antecedent, (5) proper spelling, (6) proper sentence order and words are neither omitted nor jumbled, and (7) proper citation given when others' work is used. Altogether, even in a STEM field, one should strive to be a great writer because effective writing can make one's arguments stronger and more forceful.

STARTING AT THE BEGINNING

As you continue your study of chemistry in the following chapters, you will be introduced to many scientific concepts. As you read and learn, it is important to keep in mind several key threads in modern science: inclusion, sustainability, and clarity. When you see the references included, ask yourself, "Who is not included here but should be?" When you learn about a new topic, ask yourself, "How does this topic lend itself, or not, to making a more sustainable future?" And finally, as you learn a new concept, ask yourself, "Could I clearly and concisely explain this to a peer?" The answers to these questions will enrich, enlarge, and reinforce your study of chemistry.

NOTES

1. This section is modified from: Tucker, W.B. *Chemistry: Energy, Matter, and Change.* CRC Press: Boca Raton, **2024**.
2. i) Reisman, S.E.; Sarpong, R.; Sigman, M.S.; Yoon, T.P. Organic Chemistry: A Call to Action for Diversity and Inclusion. *J. Org. Chem.*, **2020**, *85* (16), 10287–10292. DOI: 10.1021/acs.joc.0c01607.
   ii) Sanford, M.S. Equity and Inclusion in the Chemical Sciences Requires Actions not Just Words. *J. Am. Chem. Soc.*, **2020**, *142* (26), 11317–11318. DOI: 10.1021/jacs.0c06482.
   iii) Ruck, R.T.; Faul, M.M. Update to Editorial "Gender Diversity in Process Chemistry". *Org. Process Res. Dev.*, **2021**, *25* (3), 349–353. DOI: 10.1021/acs.oprd.0c00471.
3. Dunn, A.L.; Decker, D.M.; Cartaya-Marin, C.P.; Cooley, J.; Finster, D.C.; Hunter, K.P.; Jacques, D.R.N.; Kimble-Hill, A.; Maclachlan, J.L.; Redden, P.; Sigmann, S.B.; Situma, C. Reducing Risk: Strategies to Advanced Laboratory Safety through Diversity, Equity, Inclusion, and Respect. *J. Am. Chem. Soc.*, **2023**, *145*, 21, 11468–11471. DOI: 10.1021/jacs.3c03627.
4. Matlin, S.; Mehta, G.; Cornell, S.E.; Krief, A.; Hopf, H. Chemistry and Pathways to Net Zero for Sustainability. *RSC Sustain.*, **2023**, *1*, 1704–1721. DOI: 10.1039/D3SU00125C.
5. These are a condensed representation of the 12 Principles of Green Chemistry from: Anastas, P.T.; Warner, J.C. *Green Chemistry: Theory and Practice.* Oxford University Press: Oxford, **1998**.
6. For a clear example of what can happen when ambiguity comes into play, you should consider the debacle of the 1999 Mars Climate Orbiter.

2 Matter and Change

Chemistry is the study of matter, material substances, and change, modification of those substances. Matter can be classified by its composition, its state, and its properties, and change can be either chemical or physical, depending on whether the makeup of substances changes. In this chapter, we will provide a general overview of these concepts as we establish common terminology that is used throughout chemistry and throughout this book.

MATTER

We will begin by considering matter. A general definition of matter is anything that has mass and takes up space. From the point of view of chemistry, the definition of matter is more specific: matter is anything that is made up of atoms. Synonyms for matter are substance, chemical, or chemical species. We can classify chemicals based on their composition – the number and type of particles that make up the substance – as an element, a compound, or a mixture.

Problem 2.1. For each of the following items identify whether they are matter (have mass and take up space).

a. Wood
b. Gold
c. Heat
d. A cell phone
e. Light
f. Water
g. Air

MATTER CLASSIFICATION BY COMPOSITION

When we classify matter by its composition, we mean what type of particle or particles make up the chemical. If there is only one type of particle (atoms, ions, or molecules) that composes a material we call those chemicals pure substances (elements and compounds). If there is more than one type of particle that a material is composed of, then we call those chemicals mixtures.

ELEMENTS

Elements are pure substances that are composed of atoms that all contain the same number of protons. Protons are elementary particles that define an atom's identity. They are found in the nucleus and are positively charged (Chapter 6). For example, a silver ring is composed of silver atoms (47 protons in each atom) and a sheet of aluminium foil is composed of aluminium atoms (13 protons in each atom). The number of protons an atom has is called the atomic number and is represented by the symbol Z.

Chemistry is a field of science that extensively uses symbols. While these symbolic systems can initially seem daunting, they serve an important purpose. There can be a large amount of information involved in chemistry. The use of symbols allows us to clarify, streamline, and minimize

DOI: 10.1201/9781003487210-2

information overload. The first set of symbols we will consider – atomic symbols – is used to represent each element. While some atomic symbols are one capital letter, like C for carbon, most atomic symbols are one capital letter and one lowercase letter, like He for helium.[1] The atomic symbol for each element is shown in Table 2.1.

TABLE 2.1
Alphabetical List of Atomic Symbols, Names, and Atomic Numbers for Each Element

Symbol	Name (Latin name)	Atomic number (*Z*)	Symbol	Name (Latin name)	Atomic number (*Z*)
Ac	actinium	89	H	hydrogen	1
Ag	silver (argentum)	47	He	helium	2
Al	aluminium	13	Hf	hafnium	72
Am	americium	95	Hg	mercury (hydrargyrum)	80
Ar	argon	18	Ho	holmium	67
As	arsenic	33	Hs	hassium	108
At	astatine	85	I	iodine	53
Au	gold (aurum)	79	In	indium	49
B	boron	5	Ir	iridium	77
Ba	barium	56	K	potassium (kalium)	19
Be	beryllium	4	Kr	krypton	36
Bh	bohrium	107	La	lanthanum	57
Bi	bismuth	83	Li	lithium	3
Bk	berkelium	97	Lr	lawrencium	103
Br	bromine	35	Lu	lutetium	71
C	carbon	6	Lv	livermorium	116
Ca	calcium	20	Mc	moscovium	115
Cd	cadmium	48	Md	mendelevium	101
Ce	cerium	58	Mg	magnesium	12
Cf	californium	98	Mn	manganese	25
Cl	chlorine	17	Mo	molybdenum	42
Cm	curium	96	Mt	meitnerium	109
Cn	copernicium	112	N	nitrogen	7
Co	cobalt	27	Na	sodium (natrium)	11
Cr	chromium	24	Nb	niobium	41
Cs	caesium	55	Nd	neodymium	60
Cu	copper (cuprum)	29	Ne	neon	10
Db	dubnium	105	Nh	nihonium	113
Ds	darmstadtium	110	Ni	nickel	28
Dy	dysprosium	66	No	nobelium	102
Er	erbium	68	Np	neptunium	93
Es	einsteinium	99	O	oxygen	8
Eu	europium	63	Og	oganesson	118
F	fluorine	9	Os	osmium	76
Fe	iron (ferrum)	26	P	phosphorus	15
Fl	flerovium	114	Pa	protactinium	91
Fm	fermium	100	Pb	lead (plumbum)	82
Fr	francium	87	Pd	palladium	46
Ga	gallium	31	Pm	promethium	61
Gd	gadolinium	64	Po	polonium	84
Ge	germanium	32	Pr	praseodymium	59

(*Continued*)

TABLE 2.1 (CONTINUED)
Alphabetical List of Atomic Symbols, Names, and Atomic Numbers for Each Element

Symbol	Name (Latin name)	Atomic number (*Z*)	Symbol	Name (Latin name)	Atomic number (*Z*)
Pt	platinum	78	Sr	strontium	38
Pu	plutonium	94	Ta	tantalum	73
Ra	radium	88	Tb	terbium	65
Rb	rubidium	37	Tc	technetium	43
Re	rhenium	75	Te	tellurium	52
Rf	rutherfordium	104	Th	thorium	90
Rg	roentgenium	111	Ti	titanium	22
Rh	rhodium	45	Tl	thallium	81
Rn	radon	86	Tm	thulium	69
Ru	ruthenium	44	Ts	tennessine	117
S	sulfur	16	U	uranium	92
Sb	antimony (stibium)	51	V	vanadium	23
Sc	scandium	21	W	tungsten	74
Se	selenium	34	Xe	xenon	54
Sg	seaborgium	106	Y	yttrium	39
Si	silicon	14	Yb	ytterbium	70
Sm	samarium	62	Zn	zinc	30
Sn	tin (stannum)	50	Zr	zirconium	40

For some elements the atomic symbol stems from an older Latin name, which is included in parentheses for context. Tungsten's atomic symbol (W) comes from the German name for the element (wolfram).

Elements on the periodic table are in order of atomic number from 1 (top left) to 118 (bottom right). On periodic tables, some information is always shown. The first is the atomic number, which is the whole number typically shown above the atomic symbol for an element (Figure 2.1).

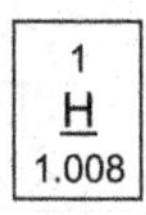

FIGURE 2.1 Example cell from the periodic table for hydrogen, with an atomic number (*Z*) of 1 and a relative atomic mass of 1.008.

The number below an element (1.008 for hydrogen, Figure 2.1) is the recommended relative atomic mass ($A_r°$) of that element, which is a decimal value. The importance of relative atomic mass will be considered in more detail in Chapter 4, but for now it is relevant to know that the value is the mass of each atom relative to the mass of a carbon-12 atom. Carbon-12, a specific isotope (Chapter 6) of carbon, is defined as having an atomic mass of *exactly* 12 unified atomic mass units (12 u) or, equivalently, 12 daltons (12 Da). All the currently known elements are arranged on the periodic table of elements (Figure 2.2). The reasons for the structure of our periodic table will become clearer as we proceed through our study of chemistry.

Each element also has a specific position on the periodic table in terms of its group (vertical column) and period (horizontal row). For example, hydrogen is in the first period (first row) and in group 1 (column 1). The periodic table has 7 periods (rows) and 18 numbered groups (columns), with numbers shown in bold. Elements 57–71 and 89–103 are shown below the periodic table (Figure 2.2)

1	2	3	4	5	6	7	8	9	10	11	12	13	14	15	16	17	18
1 H 1.008																	2 He 4.002
3 Li 6.94	4 Be 9.012											5 B 10.81	6 C 12.01	7 N 14.01	8 O 16.00	9 F 19.00	10 Ne 20.18
11 Na 22.99	12 Mg 24.31											13 Al 26.98	14 Si 28.09	15 P 30.97	16 S 32.06	17 Cl 35.45	18 Ar 39.95
19 K 39.10	20 Ca 40.08	21 Sc 44.96	22 Ti 47.87	23 V 50.94	24 Cr 52.00	25 Mn 54.94	26 Fe 55.85	27 Co 58.93	28 Ni 58.69	29 Cu 63.55	30 Zn 65.38	31 Ga 69.72	32 Ge 72.63	33 As 74.92	34 Se 78.97	35 Br 79.90	36 Kr 83.80
37 Rb 85.47	38 Sr 87.62	39 Y 88.91	40 Zr 91.22	41 Nb 92.91	42 Mo 95.95	43 Tc –	44 Ru 101.1	45 Rh 102.9	46 Pd 106.4	47 Ag 107.9	48 Cd 112.4	49 In 114.8	50 Sn 118.7	51 Sb 121.8	52 Te 127.6	53 I 126.9	54 Xe 131.3
55 Cs 132.9	56 Ba 137.3	*	72 Hf 178.5	73 Ta 181.0	74 W 183.8	75 Re 186.2	76 Os 190.2	77 Ir 192.2	78 Pt 195.1	79 Au 197.0	80 Hg 200.6	81 Tl 204.4	82 Pb 207.2	83 Bi 209.0	84 Po –	85 At –	86 Rn –
87 Fr –	88 Ra –	**	104 Rf –	105 Db –	106 Sg –	107 Bh –	108 Hs –	109 Mt –	110 Ds –	111 Rg –	112 Cn –	113 Nh –	114 Fl –	115 Mc –	116 Lv –	117 Ts –	118 Og –

*	57 La 138.9	58 Ce 140.1	59 Pr 140.9	60 Nd 144.2	61 Pm –	62 Sm 150.4	63 Eu 152.0	64 Gd 157.3	65 Tb 158.9	66 Dy 162.5	67 Ho 164.9	68 Er 167.4	69 Tm 168.9	70 Yb 173.1	71 Lu 175.0
**	89 Ac –	90 Th 232.0	91 Pa 231.0	92 U 238.0	93 Np –	94 Pu –	95 Am –	96 Cm –	97 Bk –	98 Cf –	99 Es –	100 Fm –	101 Md –	102 No –	103 Lr –

FIGURE 2.2 Periodic table of elements showing the recommended, abridged, relative atomic mass of each element. The phase of each element at standard temperature and pressure (1×10^5 Pa and 273.15 K) is indicated by the font: **solid**, liquid, and gas.

solely for the sake of legibility. If all elements are shown in one contiguous table (Figure 2.3) the font must be very small to accommodate everything.

1	2															3	4	5	6	7	8	9	10	11	12	13	14	15	16	17	18
H																															He
Li	Be																									B	C	N	O	F	Ne
Na	Mg																									Al	Si	P	S	Cl	Ar
K	Ca															Sc	Ti	V	Cr	Mn	Fe	Co	Ni	Cu	Zn	Ga	Ge	As	Se	Br	Kr
Rb	Sr															Y	Zr	Nb	Mo	Tc	Ru	Rh	Pd	Ag	Cd	In	Sn	Sb	Te	I	Xe
Cs	Ba	La	Ce	Pr	Nd	Pm	Sm	Eu	Gd	Tb	Dy	Ho	Er	Tm	Yb	Lu	Hf	Ta	W	Re	Os	Ir	Pt	Au	Hg	Tl	Pb	Bi	Po	At	Rn
Fr	Ra	Ac	Th	Pa	U	Np	Pu	Am	Cm	Bk	Cf	Es	Fm	Md	No	Lr	Rf	Db	Sg	Bh	Hs	Mt	Ds	Rg	Cn	Nh	Fl	Mc	Lv	Ts	Og

FIGURE 2.3 A 32-column periodic table showing the correct placement of the rare earth elements within the larger periodic framework. Note that the exact composition of group 3 is currently debated as it could be Sc, Y, Lu, and Lr (as shown) or Sc, Y, La, and Ac.

The bold, zigzag line towards the right side of the table (Figure 2.2) separates elements that are metals (left) from those elements that are nonmetals (right) (Figure 2.4). Elements that are adjacent to the zigzag line are referred to as metalloids. Metals are, as you may know from experience, shiny, dense, and typically silvery in color (except for copper and gold). Metals are also, except for liquid mercury, all solids. Metals are good conductors and are both malleable (can be flattened into sheets) and ductile (can be pulled into wires). In contrast, nonmetals are typically not shiny and are typically poor conductors. Nonmetals also have a low density and can range from solid to liquid to gas. Metalloids have intermediate properties. They are often shiny but have lower density than metals and are brittle rather than malleable or ductile.

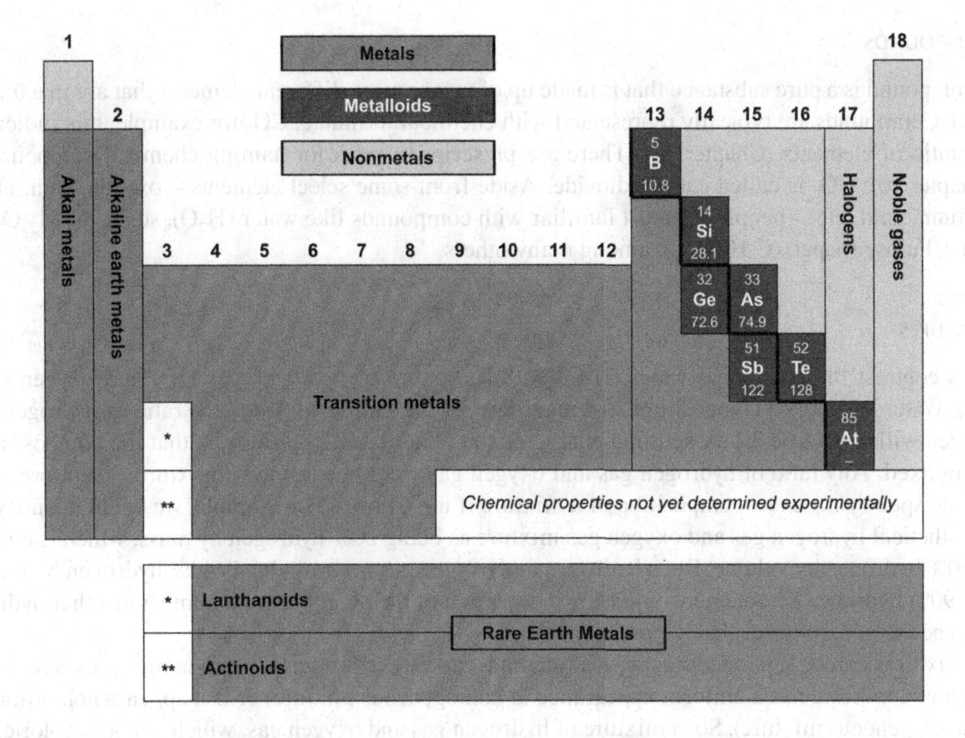

FIGURE 2.4 Designations of elements in the periodic table: black-gray are metalloids, dark gray are metals, and light gray are nonmetals.

Some groups and blocks of the periodic table have different collective names (Figure 2.4). Broadly, elements are divided into metals, metalloids,[2] and nonmetals. From there, the elements are further subdivided into those in the main group (those in groups 1, 2, 13, 14, 15, 16, 17, and 18, with group 12 elements sometimes included in the set of main group elements), the transition metals (those elements in groups 3–12), and the lanthanoids and actinoids, which are collectively called inner transition metals and, along with the elements in group 3, rare earth metals.

Problem 2.2. Using Table 2.1, Figure 2.2, and Figure 2.4 answer the following questions.

a. What is the name of the element with the atomic symbol Na?
b. What is the atomic symbol for the element sulfur?
c. Halogens are the name given to the elements in which group?
d. What element is in period 4, group 8?
e. What is the atomic number of carbon?
f. Gold is in which period and in which group?
g. What type of element (metal, nonmetal, metalloid) is W?
h. What is the state of matter for Hg?
i. What is the name of element Kr?
j. Which element has the atomic number 94?
k. Identify the elements that are gases at standard temperature and pressure.
l. Identify the element, other than mercury, that is a liquid at standard temperature and pressure.
m. Identify the lanthanoid named for the mythological titan who stole fire from Mount Olympus.
n. What type of element is samarium? Sodium? Silicon? Sulfur?

Compounds

A compound is a pure substance that is made up of two or more different elements that are in a fixed ratio. Compounds are typically represented with chemical formulae, CO_2 for example, that indicate the ratio of elements (Chapter 10). There are prescriptive rules for naming chemical compounds (Chapter 10): CO_2 is called carbon dioxide. Aside from some select elements – oxygen, iron, aluminium, gold, etc. – people are most familiar with compounds like water (H_2O), sugar ($C_{12}H_{22}O_{11}$), and cellulose (paper, $(C_6H_{10}O_5)_n$), among many others.

Mixtures

Let's contrast the compound water (H_2O) with a mixture of hydrogen gas (H_2) and oxygen gas (O_2). Water has a fixed composition, and regardless of the sample of water, the ratio of hydrogen to oxygen will always be 2:1. A defining characteristic of a mixture, however, is that the composition is not fixed. Any ratio of hydrogen gas and oxygen gas would be termed a mixture. Therefore, we tend to specify the exact composition of a mixture if it is known. For example, we could quantify a hypothetical hydrogen gas and oxygen gas mixture as being 20% hydrogen by mass, which is equal to 80% hydrogen by volume. But a hydrogen and oxygen gas mixture that is 40% hydrogen by mass and 90% hydrogen by volume would be just one more of the nearly infinite number of other hydrogen and oxygen mixtures that is possible.

Mixtures can be separated by physical methods, and we distinguish between mixtures based on whether a mixture has a uniform appearance (a homogeneous mixture) or if it appears nonuniform (a heterogeneous mixture). So, a mixture of hydrogen gas and oxygen gas, which are both colorless, would be a homogeneous mixture. In contrast, smoke rising off a fire, where the particulate matter is concentrated, would be a heterogeneous mixture because we can see differences between the smoke and the surrounding air. There are many examples of both heterogeneous and homogeneous mixtures. Some examples of homogeneous mixtures are those of gases (like air), metals (alloys like steel), miscible liquids (vinegar is a mixture of acetic acid and water), and solutions of solids and liquids (like sugar in water). There are many examples of heterogeneous mixtures: salad dressing, concrete, the book (or digital device) you are reading this on, and many others. The important thing to remember is that the designation of a mixture as homogeneous or heterogeneous is based on our ability to see or perceive differences in the components of the mixture.

Problem 2.3. Identify each of the following as an element, compound, homogeneous mixture, or heterogeneous mixture.

a. Ocean water
b. Salt water
c. Table salt (sodium chloride, NaCl)
d. Water (H_2O)
e. Sodium metal (Na)
f. Chlorine gas (Cl_2)
g. Chocolate chip cookies
h. Sugar cookies
i. Laptop
j. Black ink
k. A cell
l. M&M's candies
m. Aluminium (Al) foil
n. Gold ring

o. Paint thinner (toluene, C_7H_8)
p. A brick
q. Stainless steel, an alloy of iron (Fe), carbon (C), and chromium (Cr)
r. A cupcake
s. A block of wood
t. Borosilicate glass (containing SiO_2 and B_2O_3)

GREEN CHEMISTRY: CHEMICAL SOURCES

Everything around you as you are reading this is composed of chemicals. Let's focus on what you are reading. If you are reading this on a digital device, that device contains many elements, including silicon, copper, gold, silver, lithium, yttrium, terbium, lanthanum, neodymium, dysprosium, hydrogen, carbon, nickel, indium, tin, oxygen, sodium, potassium, phosphorus, gallium, cobalt, bromine, magnesium, arsenic, antimony, lead, aluminium, europium, and dysprosium, among others. As can be seen in Figure 2.5, some of these elements are abundant in the Earth's crust and some of them are quite rare. Regardless of their abundance, the extraction of these elements (often through mining) and purification (requiring heat, electricity, and chemical methods) can be materially and energetically intensive. As such, the recycling and reuse of these materials and the development of new chemistry that may not necessitate using rare elements are imperative to a sustainable future.

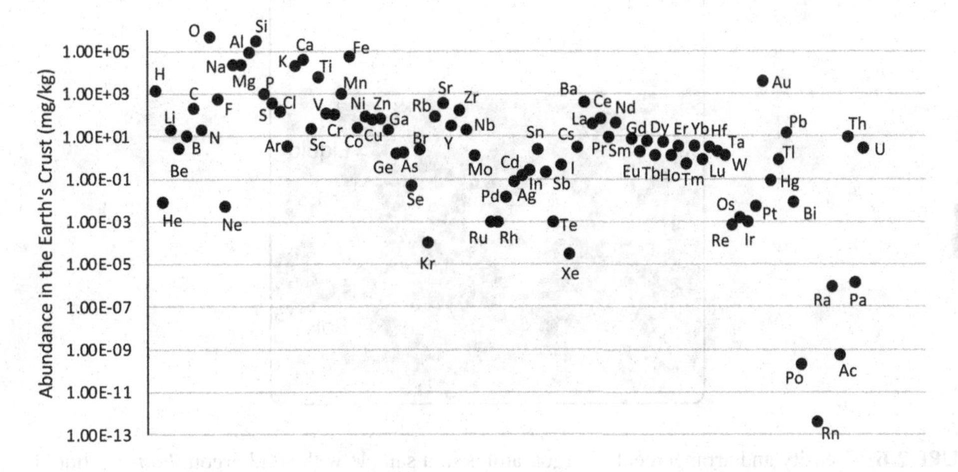

FIGURE 2.5 Abundance of naturally occurring elements in the Earth's crust. Note that the *y*-axis is a log scale.

If you are reading this as a traditional, physical textbook, then the predominant chemicals are cellulose (a polymer of glucose derived from plants) and ink (dyes from hydrocarbons). Cellulose-based paper comes from wood pulp, which can lead to deforestation if not done sustainably and if paper is not recycled. Hydrocarbons typically come from the extraction of nonrenewable feedstocks such as coal, petroleum, and natural gas. Possible renewable sources of hydrocarbons include algae, biomass, or using lignocellulosic biomass directly.

Finally, the energy (Chapter 5) necessary to extract, purify, synthesize, process, and ship whatever product you are reading, and power the device if it is electronic, likely derives from burning fossil fuels. Burning fossil fuels uses a nonrenewable resource and produces carbon dioxide, a greenhouse gas. At present, society still uses enormous amounts of fossil fuels, and the world (as of 2023) emits over 30 billion metric tons of CO_2 per year. Renewable energy sources – solar, wind, and hydroelectric – are being used to ever greater extents, but significant work is needed to make society's energy use sustainable.

As you can see, regardless of the medium, simply reading this information requires a significant amount of material and energy. A sustainable future requires that we recycle and reuse materials and derive energy from renewable sources that are, at least, net zero in terms of carbon dioxide pollution.

STATES OF MATTER

So far, we have considered how matter is classified according to its composition. Another important method of classification of matter is based on its state of aggregation. The three most common states are solids, liquids, and gases. There are other states of matter, like plasma, but they are less common and will not be considered here. We can understand the differences between solids, liquids, and gases in terms of the density of particle packing (Figure 2.6), the arrangement and degree of order, and the type(s) of particle motion.

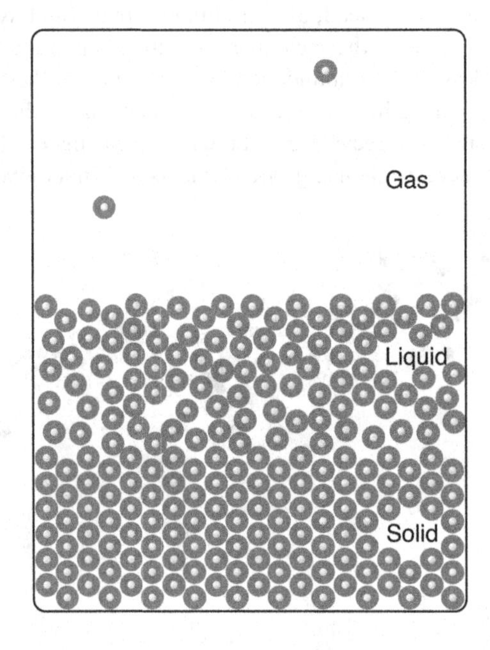

FIGURE 2.6 Density and arrangement of argon atoms in a sample with solid argon (*bottom*), liquid argon (*middle*), and gaseous argon (*top*).

Both solids and liquids are condensed phases of matter, which means they cannot be easily compressed, and there is very little space between particles (Figure 2.6). In contrast, gases are dispersed phases with significant distance between the particles, which makes gases compressible. Unlike liquids and gases, solids are distinguished by having a repeating pattern in the arrangement of their particles. The particles in a solid vibrate (move around a fixed point) but neither rotate (spin around their center of mass) nor translate (move through three-dimensional space). In addition, the close packing and limited movement of solid particles give rise to a regular repeating pattern. In contrast, liquids and gases are fluid phases that take on the shapes of their containers as the particles not only vibrate but also rotate and translate. The greater extent of motion in liquids and gases is observable in terms of the lack of a repeating pattern of the particles (Figure 2.6). The phase that a given chemical has at normal temperature and pressure (20.0 °C, 101 kPa) is dictated by the strength of the intermolecular forces (Chapter 13) present in the compound.

To indicate the phase[3] of a substance, chemical formulae utilize abbreviations: (s) for solid, (l) for liquid, and (g) for gas. Another common notation is aqueous (aq), which is used to indicate that a substance is dissolved in water as part of a homogeneous mixture.

Problem 2.4. Smell is dependent upon molecules reaching olfactory receptors in our noses. Provide an explanation why hydrogen sulfide ($H_2S(g)$) is intensely smelly even in small amounts, but phenol ($C_6H_5OH(s)$) is only intensely smelly in large amounts (or when heated).

PROPERTIES OF MATTER

Lastly, we will consider the properties of matter. A property is any observable or quantifiable attribute of a substance. There are two types of properties: physical properties and chemical properties.

A physical property is any attribute of a substance that can be determined using one's senses or a tool (thermometer, ruler, balance, graduated cylinder, barometer, and so on). Examples of physical properties are mass, length, width, height, volume, density, state, color, smell, taste, luster, and texture.

In contrast, a chemical property is determined through the interaction of two different chemicals. That is, with a chemical property, we are identifying the likelihood of a chemical to interact (or not) with another. If a chemical does not interact with others, we call it inert. There are specific terms for other chemical properties: flammability (the tendency to combine with oxygen to produce heat and light),[4] corrosiveness (a tendency to damage or destroy other chemicals), caustic (a tendency to damage or destroy living tissue), shock sensitive (a tendency to undergo chemical reactions after mechanical stress), water reactive (a tendency to react, usually violently, with water), oxidizer (a tendency to remove electrons from other chemicals), reducing agent (a tendency to add electrons to other chemicals), acid (a tendency to add H^+ to other chemicals), and base (a tendency to remove H^+ from other chemicals).

Problem 2.5. Identify each of the following as a chemical property or a physical property.

a. Blue color
b. Density (mass per volume)
c. Flammability
d. Solubility
e. State (solid, liquid, or gas)
f. Reacts with acid
g. Sour taste
h. Boiling point
i. Odor
j. Reacts with water

CHANGE

As mentioned, chemistry is a branch of science that focuses on the study of matter and change. So far, we have considered matter, which will be explored in more detail in subsequent chapters. Before this chapter concludes, a brief introduction to change will be presented with a short discussion of physical change (discussed in more detail in Chapter 14) and chemical change (discussed in more detail in Chapters 15 and 16).

A physical change is any change that does not affect the composition, but there is a change in the state of matter or the physical appearance of the material. In terms of changes in state, there are changes that result from adding thermal energy: melting (solid to liquid), vaporization (liquid to gas), and sublimation (solid to gas); other state changes are caused by removing thermal energy: condensation (gas to liquid), crystallization (liquid to solid), and deposition (gas to solid). Any change to the appearance, size, or shape of a material is also a physical change. For example, if a substance is broken, cut, abraded, bent, reshaped, or separated into pieces are all examples that are not state changes but are still physical changes.

A chemical change, in contrast to a physical change, is a change in the composition of a substance. We tend to call a chemical change a chemical reaction or we say that the chemicals react together. Unlike physical changes, we cannot directly see changes in chemical composition. Instead, we look for evidence of chemical change, which can include a change in color, bubbling (without added heat), the appearance of a solid (without cooling), the appearance of light, changes in smell, changes in taste, changes in texture or luster without external forces, and temperature changes.

A brief note will be made here about dissolving (Chapter 14), specifically the dissolution of a solid into water. This is an obvious change as a solid and a liquid combine to produce a new homogeneous, liquid-phase mixture. Dissolution involves the physical dispersal of the particles that make up a substance throughout the liquid particles. In terms of the type of change, this is typically a physical change when covalent compounds (compounds involving only nonmetal elements) dissolve in water. When ionic compounds (compounds that contain metal and nonmetal elements) dissolve in water, however, this is typically a chemical change. This is particularly evident when transition metal compounds dissolve in water and are accompanied by color changes.

Problem 2.6. Identify each of the following as an example of physical change or evidence of chemical change.

a. Baking soda ($NaHCO_3$) bubbles with vinegar
b. Table salt ($NaCl$) dissolves in water
c. Milk sours
d. Grass grows
e. Iron (Fe) rusts
f. Sugar ($C_{12}H_{22}O_{11}$) caramelizes
g. An apple is cut
h. Wood rots
i. Heat converts water (H_2O) to steam
j. A tire is inflated
k. Alcohol (CH_3CH_2OH) evaporates
l. Food is digested
m. Pancakes cook
n. Ice melts
o. Silver (Ag) tarnishes
p. A paper towel absorbs water
q. Two chemicals are combined, and gas bubbles form
r. A solid is crushed into a powder
s. Mixing salt and pepper
t. A marshmallow is cut in half
u. A marshmallow is toasted over a fire

THE STUDY OF MATTER AND CHANGE

Throughout this chapter, matter and change have been briefly introduced and some common language has been established. Over the course of the following chapters, the ideas introduced here will be given greater attention. In this work, your focus should be on understanding the explanation for these phenomena.

NOTES

1. Berzelius, J.J.; Article V. Essay on the Cause of Chemical Proportions, and on Some Circumstances Relating to Them: Together with a Short and Easy Method of Expressing Them. III. On the Chemical Signs, and the Method of Employing Them in Express Chemical Proportions. *Ann. Philos.*, **1814**, *3*, 51–52. https://www.biodiversitylibrary.org/page/15909112 (Date Accessed 09/24/23).
2. Note that there is no formal definition for a metalloid, and so the identification shown here, while broadly in line with most sources, is based on the working definition from 1. Masterston, W.L.; Slowinski, E.J. *Chemical Principles.* W.B. Saunders Company: Philadelphia, **1977**. 2. Vernon, R.E. Which Elements are Metalloids? *J. Chem. Educ.*, **2013**, *90* (12), 1703–1707. DOI: 10.1021/ed3008457.
3. Phase denotes a material that has a specific state of aggregation and is uniform in chemical composition, while state refers only to solid, liquid, and gas but not to the chemical composition.
4. Note that flammability and inflammability mean the same thing.

3 Measurement and Significant Figures

Chemistry is unique – among the traditional triad of biology, chemistry, and physics – in that the objects we study (atoms and molecules) cannot be directly observed. While there are now tools that allow us to visualize the atomic scale, its scale, direct observation is obscured by the significant difference in scale between our macroscopic world and the nanoscopic world of atoms and molecules. As such, chemistry is intensely dependent upon proxies, that is, those things that we can observe and measure, from which we can infer an understanding of the atoms and molecules that we are interested in.

SCIENTIFIC NOTATION AND E NOTATION

Before discussing measurement, it is worth reviewing the basics of scientific notation. Scientific notation is particularly important in chemistry because the fundamental particles of chemistry are very small (a hydrogen atom has a radius of 0.000 000 000 120 m), which means that numbers we consider can be both very small and very large. In general, numbers in scientific notation take the form of $p \times 10^y$. In this notation, p is the significand, which shows the value of the number. The y term is an exponent, which with $\times 10^y$ provides the appropriate place (10s, 1s, 10ths, 100ths, etc.) of the value.

We will first consider numbers less than one and how to think of scientific notation conceptually before thinking about them practically. Let's reconsider the hydrogen atom radius $\underline{0.000\,000\,000}\,\mathbf{120}$ m. The leading zeros, underlined, do not tell us anything about the value of the number. Instead, those leading zeros only tell us that the 1 is in the ten billionths place. The significand is the value, the important digits, which is shown in bold. In normalized scientific notation, we write the significand with one nonzero digit and all other digits after the decimal place. So, the significand, p, is 1.20, but alone we don't know the places of those three digits, which is where $\times 10^y$ comes in. If we look at the ratio $\frac{1.20}{0.000\,000\,000\,120}$, that ratio is 10 000 000 000 (or our significand is 10 billion times larger than the actual number). To show the place appropriately, 1.20 would need to be divided by 10 000 000 000 or $1.20 \div 10^{10}$. By convention, however, scientific notation is always shown as a multiplicative process and so dividing by a positive exponent (10^{+10}) is the same as multiplying by a negative exponent (10^{-10}). Altogether, then, the scientific notation expression for 0.000 000 000 120 m is 1.20×10^{-10} m.

Now a question might be: Wait, does this mean I have to think about ratios and places every time I need to use scientific notation? No, the above explanation provides a framework for thinking about scientific notation. If we are looking at 0.000 000 000 120 m, there is always implicitly $a \times 10^0$ after any number. The practical thing to remember is that when we move a decimal point to the right, the exponent decreases by 1, which gives the following series of equivalent ways of representing the number.

$$0.000\,000\,000\,120 \times 10^{0} \text{ m}$$

$$0.000\,000\,001\,20 \times 10^{-1} \text{ m}$$

$$0.000\,000\,012\,0 \times 10^{-2} \text{ m}$$

DOI: 10.1201/9781003487210-3

$$0.000\,000\,120 \times 10^{-3}\ \text{m}$$

$$0.000\,001\,20 \times 10^{-4}\ \text{m}$$

$$0.000\,012\,0 \times 10^{-5}\ \text{m}$$

$$0.000\,120 \times 10^{-6}\ \text{m}$$

$$0.001\,20 \times 10^{-7}\ \text{m}$$

$$0.0120 \times 10^{-8}\ \text{m}$$

$$0.120 \times 10^{-9}\ \text{m}$$

$$1.20 \times 10^{-10}\ \text{m}$$

$$12.0 \times 10^{-11}\ \text{m}$$

$$120. \times 10^{-12}\ \text{m}$$

Problem 3.1. Rewrite each of the following numbers that are less than one in scientific notation.

a. 0.0050 ____________________
b. 0.000 028 7 ____________________
c. 0.000 000 000 000 000 000 000 000 000 000 000 662 6 ____________________

Problem 3.2. Rewrite each number in scientific notation in general form.

a. 1.38×10^{-23} ____________________
b. 2.54×10^{-2} ____________________
c. 1.3×10^{-1} ____________________

Now let's consider numbers larger than one. We will again start conceptually before looking at this practically. For example, an important constant is the Avogadro constant: **602 214 076** 000 000 000 000 000. Like the leading zeros in 0.000 000 000 120 m, the trailing zeros tell us nothing about the value but only help to identify the place of the significand (in bold), that is, that the first 6 is in the hundred sextillion place. In normalized scientific notation, the significand is 6.022 140 76. This value is 100 sextillion times smaller than the actual value ($\frac{6.022\,140\,76}{602\,214\,076\,000\,000\,000\,000\,000} =$ 0.000 000 000 000 000 000 000 01) and so to show the appropriate place for our significand, we would need to multiply it by 100 sextillion (10^{23}). Altogether, then, the Avogadro constant in scientific notation would be $6.022\,140\,76 \times 10^{23}$. The practical thing to remember is that when we move a decimal point to the left, the exponent increases by 1, which gives this series of equivalent ways of representing the Avogadro constant.

$$602\ 214\ 076\ 000\ 000\ 000\ 000\ 000 \times 10^{0}$$

$$60\ 221\ 407\ 600\ 000\ 000\ 000\ 000 \times 10^{1}$$

$$6\ 022\ 140\ 760\ 000\ 000\ 000\ 000 \times 10^{2}$$

$$602\ 214\ 076\ 000\ 000\ 000\ 000 \times 10^{3}$$

$$60\ 221\ 407\ 600\ 000\ 000\ 000 \times 10^{4}$$

$$6\ 022\ 140\ 760\ 000\ 000\ 000 \times 10^{5}$$

$$602\ 214\ 076\ 000\ 000\ 000 \times 10^{6}$$

$$60\,221\,407\,600\,000\,000 \times 10^{7}$$

$$6\,022\,140\,760\,000\,000 \times 10^{8}$$

$$602\,214\,076\,000\,000 \times 10^{9}$$

$$60\,221\,407\,600\,000 \times 10^{10}$$

$$6\,022\,140\,760\,000 \times 10^{11}$$

$$602\,214\,076\,000 \times 10^{12}$$

$$60\,221\,407\,600 \times 10^{13}$$

$$6\,022\,140\,760 \times 10^{14}$$

$$602\,214\,076 \times 10^{15}$$

$$60\,221\,407.6 \times 10^{16}$$

$$6\,022\,140.76 \times 10^{17}$$

$$602\,214.076 \times 10^{18}$$

$$60\,221.4076 \times 10^{19}$$

$$6022.140\,76 \times 10^{20}$$

$$602.214\,076 \times 10^{21}$$

$$60.221\,407\,6 \times 10^{22}$$

$$6.022\,140\,76 \times 10^{23}$$

Problem 3.3. Rewrite each of the following numbers that are greater than one in scientific notation.

a. 5280 ____________________
b. 10973731 ____________________
c. 100 000 000 000 000 ____________________

Problem 3.4. Rewrite each number in scientific notation in general form.

d. 9.6485×10^{4} ____________________
e. 3.00×10^{8} ____________________
f. 5.5×10^{1} ____________________

Finally, scientific notation, as we have seen so far, is common in writing, but calculators and computers use a slightly different E notation. In this form, the ×10 is replaced by an E, and the exponent is written in normal sentence case. The number 1.20×10^{-10} m would be written as 1.20E–10 m, and $6.022\,140\,76 \times 10^{23}$ would be written as 6.022 140 76E23.

MEASUREMENT

Measurement involves the use of a standardized tool, or device, from which we can obtain a numerical value that quantifies a dimension of an object. While digital tools are now common, they are not universal and they obscure some aspects of how measurement works. For our discussion of measurement, consider the black object and ruler in Figure 3.1. Using the ruler in Figure 3.1, we can see that the object is less than 10 cm – how much less is up to the estimation of the viewer – and one

FIGURE 3.1 Rectangular object with a ruler that denotes every 10 cm.

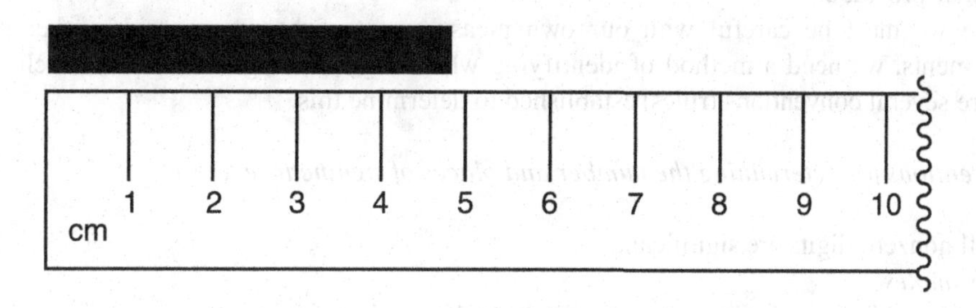

FIGURE 3.2 Rectangular object with a ruler that denotes every 1 cm.

could argue that it is 5 cm (4 cm, 5 cm, or 6 cm would all be reasonable). That is, we can estimate one more place (here the 1s) than our nondigital tool shows (here the 10s).

Now consider the same object with a better ruler in Figure 3.2. Our ruler now denotes every 1 cm. We can see that the object is more than 4 cm but less than 5 cm, and so a reasonable measurement would be 4.8 cm (any value between 4.7 cm and 4.9 cm would all be reasonable). As the tool becomes better, the magnitude of estimation becomes smaller, and measurements will become more precise. Precision is getting similar measurements when an object is measured. This contrasts with accuracy, which is how close a measurement is to the true value.

Finally, consider the object and an even better ruler in Figure 3.3. The object is between 4.8 cm and 4.9 cm, but again where in between is up to the reader. Any value between 4.80 cm and 4.85 cm is plausible. To the author it looks like 4.84 cm.

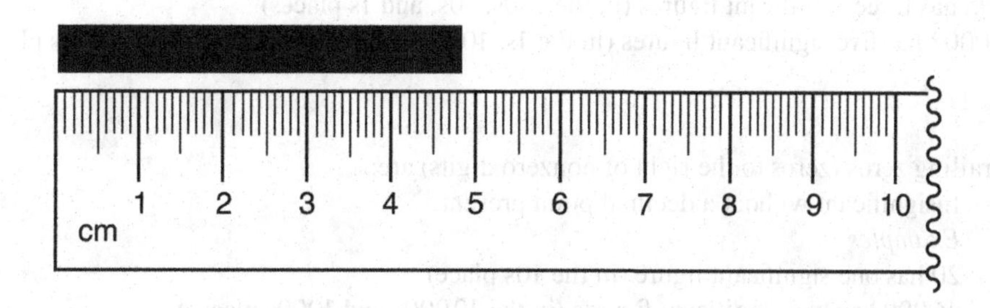

FIGURE 3.3 Rectangular object with a ruler that denotes every 0.1 cm.

In each case with the rulers in Figures 3.1–3.3, the measurement reported includes an estimation of the final digit, which means that there is some uncertainty in the measurements and the values reported. For a digital tool, it feels as though we avoid the whole problem of estimation and uncertainty, but the tool is still reporting a value that involves uncertainty and this is stated by the manufacturer. The next time you are in a laboratory, you should ask the instructor for the manufacturer specifications on any digital instruments you are using to see what the reported precision of the device is.

SIGNIFICANT FIGURES

When we say, for example, "there are 7 people in the room," there is no ambiguity in the number of people. In contrast, a measurement produces an inexact number. Therefore, we must take the inexact nature of measurements into account when we do math with measured values.[1] First, when we consider a measured value, we must determine how many of the digits are significant. The term "significant" here should be thought of as the reasonableness of the number based on the tool used. For example, 6 cm is a reasonable measurement of the length of the unknown in Figure 3.1 based on the ruler provided.

While we must be careful with our own measurements, when we consider other people's measurements, we need a method of identifying which digits are significant and which are not. There are several conventions (rules) established to determine this.

Conventions for determining the number and places of significance

(1) All nonzero digits are significant.
Examples
824 has three significant figures (in the 100s, 10s, and 1s places)
9.1 has two significant figures (in the 1s and 10ths places)

(2) All leading zeros (zeros to the left of nonzero digits) are insignificant.
Examples
0.25 has two significant figures (in the 10ths and 100ths places)
0.000 874 8 has four significant figures (in the 10 000ths, 100 000ths, 1 000 000ths, and 10 000 000ths places)

(3) All sandwiched zeros (zeros between nonzero digits) are significant.
Examples
405 has three significant figures (in the 100s, 10s, and 1s places)
7.0002 has five significant figures (in the 1s, 10ths, 100ths, 1000ths, and 10 000ths places)

(4) Trailing zeros (zeros to the right of nonzero digits) are:

a. Insignificant without a decimal point present.
Examples
20 has one significant figure (in the 10s place)
46 000 has two significant figures (in the 10 000s and 1000s places)

b. Significant with a decimal point present.
Examples
20. has two significant figures (in the 10s and 1s places)
20.0 has three significant figures (in the 10s, 1s, and 10ths places)
0.070 70 has four significant figures (in the 100ths, 1000ths, 10 000ths, and 100 000ths places)

Problem 3.5. Which of the following are exact numbers or should be treated as exact numbers? If they are not exact, how many significant figures does each have?

a. 7 computers ______
b. 12.5 gal gasoline ______
c. The atomic mass of any element ______
d. 10 g sugar ______
e. 1 mi = 5280 ft ______

Problem 3.6. Identify the number of significant figures for each of the following.

a. 6.7540×10^{-3} ______
b. 0.0204 ______
c. 124 people ______
d. 12 in/1 ft ______
e. 28.5 °C ______
f. 20 ______
g. 20 paper clips ______
h. 900. ______
i. 5723.090 ft ______
j. 0.000 357 g ______
k. 907.1 lb ______
l. 7.92×10^{-4} L ______
m. 3.141 59 in ______
n. 200.000 m ______
o. 0.00065×10^{3} s ______
p. 1.0065×10^{3} h ______
q. 6.022×10^{23} atoms ______

Problem 3.7. Rewrite each of the following numbers with three significant figures.

a. 100.000 ______
b. 8854.05 ______
c. 0.005000 ______
d. 5×10^{-3} ______
e. 73000 ______

Given the limitations of inexact, measured values, it is important that when we manipulate these numbers mathematically, we report results that are consistent with the precision of the tools used. The following conventions ensure that our results maintain the significance of any measured values and that we are not reporting insignificant digits (which would be comparable to reporting a value of 4.7 cm with the ruler in Figure 3.1).

In multiplication and division, the number of significant figures in the answer is determined by the number that has the fewest significant figures.

$4.50 \times 3.050 \times 8.004 = 110.$
(4.50 has the fewest number of significant figures (three), and so the final answer is rounded to three significant figures.)

2424/51.0 = 47.5
(51.0 has the fewest number of significant figures (three), and so the final answer is rounded to three significant figures.)

In addition and subtraction, the final answer is determined by rounding to the same place (tens, ones, tenths, hundredths, etc.) as the least precise number, e.g.:

1.50
4.82
+3.875
10.20 (rounded to the 100ths because the least precise numbers 1.50 and 4.82 have no significant figures in the 1000ths place)

82 000
100
− 42.22
82 000 (the least precise number (82 000) has no places of significance past the 1000s place, and so the final answer is rounded to the 1000s place)

Problem 3.8. Complete each of the following calculations and express the answer with the correct number of significant figures.

a. $321.55 - \frac{6104.5}{2.3} =$ ______________________
b. (0.000 45 × 20 000.0) + (2813 × 12) = ______________________
c. 863 [1255 − (3.45 × 108)] = ______________________
d. $2.823{\times}10^5 - 1.220 \times 10^3 =$ ______________________

Problem 3.9. Rewrite each of the following numbers without either scientific notation.

a. 1.30×10^6 g = ______________________
b. 4.4×10^{-6} g = ______________________
c. 1.1×10^{-4} L = ______________________
d. 1.9×10^2 J = ______________________
e. 7.41×10^{-10} s = ______________________

THE INTERNATIONAL SYSTEM OF UNITS: UNITS AND PREFIXES

On the topic of measurements, one of the single most important requirements for clarity is that every number has an appropriate unit label (Table 3.1). In the absence of a unit label, significant confusion can result. It is also important to differentiate between a number with a unit label and a coefficient and variable. There is always a space between the number and its unit label. This is for two purposes. First, when we write 5 g, it is a translation of the words five grams. Writing 5g would imply the word fivegrams. Second, 5 g is a measurement with an appropriate unit label, while 5*g* is a coefficient and a variable (variables are always in italicized typeface and unit labels are in Roman typeface). One of the most frequent sources of confusion among students is whether something is a measurement and unit label or a coefficient and variable. The space between the number and label (or lack thereof) and the typeface (Roman or *italicized*) is a key identifier as to the type of number you are looking at.

Chemistry, and science generally, uses the International System of Units (SI) for measurement. This system originated during the French Revolution (1789) as a more rational system of units than the previous myriad of customary measurement systems used in France and in Europe. Today, SI is the fundamental system of measurement in science, and all other measurement systems, including the US customary system, are defined by the SI base units. The standard SI units, unit labels, and their associated quantities are shown in Table 3.1.

Finally, in terms of measurement, in the metric system the magnitude of a number is frequently indicated with SI prefixes. The most used prefixes in chemistry are those shown in Table 3.2. While

TABLE 3.1
List of SI Standard Units, Unit Labels, and Related Quantity

Unit (unit label)	Unit of (unit symbol)
meter (m)	length (l)
cubic meter (m^3)	volume (V)
kilogram (kg)	mass (m)
mole (mol)	amount (n)
second (s)	time (t)
kelvin (K)	temperature (T)
joule (J)	energy (E)
pascal (Pa)	pressure (p)
coulomb (C)	charge (Q)
volt (V)	electric potential (Φ)
newton (N)	force (F)

TABLE 3.2
SI Prefixes, Orders of Magnitude, and Symbols

Prefix	Order of magnitude	Symbol
giga-	$\times 10^9$	G
mega-	$\times 10^6$	M
kilo-	$\times 10^3$	k
hecto-	$\times 10^2$	h
deca-	$\times 10^1$	da
–	$\times 10^0$	–
deci-	$\times 10^{-1}$	d
centi-	$\times 10^{-2}$	c
milli-	$\times 10^{-3}$	m
micro-	$\times 10^{-6}$	μ
nano-	$\times 10^{-9}$	n

one may be familiar with these as a reader, what is important to stress here is that the SI prefixes imply a scientific notation. That is, 500 nm implies 500×10^{-9} m.

Problem 3.10. Rewrite each of the following numbers by using an appropriate SI prefix instead of scientific notation.

f. 1.30×10^{6} g = __________
g. 4.4×10^{-6} g = __________
h. 1.1×10^{-4} L = __________
i. 1.9×10^{2} J = __________
j. 7.41×10^{-10} s = __________

Problem 3.11. Conduct the following mathematical operations. Round your final answer to the appropriate number of significant figures.

a. 1.2 mL + 2.7×10^{-4} L = __________
b. $\frac{23.1000\,\text{g} - 22.0000\,\text{g}}{25.10\,\text{mL} - 25.00\,\text{mL}}$ = __________
c. 500 032.1 cm + 3 cm = __________
d. 500 032.1 cm + 3.00 cm = __________
e. 7001 g + 6.001 kg = __________
f. 1459.3 Å + 9.77 Å + 4.32 Å = __________
g. 4.1(6.022×10^{23} atom) = __________
h. (1206.7 mm – 0.904 mm) 89 mm = __________
i. 3.8×10^{5} nm – 8.45×10^{4} nm = __________
j. $\frac{9.2\times10^{24}\ \text{atom}}{6.022\times10^{23}\ \text{atom}}$ = __________
k. $\frac{4.55\,\text{g}}{407\ 859\ \text{mL}}$ + 1.00098 g/mL = __________

Problem 3.12. Rewrite the following numbers in scientific notation and in E notation.

a. Altitude of Concord Academy Science Office: 52.98 m = __________
b. Altitude of the summit of Mount Washington: 1917 m = __________
c. Wavelength of green light: 0.000 000 558 m = __________
d. Number of galaxies in the universe: 100 billion galaxies = __________
e. Volume of an H atom: 0.000 000 000 000 000 000 000 000 000 621 L = __________

Problem 3.13. Rewrite each of the following (from Problem 3.12) using the indicated SI prefix.

a. The altitude of Concord Academy Science Office in km = __________
b. The altitude at the summit of Mt. Washington in km = __________
c. The wavelength of green light in nanometers = __________
d. The volume of an H atom in quectoliters (1 qL = 1×10^{-30} L) = __________

Problem 3.14. Carry out each of the following calculations using scientific notation.

a. Ratio of the mass of the Earth to the mass of the Moon:
$$\frac{5.974\times10^{27}\text{ g}}{7.348\times10^{25}\text{ g}} = \underline{\hspace{6cm}}$$
b. Difference between the mass of the Earth and the mass of the Moon:
$5.974\times10^{27}\text{ g} - 7.348\times10^{25}\text{ g} =$ ____________________
c. Ratio of the mass of a proton to the mass of an electron:
$$\frac{1.672\ 621\ 9\times10^{-24}\text{g}}{9.109\ 389\ 7\times10^{-28}\text{ g}} = \underline{\hspace{6cm}}$$
d. Difference between the mass of an ^{1}H atom and the mass of an electron:
$1.673\ 557\ 5\times10^{-24}\text{ g} - 9.109\ 389\ 7\times10^{-28}\text{ g} =$ ____________________
e. Difference between the mass of an ^{1}H atom and the mass of a proton:
$1.673\ 557\ 5\times10^{-24}\text{ g} - 1.672\ 621\ 9\times10^{-24}\text{ g} =$ ____________________

MEASUREMENT AND SIGNIFICANCE

Measurement is a key part of science and chemistry, where we cannot routinely see the particles that are the objects of our study. In this chapter, we have discussed how to properly use tools and report significance in our measurements, and then we investigated how to properly carry that significance through resulting calculations. As you continue your study of chemistry, proper significant figure work will not only be incumbent upon you to maintain, but it will also be expected in all of the work that you do.

NOTE

1. i) The Oxford English Dictionary (*Oxford English Dictionary*, s.v. "significant, adj., sense 2.b," July 2023. DOI: 10.1093/OED/1149173983) cites William Bedwell's 1614 work as the earliest English source to use the term. Bedwell, W. *De Numeris Geometricis*, **1614**.

   ii) Gauss gives an extensive treatment of the importance of precise and imprecise numbers and the impact on error in calculations: Gauss, C.F. *Theory of the Motion of the Heavenly Bodies Moving About the Sun in Conic Sections, a Translation of Gauss's "Theoria Motus"*. Translated by Charles Henry Davis. Little, Brown and Company: Boston, **1857**.

   iii) The earliest complete outline of the rules and conventions presented here that the author could find comes from: Holman, S.W. *Discussion of the Precision of Measurements with Examples Taken Mainly from Physics and Electrical Engineering*. Kegan Paul, Trench, Trübner, and Co., Ltd: London, **1892**.

4 Conversion Factors and Chemical Amounts

Let's imagine you were driving from Boston Common to Central Park in New York, a driving distance of 330 km. Now imagine further that there is no traffic along the way and the speed limit is 110 km/h. You could then estimate that it would take you three hours to drive from Boston Common to Central Park. Here you have converted from the unit of distance (a physical dimension) to the unit of time (the temporal dimension). We call this type of calculation unit conversion or dimensional analysis.[1] We have experience with this in our everyday life in terms of travel, and with purchases: it costs $15 to see a movie and you want to go with three friends so the four tickets will cost $60. In this chapter, we will look at making this process more explicit by using conversion factors and the factor-label method. After establishing the basic structure of unit cancellation, we will look at chemistry-specific problems.

CONVERSION FACTORS

Conversion factors relate different measured units to one another. The first type that we will consider is the relationship between different units for the same dimension. Consider the SI and US customary rulers shown in Figure 4.1. We can see that 1 cm does not equal 1 in, which means that an object measured in centimeters will have a very different value for length than an object measured in inches.

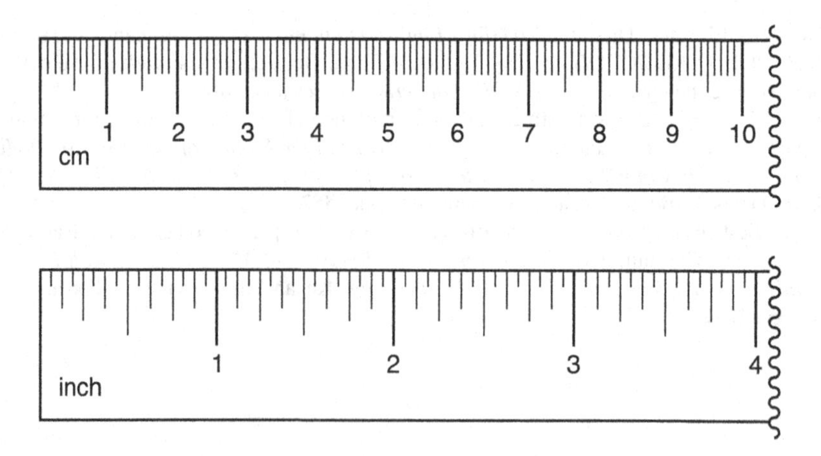

FIGURE 4.1 An SI ruler showing centimeters (*above*) and a US customary ruler showing inches (*below*).

The relationship between centimeters and inches is defined as one inch exactly equals 2.54 centimeters. Mathematically, this can be expressed as shown in Equation 4.1.

$$1 \text{ in} = 2.54 \text{ cm (exact)} \tag{4.1}$$

Relationships between units are frequently shown as equalities (Equation 4.1), but this formulation obscures the mathematical utility of conversion factors. For example, if we divide both sides by 1 in

DOI: 10.1201/9781003487210-4

(Equation 4.2), then we can arrive at a useful mathematical ratio that we will employ in unit conversions (*below*). In addition, we can see that this ratio $\left(\frac{2.54\,\text{cm}}{1\,\text{in}}\right)$ is equal to 1, which means that we can multiply or divide by this ratio, and it is the same as multiplying or dividing by 1.

$$1 = \frac{2.54\,\text{cm}}{1\,\text{in}}(\text{exact}) \tag{4.2}$$

It is worth mentioning that we could have also divided both sides by 2.54 cm to arrive at an equivalent relationship, but with centimeters in the denominator rather than inches (Equation 4.3).

$$\frac{1\,\text{in}}{2.54\,\text{cm}} = 1(\text{exact}) \tag{4.3}$$

Table 4.1 provides a list of conversion factors that relate dimensions (length, time, mass, volume, pressure, and energy) from different systems of measurement to the International System of Units standard unit for that dimension.

The second type of conversion factor relates units of different dimensions. As we saw in the opening paragraph, velocity relates distance (miles or kilometers) and time (hour). While the conversion factors we see in Table 4.1 are a way of relating different unit systems for the same dimension to one another, conversion factors like that for velocity come from measuring different dimensions of the same object. For example, if a sample of gold was found to be a cubic meter (m^3) in volume and have

TABLE 4.1
Conversion Factors Relating Dimensions from Other Units to the SI Standard Unit

Dimension	Units related together	Conversion factor
Length	inch to meter	1 in = 0.0254 m
	foot to meter	1 ft = 0.3048 m
	yard to meter	1 yd = 0.9144 m
	mile to meter	1 mi = 1609.344 m
	astronomical unit to meter	1 au = 149 597 870 700 m
	angstrom to meter	1 Å = 1×10^{-10} m
Time	minute to second	1 min = 60 s
	hour to second	1 h = 3600 s
	day to second	1 d = 86 400 s
	year (annum) to second	1 a = 31 557 600 s
Mass	ounce to kilogram	1 oz = 0.028 349 52 kg
	pound to kilogram	1 lb = 0.453 592 37 kg
	unified atomic mass unit to kilogram	1 u = $1.660\,539\,066\,60 \times 10^{-27}$ kg (inexact)
Volume	gallon to m^3	1 gal = 0.003 785 411 784 m^3
	liter to m^3	1 L = 0.001 m^3
Pressure	atmosphere to pascal	1 atm = 101 325 Pa
	bar to pascal	1 bar = 100 000 Pa
	millimeter mercury to pascal	1 mmHg = 133.322 Pa
	pound per square inch to pascal	1 psi = 6894.757 Pa (inexact)
Energy	calorie to joule	1 cal = 4.184 J
	kilowatt-hour to joule	1 kWh = 3 600 000 J
	megaton of TNT equivalent to joule	1 Mt = 4.184×10^{15} J
	electronvolt to joule	1 eV = $1.602\,176\,634 \times 10^{-19}$ J

Note that all conversion factors are exact unless otherwise noted.

TABLE 4.2
Selected Fundamental Physical Constants

Related dimensions	Fundamental physical constant	Conversion factor
Length to time	Speed of light in a vacuum	299 792 458 m = 1 s
Energy to temperature	Boltzmann constant	$1.380\,649 \times 10^{-23}$ J = 1 K
Energy to frequency	Planck constant	$6.626\,070\,15 \times 10^{-34}$ J = 1 Hz
Charge to mass	Electron charge-to-mass quotient	$-1.758\,820\,010\,76 \times 10^{11}$ C = 1 kg
Charge to amount	Faraday constant	96 485.332 12 C = 1 mol

These are examples of conversion factors that relate different dimensions, with some providing the basis of the SI system of units.

a mass of 19 300 kg, then we could determine its density (a conversion factor relating volume and mass) is 1 m^3 = 19 300 kg. Fundamental physical constants, like the speed of light (Table 4.2), are examples of conversion factors that relate different dimensions. Some of these physical constants provide the basis of the units in the SI system itself.

UNIT CONVERSION: THE FACTOR-LABEL METHOD

We have discussed conversion factors, that is, relationships among and between dimensional quantities. We will now turn our attention to why these quantities are useful and how we use them in problem solving. The system presented here is called the factor-label method. Let's investigate this problem using an example that we have already seen finding how long it will take to drive 330 km at 110 km/h.

In any unit-conversion problem, the first step is to find the initial, measured quantity. In the problem there are two values 330 km and 110 km/h. The first (330 km) is a measured quantity of distance; the second is a conversion factor relating distance and time (110 km = 1 h or $\frac{110\,\text{km}}{\text{h}}$ or $\frac{\text{h}}{110\,\text{km}}$). We will start with 330 km:

330 km

Now we will multiply by our conversion factor. We wish to go from the unit kilometer to the unit hour, which means that we want the unit kilometer to cancel. If we use our conversion factor in the form $\frac{\text{h}}{110\,\text{km}}$, the unit kilometer is in the denominator, which means that it will cancel with any unit in the numerator (any unit that we started with). We are creating, in effect, a fraction expression. The 330 km is, implicitly, 330 km/1, and we then write in our conversion factor with hour in the numerator and 110 km in the denominator:

330 km	h
	110 km

Now, we can see that the unit of kilometer is in both the numerator and the denominator, which means that it will cancel. The remaining unit hour (h) is the unit that we are looking for in this problem.

330 ~~km~~	h
	110 ~~km~~

Mathematically, then we have 330 multiplied by 1 h in the numerator and 1 multiplied by 110 in the denominator. Considering that our final answer should have two significant figures, the final answer is 3.0 h:

$$330\ \cancel{\text{km}} \times \frac{\text{h}}{110\ \cancel{\text{km}}} = 3.0\ \text{h}$$

This is the general approach to the factor label method. The key points to remember are: (1) always start with a measured quantity (not a conversion factor) and (2) each conversion factor should be written such that the units in the numerator of the prior term cancel with the units in the denominator of the next term.

Let's look at a slightly longer example before turning to some practice. If we have 1.0 L of water, what is the mass in kilograms, given that the density of water is 1.0 g/mL? The initial measured quantity is 1.0 L. Density provides us with a way of converting from gram to milliliter (1.0 g/mL). To use this conversion factor, we will first have to convert 1.0 liter to milliliter, using our knowledge of the SI prefixes that the prefix milli means $\times 10^{-3}$:

$$1.0\ \text{L} \times \frac{1\ \text{mL}}{1 \times 10^{-3}\ \text{L}}$$

Now the unit liter cancels, and we can use density to convert the unit milliliter to gram:

$$1.0\ \cancel{\text{L}} \times \frac{1\ \text{mL}}{1 \times 10^{-3}\ \cancel{\text{L}}} \times \frac{1.0\ \text{g}}{\text{mL}}$$

And now that the unit milliliter cancels, we again use our knowledge of the SI prefixes (kilo means $\times 10^{3}$) to convert gram to kilogram:

$$1.0\ \cancel{\text{L}} \times \frac{1\ \cancel{\text{mL}}}{1 \times 10^{-3}\ \cancel{\text{L}}} \times \frac{1.0\ \text{g}}{\cancel{\text{mL}}} \times \frac{1\ \text{kg}}{1 \times 10^{3}\ \text{g}}$$

With that our final unit is kilogram, the desired unit. We can carry out the arithmetic to find that the final answer (with two significant figures) is 1.0 kg:

$$1.0\ \cancel{\text{L}} \times \frac{1\ \cancel{\text{mL}}}{1 \times 10^{-3}\ \cancel{\text{L}}} \times \frac{1.0\ \cancel{\text{g}}}{\cancel{\text{mL}}} \times \frac{1\ \text{kg}}{1 \times 10^{3}\ \cancel{\text{g}}} = 1.0\ \text{kg}$$

Now try the practice problems (Problem 4.1), making sure to let the units be your guide and that at each step the units appropriately cancel.

Problem 4.1. Carry out each of the following unit conversion problems using Table 4.1 and any conversion factors that are provided in the problems.

a. Convert 675 cal to J.

b. Convert 0.984 atm to mmHg.

c. Convert 764 mmHg to kPa.

d. How many milligrams are equal to 0.542 kg?

e. If you decide to switch from drinking soda to drinking milk at every meal, how many gallons of milk will you drink if you drink 2.5 L of soda every week?

f. You see a bumper sticker for 100.0 km, implying the owner of the car has run a 100 km race. 100.0 km corresponds to how many miles?

g. A red blood cell has a diameter of 8 μm.
   i. What is this diameter in mm?

   ii. Your body makes 17.500 million new red blood cells every second. How many new red blood cells do you make in a week?

h. What is the volume (in liters) of 10. kg of bismuth (density is 9.747 g/mL)?

i. The estimated volume of the world's oceans is 1.5×10^{21} L. If 1 mL of seawater contains 4.0×10^{-12} g gold and gold is valued at nearly $2000 per ounce, how much is the gold in the ocean worth? (1 lb = 16 oz).

j. An average home in Concord, MA uses 28 kWh of electricity every day. How much energy (kJ) does an average home in Concord use every day?

k. During Operation Sandblast, the first completely submerged circumnavigation of the ocean by a nuclear-powered submarine, the crew completed a trip of 49491 km. If the average speed was 33 km/h, how many days was the trip?

l. Your heart beats 80 beats per minute and pumps 80 mL per beat. Over 24 hours, what volume (mL) of blood does the heart circulate?

USING CONVERSION FACTORS THAT RELATE MULTIPLE QUANTITIES

So far, we have only considered conversion factors that relate two quantities. There are also conversion factors that relate three or more quantities. There are two common examples in chemistry. The first is specific heat capacity, which relates energy, mass, and temperature (J/(g °C)). The second is the gas constant, which relates volume, pressure, amount, and temperature $\left(\frac{\text{L atm}}{\text{mol K}}\right)$.

To see how to work with conversion factors that relate more than two quantities, we will work with a more tangible example: building a house. To build a moderately sized house it takes 2000 people hours, that is, we have the conversion factor $\frac{2000\text{ people hours}}{\text{house}}$. If two houses had to be built in 160 hours (a month of 40-hour work weeks), how many people would need to be part of the construction crew to complete the houses in 160 hours?

First, when working with conversion factors, always make sure that the desired unit is in the numerator. In this problem, we want to find the number of people, which is in the numerator. We are then going to multiply and divide in individual steps to cancel the other units. Here, we will start by canceling the unit house:

$$\frac{2\ \cancel{\text{houses}}}{} \left| \frac{2000\text{ people hours}}{\cancel{\text{house}}} \right. = 4000\text{ people hours}$$

Now we have the complex unit of 4000 people hours. We want to find the number of people, and so to cancel the unit of hours, we must divide by the 160 hours given in the problem. This tells us that 25 people need to work on this two-house project.

$$\frac{4000\text{ people }\cancel{\text{hours}}}{} \left| \frac{1}{160\ \cancel{\text{hours}}} \right. = 25\text{ people}$$

Unlike the two-quantity, conversion-factor problems seen so far, conversion factors relating multiple quantities problems should always be approached through separated, individual steps to successively cancel each unit. Problem 4.2 provides a conversion-factor problem using specific heat capacity, which relates energy, mass, and temperature.

Problem 4.2. The specific heat capacity of water is 4.184 J/(g °C). How much energy (J) does it take to warm up 155 kg of water (mass of water in a bathtub) from 25 °C to 38 °C? Note that in this problem, the change in temperature (final temperature minus initial temperature) is used to cancel the unit of temperature (°C).

CHEMICAL AMOUNTS (MOLE)

With the recognition that atoms are the basic unit of chemistry (Chapter 6), an important challenge arises in the practice of chemistry. That is, chemists measure and observe macroscopic quantities of chemicals, from which we make inferences about the particles (atoms and molecules) we are studying. This necessitates a clearly defined way of linking the macroscopic to the nanoscopic (Figure 4.2), which is called the amount of substance and is given the unit mole.[2] We are familiar with units of amount like dozen (12 items), score (20 items), and gross (144 items). That is each of these units (mole, dozen, score, and gross) is just a set number of things. A mole is similar to the units dozen, score, and gross in that it is a set number of countable entities, but there is also a significant difference between a mole and a dozen, score, or gross. A mole is a much larger number:

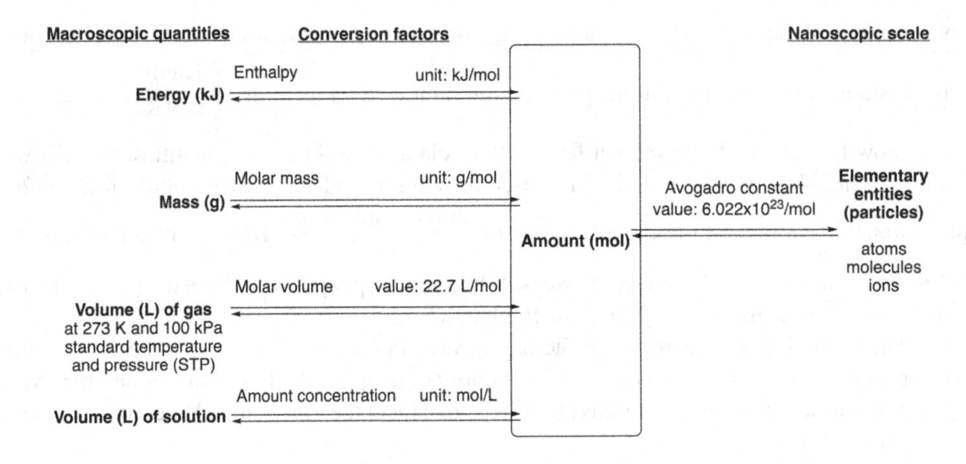

FIGURE 4.2 A conversion factor diagram showing the centrality of the amount (mol) of a substance as the link between the macroscopic and the nanoscopic (elementary entity) scales.

roughly a trillion trillion. Specifically, a mole is defined as exactly $6.022\,140\,76 \times 10^{23}$ particles (elementary entities). Just to highlight how large this number is, if there were $6.022\,140\,76 \times 10^{23}$ marshmallows, this amount of marshmallows would cover the surface of the Earth in a layer about 10-km thick.

The importance of the mole unit is that it connects the macroscopic and nanoscopic worlds (Figure 4.2). This means, for example, that we can measure the mass of an element and know how many particles are present. The relationship between mass and amount of an element is called the molar mass. The molar mass of carbon is 0.012 011 kg/mol or, equivalently, 12.011 g/mol. The molar mass of each element in g/mol is equivalent to its recommended relative atomic mass (A_r°) (Table 4.3).[3] If we had 12 g of carbon and 56 g of iron then we would know that there is an equal amount, one mole, in each sample and that means that there are equal numbers of atoms in the two samples.

For a compound, the molar mass is the sum of the recommended relative atomic mass value of each element in the chemical formula. Sodium chloride (NaCl) has a molar mass of 58.44 g/mol (the sum of 22.990 g/mol for sodium and 35.45 g/mol for chlorine), and water (H_2O) has a molar

TABLE 4.3
List of Elements and Their Abridged Recommended Relative Atomic Mass Values (A_r°)

Atomic number	Name	Symbol	A_r°	Atomic number	Name	Symbol	A_r°
1	hydrogen	H	1.0080	13	aluminium	Al	26.982
2	helium	He	4.0026	14	silicon	Si	28.085
3	lithium	Li	6.94	15	phosphorus	P	30.974
4	beryllium	Be	9.0122	16	sulfur	S	32.06
5	boron	B	10.81	17	chlorine	Cl	35.45
6	carbon	C	12.011	18	argon	Ar	39.95
7	nitrogen	N	14.007	19	potassium	K	39.098
8	oxygen	O	15.999	20	calcium	Ca	40.078
9	fluorine	F	18.998	21	scandium	Sc	44.956
10	neon	Ne	20.180	22	titanium	Ti	47.867
11	sodium	Na	22.990	23	vanadium	V	50.942
12	magnesium	Mg	24.305	24	chromium	Cr	51.996

(Continued)

TABLE 4.3 CONTINUED

List of Elements and Their Abridged Recommended Relative Atomic Mass Values (Ar°)

Atomic number	Name	Symbol	$A_r°$	Atomic number	Name	Symbol	$A_r°$
25	manganese	Mn	54.938	72	hafnium	Hf	178.49
26	iron	Fe	55.845	73	tantalum	Ta	180.95
27	cobalt	Co	58.933	74	tungsten	W	183.84
28	nickel	Ni	58.693	75	rhenium	Re	186.21
29	copper	Cu	63.546	76	osmium	Os	190.23
30	zinc	Zn	65.38	77	iridium	Ir	192.22
31	gallium	Ga	69.723	78	platinum	Pt	195.08
32	germanium	Ge	72.630	79	gold	Au	196.97
33	arsenic	As	74.922	80	mercury	Hg	200.59
34	selenium	Se	78.971	81	thallium	Tl	204.38
35	bromine	Br	79.904	82	lead	Pb	207.2
36	krypton	Kr	83.798	83	bismuth	Bi	208.98
37	rubidium	Rb	85.468	84	polonium	Po	–
38	strontium	Sr	87.62	85	astatine	At	–
39	yttrium	Y	88.906	86	radon	Rn	–
40	zirconium	Zr	91.224	87	francium	Fr	–
41	niobium	Nb	92.906	88	radium	Ra	–
42	molybdenum	Mo	95.95	89	actinium	Ac	–
43	technetium	Tc	–	90	thorium	Th	232.04
44	ruthenium	Ru	101.07	91	protactinium	Pa	231.04
45	rhodium	Rh	102.91	92	uranium	U	238.03
46	palladium	Pd	106.42	93	neptunium	Np	–
47	silver	Ag	107.87	94	plutonium	Pu	–
48	cadmium	Cd	112.41	95	americium	Am	–
49	indium	In	114.82	96	curium	Cm	–
50	tin	Sn	118.71	97	berkelium	Bk	–
51	antimony	Sb	121.76	98	californium	Cf	–
52	tellurium	Te	127.60	99	einsteinium	Es	–
53	iodine	I	126.90	100	fermium	Fm	–
54	xenon	Xe	131.29	101	mendelevium	Md	–
55	caesium	Cs	132.91	102	nobelium	No	–
56	barium	Ba	137.33	103	lawrencium	Lr	–
57	lanthanum	La	138.91	104	rutherfordium	Rf	–
58	cerium	Ce	140.12	105	dubnium	Db	–
59	praseodymium	Pr	140.91	106	seaborgium	Sg	–
60	neodymium	Nd	144.24	107	bohrium	Bh	–
61	promethium	Pm	–	108	hassium	Hs	–
62	samarium	Sm	150.36	109	meitnerium	Mt	–
63	europium	Eu	151.96	110	darmstadtium	Ds	–
64	gadolinium	Gd	157.25	111	roentgenium	Rg	–
65	terbium	Tb	158.93	112	copernicium	Cn	–
66	dysprosium	Dy	162.50	113	nihonium	Nh	–
67	holmium	Ho	164.93	114	flerovium	Fl	–
68	erbium	Er	167.36	115	moscovium	Mc	–
69	thulium	Tm	168.93	116	livermorium	Lv	–
70	ytterbium	Yb	173.05	117	tennessine	Ts	–
71	lutetium	Lu	174.97	118	oganesson	Og	–

mass of 18.015 g/mol (the sum of 2(1.0080 g/mol) for the two equivalents of hydrogen and 15.999 g/mol for oxygen).

As can be seen in Figure 4.2, we can also relate amount (mol) to the volume (L) of a gas at standard temperature, to energy (kJ), and to the volume of a solution. The volume of a gas at STP has a set conversion factor (22.7 L/mol), but the conversion factors for energy and volume of a solution must be determined empirically (see section "Deriving Conversion Factors" below) or will be provided as given information. A final note on amount conversion factors concerns the conversion factor amount concentration, which relates the amount of a chemical to the volume of solution. Amount concentration is frequently called molarity and represented with the unit label M, which obscures that it is a conversion factor. For example, it is recommended by IUPAC that 0.900 M NaCl be represented as 0.900 mol/L NaCl.

Problem 4.3. Use Figure 4.2 and Table 4.3 to answer the following questions.

a. What is the mass (in kg) of 5.0×10^{25} carbon atoms?

b. What volume (L at STP) does 1.23 g of helium gas occupy?

c. If a solution is 0.900 mol/L NaCl, how many grams NaCl are in 500.0 mL?

d. What is the mass (kg) of 5.0×10^{25} uranium atoms?

e. If a solution is 0.014 mol/L HI, what mass of HI is in 5.500 L?

f. If burning carbon (coal) produces 418 kJ/mol, how much energy (in kJ) is produced by a coal power plant that burns 9.00×10^{9} g of coal? (This is the amount of coal typically combusted in a single day)?

Problem 4.4. The gas constant is $\frac{0.082\ 06\ \text{L atm}}{\text{mol K}}$, what is the volume (L) of 1.00 mol of water vapor at 373 K (100 °C) and 0.987 atm?

DERIVING CONVERSION FACTORS

Finally, we conclude this chapter by considering how conversion factors are derived. Recall that a conversion factor relating different types of units comes from measuring different dimensions of the same object. These two dimensions are then related to one another to give a conversion factor.

A common conversion factor that is a physical property of a substance is density. Density is a measure of the mass of material per unit volume. In the SI system, this can be expressed as kg/

m^3 (kilogram per cubic meter). In chemistry, it is far more common to see density shown as g/cm^3 (gram per cubic centimeter), which is equivalent to g/mL, gram per milliliter.

Problem 4.5. You are tasked with determining the density (g/mL) of an unknown metal. You place a piece of metal onto a balance and find that it weighs 45.001 g. To determine the volume, you place the cube into a graduated cylinder of water. Before adding the cube, you find the volume of water to be 22.5 mL. After adding the piece of metal to the cylinder you find the water level to now be 24.5 mL. What is the density (g/mL) of the unknown metal?

Amount concentration is a measure of the amount of a substance dissolved in a specific volume of solution. In the SI system, amount concentration is calculated as mol/L.

Problem 4.6. You come across a solution of sodium carbonate (Na_2CO_3) in a chemistry supply room. The label is missing the amount concentration of the solution, but there is an ingredient list. Contents: 1.52 g of sodium carbonate mixed with enough water to make a 250 mL solution.

a. What amount (mol) of sodium carbonate is dissolved in solution?

b. What is the volume of the solution in liters?

c. Using your answers to (a) and (b), what is the amount concentration (mol/L) of the unknown solution?

Enthalpy (ΔH) is a measure of the amount of heat produced, or consumed, per amount of chemical. In the SI system of units, enthalpy is typically calculated as the kilojoules of heat energy per mole (kJ/mol). To find the amount of heat, we use the specific heat capacity of a substance, typically water. Specific heat capacity has the units of J/(g °C), and the specific heat capacity of water is 4.184 J/(g °C).

Problem 4.7. A 9.75 g sample of potassium is added to 500 g of water. After all the potassium reacted (producing a pink flame on the water's surface), the water (which was at 25.00 °C before adding the potassium) was now 36.70 °C. If water has a specific heat capacity of 4.184 J/(g °C), calculate the following.

a. What is the difference in temperature (final temperature minus initial temperature)?

b. Based on this temperature change and the heat capacity of water, how much thermal energy (kJ) does the water take in?

c. Note that numerically the water has a positive thermal energy transfer value because it took in energy. That energy came from the metal. What is the value of thermal energy transfer from the reaction (note the sign should be negative because it is energy given off)?

d. Calculate the amount of potassium (mol) given that we start with 9.75 g.

e. Determine the enthalpy (kJ/mol potassium) using your answers from (c) and (d).

DIMENSIONAL ANALYSIS AND THE MOLE

The multitude of units that are used in chemistry, and science generally, can be bewildering. In this chapter, we discussed the means of converting from one unit to another and introduced a unit central to chemistry's study: the mole. The amount of a substance (mol) provides us with a means of comparing making direct comparison between different chemicals and different reactions. The amount of a substance also provides a tangible link between the macroscopic, laboratory measurements we collect and the nanoscopic world that we are studying.

NOTES

1. Dimensional analysis evolved over many years and many people can be credited with its development. For a review of the topic:
   i) Macagno, E.O. Historico-critical Review of Dimensional Analysis. *J. Franklin Inst.*, **1971**, *292* (6), 391–402. DOI: 10.1016/0016-0032(71)90160-8.
   ii) Martins, R. De A. The Origin of Dimensional Analysis. *J. Franklin Inst.*, **1981**, *311* (5), 331–337. DOI: 10.1016/0016-0032(81)90475-0.
2. Ostwald, W. *Hand- und Hilfsbuch zur Ausführung Physiko-Chemischer Messungen.* Wilhelm Engelmann: Leipzig, **1893**.
3. For most practical purposes the values are equivalent, but the molar mass constant is not quite unity: $0.99999999965 \times 10^{-3}$ kg/mol. This difference is only significant in very high-precision work, and it will not be significant for the material covered in this book.

5 Energy

In this chapter we will discuss some of the basics of energy and thermodynamics. Energy is a core concept across science, which we will see throughout this book. Given the importance of energy, we will develop a common language and approaches to understanding and quantifying energy. First, let us start with a definition. Energy is defined as the ability to move an object (do work) or to produce heat,[1] and it comes in two forms: kinetic energy and potential energy. The SI unit for energy is joule (J), but other units with which you may be familiar are kilowatt-hour (kWh), calorie (cal), and kilocalorie (kcal or Cal). This chapter will also introduce some of the basic concepts in thermodynamics (heat, work, and temperature), which provide us with tools to understand and quantifying energy in chemistry.

KINETIC ENERGY (E_K)

When a ball is in motion and collides with a person's hand, two things can happen. Either the ball will cause the hand to move or if the hand does not move, the hand will be warmer where the ball hit it. If you have ever missed catching a ball and had it hit your person, you have experienced the increase in temperature, which you may or may not have noticed in addition to any pain, that comes with a ball's collision with your body. As we can see based on the definition of energy, the ball (a moving object) has energy. We say that an object in motion has kinetic energy (E_k), where the word "kinetic" comes from the Greek *kinesis* meaning motion.[2] We can calculate the amount of kinetic energy present using Equation 5.1.

$$E_k = \frac{1}{2}mv^2 \tag{5.1}$$

E_k is the kinetic energy (J, J = (m^2 kg)/s^2)
m is the mass of the object (kg)
v is the speed of the object (m/s)

Let's look at an example by calculating the amount of kinetic energy present in an elephant (2000 kg) running at top speed (4 m/s). First, we need to make sure that the units are kilograms (kg) for the mass and meters per second (m/s) for the speed, which they are. If the units were not those units, we would first have to do unit conversion to convert the mass and speed to the appropriate kilogram and meter per second units. We then plug the values into the equation (m = 2000 kg and v = 4 m/s). Plugging these into Equation 5.1, we find that the kinetic energy of a running elephant is 20000 (m^2 kg)/s^2, which is 20000 J.

$$E_k = \frac{1}{2}mv^2 = \frac{1}{2}(2000\text{ kg})(4\text{ m/s})^2 = 20000\text{ (m}^2\text{ kg)/s}^2 = 20000\text{ J}$$

Problem 5.1. Calculate the kinetic energy (E_k) in each of the following examples.

a. A horse (421 kg) galloping at 12 m/s.

DOI: 10.1201/9781003487210-5

b. A baseball (145 g) flying through the air at 120 km/h.

c. An electron (9.11×10^{-31} kg) moving at 3.00×10^{7} m/s.

KINETIC ENERGY–TEMPERATURE (*T*)

Now based on the introduction to kinetic energy and Problem 5.1, we must clarify that in chemistry we tend to avoid having our experimental setups move. Typically, when an experiment moves, especially if it moves through the lab, that can represent a significant safety hazard. But the nanoscopic motion of particles – how much they vibrate, rotate, and translate (move through three-dimensional space) – is an important quantity. The mean kinetic energy of particles is proportional to their temperature. For example, at 0 °C, the mean speed of particles in air is 485 m/s, and at 20 °C, this increases to 505 m/s.

We measure thermal energy and changes in thermal energy (heat) with thermometers (Figure 5.1), which provide a macroscopic measure of average particle motion. Historically, the most common temperature scale in the Anglophone world – and still the primary scale in the United States, Cayman Islands, and Liberia – is the Fahrenheit scale (°F), where the zero point (0 °F) is a mixture of ice, water, and ammonium chloride.[3] The most common laboratory thermometers record values in Celsius (°C), where the zero point (0 °C) is the freezing point of water.[4] The preferred thermodynamic temperature scale is the kelvin (K) scale, which starts at absolute zero (0 K) where molecular motion, except vibrations, ceases.[5]

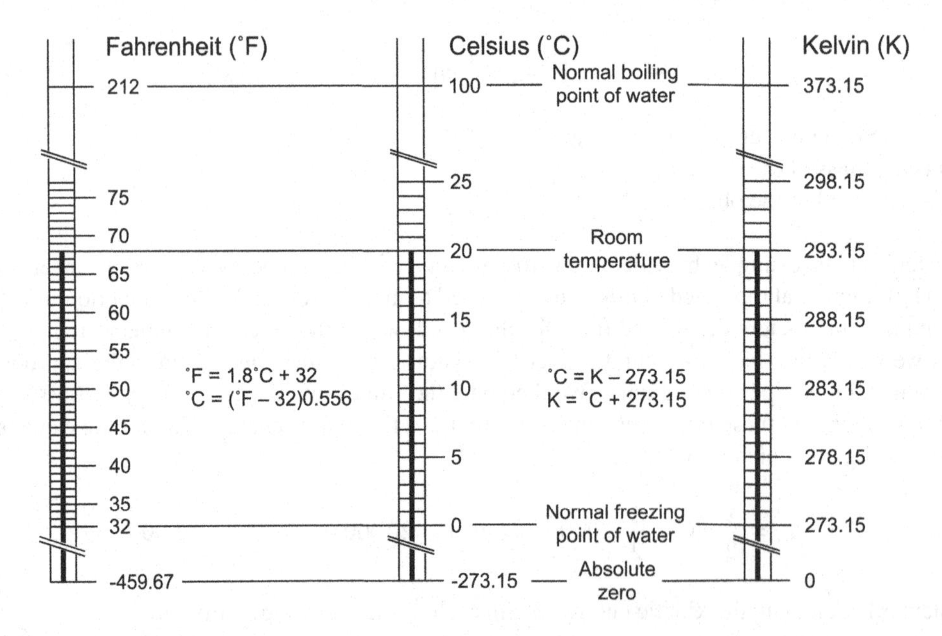

FIGURE 5.1 Comparison of Fahrenheit, Celsius, and kelvin temperature scales and their interconversions. Standard temperature is defined by the International Union of Pure and Applied Chemistry (IUPAC) as 0 °C, and room temperature is the standard defined by the National Institutes of Standards and Technology (NIST) as 20 °C.

POTENTIAL ENERGY (E_p)

Potential energy (E_p) is more nuanced than kinetic energy because we are determining the energy an object has based on its configuration.[6] Given our lived experience on Earth, we are familiar with the force due to gravity and how different configurations lead to different amounts of potential energy (different likelihoods of falling). Consider, for example, this book (whether a physical book or a digital device) sitting on the edge of your desk (or table). Sitting there the book has potential energy, which is to say that it has the potential to move. We can see this if the book slips off the table or desk and then falls (moves) to the floor. This is different from the book sitting on the floor, which has lower potential energy and cannot fall because there is no lower place for it to be. The difference between the two books is their configuration (position) in relation to the floor (the gravitational potential energy minimum in the room), which leads to differences in potential energy (Figure 5.2).

The height of an object on Earth is only one type of potential energy (gravitational potential energy). Other types of potential energy are elastic potential energy (think of what a stretched rubber band could do) and electric potential energy, which is what charged batteries possess and the amount of potential energy is measured in volts (V). A type of potential energy, more relevant to chemistry, is the energy stored in the arrangement of and interactions among atoms and molecules (chemical potential energy). Interactions among atoms and molecules may be bonds (Chapter 11) and/or intermolecular forces (Chapter 13).

POTENTIAL ENERGY DIAGRAM

Instead of a book, on a desk or the floor, let's consider the chemical sodium azide, NaN_3, which can be used in car airbags. Sodium azide has a lot of chemical potential energy, energy stored in the configuration of atoms and their electrons. When sodium azide decomposes, it becomes sodium (Na) and dinitrogen (N_2), which are lower in energy. Changes in potential energy are typically shown in potential energy diagrams like that in Figure 5.2.

Looking at Figure 5.2, we can see the difference in the potential energy of the book and its configuration in the room. Processes that freely occur are driven by a decrease in energy. The book falling from the desk to the floor will, for example, happen freely if nudged. Put another way, if the

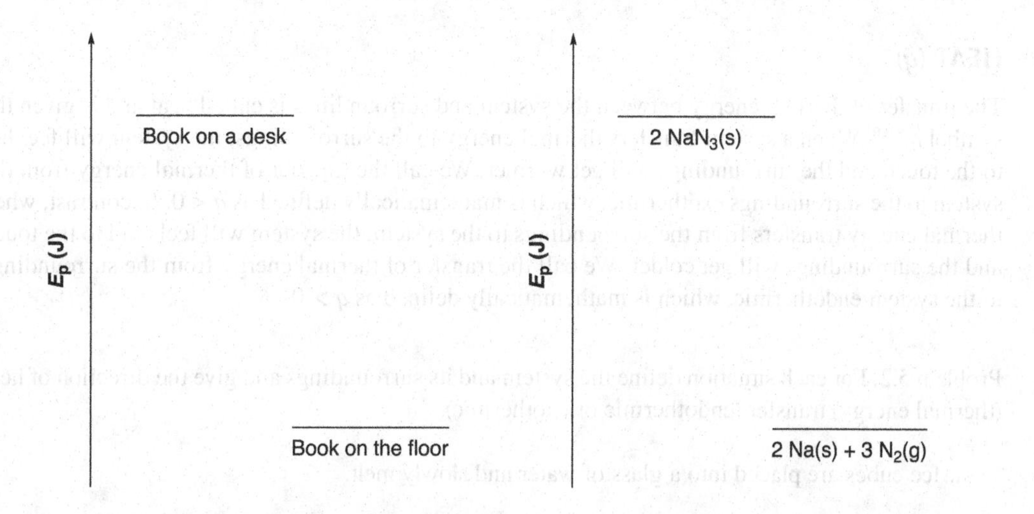

FIGURE 5.2 Potential energy diagram of a book on a desk and the floor (*left*) and a potential energy diagram of sodium azide (*right*) solid (NaN_3(s)) decomposing into sodium solid (Na(s)) and dinitrogen gas (N_2(g)). The energy scales are arbitrary.

book is pushed off the desk it will not sit there in the air, and we do not have to move it to the floor ourselves. The decomposition of sodium azide occurs freely, once provided with a nudge (a spark). This is why car airbags have used sodium azide: Once a spark was supplied, the decomposition will proceed without further help to fill the bag with dinitrogen gas to protect the occupants.

Processes that result in higher potential energy (moving the book from the floor to the desk) do not occur freely. The book, once on the floor, will stay on the floor until someone lifts it back to the desk. Energy must be added to move the book to a higher potential energy. Similarly, sodium and dinitrogen gas will not freely combine to produce sodium azide, and chemists must add energy to make this happen.

ENERGY TRANSFER: SYSTEM AND SURROUNDINGS

If we reconsider (Figure 5.2), sodium azide is higher in energy than sodium and dinitrogen. When sodium azide decomposes, then, energy must be given off. In describing energy flows, we must define our system, the object of our interest, and the surroundings, those things that are not our system. If we consider the sodium azide airbag, the system would be the sodium azide. In general, in chemistry the system is invariably a chemical or mixture of chemicals. The surroundings, then, would be everything that is inside the car other than the sodium azide. And so, the sodium azide (the system) is giving off energy to the inside of the car (to the surroundings). This would mean that $\Delta E_{\text{p}} < 0$, and this is defined as exergonic (energy going out). A more familiar term might be exothermic, which is an exergonic change where the energy is lost only as heat.

If we consider the book on the floor moving to the desk, the system (the book) has lower potential energy initially. It will have to take in energy (someone picking it up and moving it to the desk). In the language of thermodynamics, the system is taking in energy from the surroundings. That would mean that $\Delta E_{\text{p}} > 0$ and this is defined as endergonic (energy coming in). A more familiar term is endothermic, which is an endergonic change where the energy gained is only in the form of heat.

According to the First Law of Thermodynamics,[7] the energy of a closed system can only change by transferring thermal energy (heat) or by changing in volume (work). A closed system is one where matter cannot enter or leave but energy can. In the following paragraphs we will consider heat and work in turn and discuss how to quantify both heat and work.

HEAT (*q*)

The transfer of thermal energy between the system and surroundings is called heat and is given the symbol q.[8,9,10] When a system transfers thermal energy to the surroundings, the system will feel hot to the touch and the surroundings will get warmer. We call the transfer of thermal energy from the system to the surroundings exothermic, which is mathematically defined as $q < 0$. In contrast, when thermal energy transfers from the surroundings to the system, the system will feel cold to the touch and the surroundings will get colder. We call the transfer of thermal energy from the surroundings to the system endothermic, which is mathematically defined as $q > 0$.

Problem 5.2. For each situation define the system and its surroundings and give the direction of heat (thermal energy) transfer (endothermic or exothermic).

a. Ice cubes are placed into a glass of water and slowly melt.

b. Hydrogen gas in a balloon explosively combusts with oxygen, producing a bright flame.

c. During exercise, metabolism in our cells provides energy for movement.

d. Propane is burning in a Bunsen burner in the laboratory.

e. After you have a swim, water droplets on your skin evaporate.

f. Water, originally at 25 °C, is placed in a freezing compartment of a refrigerator.

g. Two chemicals are mixed in a flask on a laboratory bench. A reaction occurs and feels hot to the touch.

h. Thorium burns to produce a white flame.

i. An ice pack is placed on a sore ankle.

j. A student (with a normal 37 °C body temperature) sits in a classroom.

As we have already seen (Chapter 4), heat can be quantified using specific heat capacity, which is defined as the amount of energy (J) needed to raise 1 gram (g) of a substance by 1 kelvin (K). The units for specific heat capacity (c_p), then, are J/(g K) or, equivalently J/(g °C). The cancellation of units, as we saw in Chapter 4, is the method of calculating the amount of thermal energy transferred (q) with the specific heat capacity. We can also symbolically show the relationship between heat and specific heat capacity by using Equation 5.2.

$$q = mc_p\Delta T \tag{5.2}$$

q is the heat (J)
m is the mass (g)
c_p is the specific heat capacity (J/(g °C))
ΔT is the change (final minus initial) in temperature (°C)

As an example of using Equation 5.2, let us calculate the amount of thermal energy (q) required to raise the temperature of 1500 g water (the mass of water in an electric tea kettle) from 25.0 °C to 100.0 °C, where the specific heat capacity is 4.184 J/(g °C). Looking at the equation and the description provided, we can identify that 1500 g is the mass (m), 4.184 J/(g °C) is the specific heat capacity (c_p), and the change in temperature (ΔT) is 100.0 °C–25.0 °C, always final minus initial, or 75.0 °C. Plugging this all in, we can see that the units will cancel to leave us with energy (J) and a value of 470000 J. Since the value is positive ($q > 0$), we know that this is an endothermic process and heat is being transferred from the surroundings into the system.

$$q = (1500\,\cancel{\text{g}})\left(\frac{4.184\,\text{J}}{\cancel{\text{g}\,^\circ\text{C}}}\right)(75.0\,\cancel{^\circ\text{C}})$$

$$q = 470000\ \text{J}$$

Equations, like Equation 5.2, can helpfully supplement the unit cancellation work that was introduced in Chapter 4. The form of these symbolic equations – so long as we plug the appropriate values and units for the correct variable in the equation – allows us to know what to multiply (or divide) without necessarily needing to consider the units beforehand. Once we have plugged in all values and units, however, we want to verify that units do appropriately cancel.

Problem 5.3. For each of the following, calculate the thermal energy transfer (J) in each of the following examples, identify whether each is endothermic or exothermic.

a. 0.125 kg of hydrogen fluoride is cooled from 19.5 ˚C to 0.0 ˚C. The specific heat capacity of hydrogen fluoride is 1.456 J/(g ˚C).

b. 535 g of iron is heated from 25.2 ˚C (room temperature) to 900 ˚C (red hot). The specific heat capacity of iron is 0.451 J/(g ˚C).

c. 200.00 g of solid sulfur goes from 20.15 ˚C to 34.27 ˚C. The specific heat capacity of solid sulfur is 0.732 J/(g ˚C).

d. 10.0 g of helium gas goes from 15.20 ˚C to –35.32 ˚C. The specific heat capacity of helium gas is 5.19 J/(g ˚C).

e. 270 mg of dihydrogen gas goes from 288.35 K to 373.38 K. The specific heat capacity of dihydrogen gas is 14.31 J/(g K).

As we saw in Chapter 4, heat can also be calculated using enthalpy (ΔH). Enthalpy is the amount of heat (J) per amount (mol) for a given process. As we saw for specific heat capacity, we can also symbolically represent the calculation of heat (q) from enthalpy using Equation 5.3. While Equation 5.2 is very useful for a single chemical increasing or decreasing in temperature, for the heat of a phase change, of mixing, or of a reaction it can become computationally cumbersome because we can only indirectly measure the temperature change of these processes. Enthalpy, which ultimately relies on someone else's work using specific heat capacity, can provide a way of calculating the heat associated with these changes more directly. Note that the calculation of heat (q) using the enthalpy of a reaction requires using a balanced equation to find the amount (n), and this will be considered in detail in Chapter 15.

$$q = n\Delta H \tag{5.3}$$

q is the heat (J)
n is the amount of substance (mol)
ΔH is the enthalpy (kJ/mol)

Let us consider an example of using Equation 5.3 by considering the amount of thermal energy required to evaporate 1500 g of water, given that the enthalpy of vaporization is 43.9 kJ/mol. We are not given the value for the amount of water (n) directly, but we can convert the 1500 g given into amount using the molar mass (18.015 g/mol for water).

$$\frac{1500\ \cancel{\text{g}}\ \Big|\ \text{mol}}{\Big|\ 18.015\ \cancel{\text{g}}} = 83\ \text{mol}$$

Now we have the amount of water (n), which is 83 mol and the enthalpy (ΔH) is 43.9 kJ/mol in this example. Plugging this into Equation 5.3, we can see that the units will cancel to leave us with energy (kJ) and a value of 3700 kJ. Since the value is positive ($q > 0$), we know that this is an endothermic process and heat is being transferred from the surroundings into the system.

$$q = (83\ \cancel{\text{mol}})(43.9\ \text{kJ}/\cancel{\text{mol}}) = 3700\ \text{kJ}$$

Problem 5.4. For each of the following, calculate the thermal energy transfer (J) in each of the following examples and identify whether each is endothermic or exothermic.

a. 0.4535 kg of water freezes (enthalpy of solidification is −6.01 kJ/mol).

b. 150 g of dry ice (CO_2) sublimates (enthalpy of sublimation is 27.2 kJ/mol).

c. 52 L of dihydrogen (H_2) gas (at STP) combusts (enthalpy of combustion is −286 kJ/mol).

d. 750 mg of ethanol (CH_3CH_2OH) evaporates (enthalpy of vaporization is 42.3 kJ/mol).

e. 47.2 L of methane (CH_4) condenses into a liquid (enthalpy of condensation is −8.9 kJ/mol).

WORK (*w*)

Going back to the example of a book on the floor. If someone were to pick the book up and put it back on the desk (Figure 5.2), the book has gained potential energy, but its temperature has not changed. To cause this change, a different type of energy was added to the system: work (w). Formally, work is the product of force (in newtons, N) and distance (in meters, m). In chemistry, we will be most concerned with pressure–volume work (Equation 5.4), which occurs when the volume of a substance changes under some external pressure.

$$w = -p\Delta V \tag{5.4}$$

w is the work (L atm)
p is the external pressure (atm)
ΔV is the change (final minus initial) in volume (L)

In a laboratory setting, it is common to measure pressure in atmospheres (atm) and volume in liters (L). To convert the resulting energy unit liter-atmosphere (L atm) to the common unit joule (J), one would need to use the conversion factor: 1 L atm = 101.325 J. Note that the recommended SI units for pressure and volume are pascal (Pa) and cubic meter (m^3). While not commonly used in most chemistry laboratories, these units have the advantage that 1 pascal-cubic meter is 1 joule (1 Pa m^3 = 1 J).

The sign convention is the same as we saw for potential energy and heat. When the system expands, the system is doing work on the surroundings ($w < 0$), which can be called exoworkic. And when the system condenses, the surroundings are doing work on the system ($w > 0$), which can be called endoworkic.[11] Let us consider an example of using Equation 5.4 to calculate the work when a piston compresses from 1.5 L to 0.5 L with an external pressure of 1.15 atm. Here the pressure (p) is 1.15 atm and the change in volume (ΔV) is 0.5 L – 1.5 L, always final minus initial, or –1.0 L. Plugging these into Equation 5.4, we find that the work is 1.15 L atm.

$$w = -(1.05 \text{ atm})\ (0.5 \text{ L} - 1.5 \text{ L})$$

$$w = 1.15 \text{ L atm}$$

This process is endoworkic ($w > 0$) as the surroundings are doing work on (compressing) the system. If we wanted to convert this to joules (J), we would use the conversion factor (1 L atm = 101.325 J) and find that 117 J of work energy were transferred to the system from the surroundings.

1.15 ~~L atm~~	101.325 J	= 117 J
	~~L atm~~	

Problem 5.5. For each of the following, calculate the work (J) in each of the following examples and identify whether each is endoworkic or exoworkic.

a. A piston expands from 0.37 L to 2.41 L with an external pressure of 1.05 atm.

b. A sample of 1.5 L liquid water evaporates and becomes 2600 L of steam with an external pressure of 0.985 atm.

c. 498 mL of solid water converts to 454 mL of liquid water with an external pressure of 1.1 atm.

d. 22.7 L of ammonia gas condenses to become 25 mL of liquid ammonia with an external pressure of 8.51 atm.

e. 14 mL of white tin becomes 17 mL of gray tin with an external pressure of 0.991 atm.

CONSERVATION OF ENERGY: THE FIRST LAW OF THERMODYNAMICS

Energy, like matter, cannot be created or destroyed; rather, it can only be transferred and transformed. For a closed system, a system where the amount of matter is fixed, it has its own internal energy (U). The internal energy is the sum of the particles' thermal energy (nanoscopic kinetic energy) and the configuration of atoms, molecules, and electrons (potential energy). Because energy is conserved, the internal energy can only change (ΔU) as the result of either heat or work, that is, only through the transference of energy. This is called the First Law of Thermodynamics and is represented symbolically by Equation 5.5. As discussed, freely occurring processes are those that lower the internal energy of a system (releasing energy to the surroundings) and those that require energy are those that increase the internal energy of a system (taking energy from the surroundings).

$$\Delta U = q + w \tag{5.5}$$

ΔU is the change in internal energy (J)
q is the transfer of thermal energy, heat (J)
w is the change in volume under external pressure, work (J)

ENERGY AND CHANGE

Energy is a central, unifying concept in science, and it will be a theme that threads its way throughout this book and all of your future studies of chemistry. In this chapter, we discussed heat and work as ways that the internal energy of a system can change. We will see energy as a deep, explanatory tool to help us understand electron configurations (Chapter 8), periodic trends (Chapter 9), bonding (Chapter 11), intermolecular interactions (Chapter 13), and chemical reactions (Chapter 16).

NOTES

1. i) William Rankin originally proposed the definition of energy, which is only slightly modified in the IUPAC definition given in this text: Rankin, W.J.M. *Miscellaneous Scientific Papers Volume 2.* Ed. W.J. Millar. Charles Griffin and Company: London, **1881**.
   ii) "Energy." IUPAC. *Compendium of Chemical Terminology*, 2nd ed. (the "Gold Book"). Compiled by A.D. McNaught, A. Wilkinson. Blackwell Scientific Publications: Oxford, **1997**. Online version (2019-) created by S.J. Chalk. DOI: 10.1351/goldbook.E02101.
2. The idea that subsequently became our modern kinetic energy was first termed *vis viva* (Latin for *living force*) by Gottfried Leibnitz (Leibnitz, G.G. Specimen dynamicum pro admirandis naturae legibus circa corporum vires & mutuas actiones detegendis & adsuas causas revocandis. *Nova Acta Erud.*, **1695**, 145–157). The term kinetic energy was formally introduced by Lord Kelvin (William Thomson): Thomson, W.; Tait, P.G. *Treatise on Natural Philosophy Volume 1.* Clarendon Press: Oxford, **1867**.
3. Fahrenheit, D.G. VIII. Experimenta & observationes de congelatione aquæ in vacuo factæ a D. G. Fahrenheit, R. S. S. *Phil Trans. R. Soc.*, **1724**, *33*, 78–84. DOI: 10.1098/rstl.1724.0016.
4. The temperature scale originally published by Celsius went from 100 °C as the freezing point of water and 0 °C as the boiling point of water: Celsius, A. Observationer om twänne beständiga grader på en thermometer. *Kongl. Vetensk. Acad. Handl.*, **1742**, *3*, 171–180.
   Several people are credited with developing the modern, forward Celsius scale.
5. Thomson, W. Art. XXXIX. On an Absolute Thermometric Scale Founded on Carnot's Theory of the Motive Power of Heat*, and Calculated from Regnault's Observations†. *Philos. Mag.*, **October 1848**, 100–106.
6. Originally defined by Rankin as having the capacity to do work: Rankin, W.J.M. *Miscellaneous Scientific Papers Volume 2.* Ed. W.J. Millar. Charles Griffin and Company: London, **1881**.
7. Clausius, R. Ueber die bewegende Kraft der Wärme und die Gesetze, welche sich daraus für die Wärmelehre selbst arbeiten lassen. *Ann. Phys.* (Leipzig), **1850**, *155* (4), 500–524. DOI: 10.1002/andp.18501550306.
8. i) Carnot, S. *Réflexions sur la Puissance Motrice du Feu et sur les Machines Propres a Développer cette Puissance.* Bachelier: Paris, **1824**.
   ii) Clapeyron, E. Mémoire sur la Puissance Motrice de la Chaleur. *Journal de l'École Royale Polytechnique*, **1834**, *14*, 153–190.
9. Joule, J.P. III. On the Mechanical Equivalent of Heat. *Phil. Trans. R. Soc.*, **1850**, *140*, 61–82. DOI: 10.1098/rstl.1850.0004.
10. Clausius, R. *The Mechanical Theory of Heat.* Translated by W.R. Browne. MacMillan and Co.: London, **1879**.
11. While there is no official term for the case of $w > 0$ or $w < 0$, this text will follow the convention presented by Bhairav D. Joshi in: Joshi, B.D. The Sign of Work: Endoworkic and Exoworkic Processes. *J. Chem. Educ.*, **1983**, *60* (10), 895. DOI: 10.10.1021/ed060p895.

6 Atomic Structure

In Chapter 5, we discussed energy and its importance as a central concept in science. In this chapter, we will turn our attention to one of the core concepts of chemistry: matter. As a reminder from Chapter 2, matter is defined, generally, as anything that has mass and occupies space. In chemistry, we define matter according to atomic theory: matter is anything that is made up of atoms. In this chapter, we will consider the development of empirical atomic theory and atomic structure

DEVELOPMENT OF ATOMIC THEORY

An atomistic philosophy is present across cultures in the ancient world and is well documented in both Greek (following the philosophical tradition of Leucippus and Democritus) and Indian (following the Nyāya-Vaiśeṣika philosophical tradition) history. The modern chemistry term "atom" comes from the Greek word *atomos* meaning indivisible. In these atomistic philosophies, the core idea is that the material world is made up of undividable and eternal particles: atoms.

Philosophically, an atomic perspective postulates that everything that we observe is the result of a small number of elementary particles combining to produce all materials. The development of an empirical, rather than philosophical, atomic theory required the establishment of several key principles, with the chemical identification of elements being foundational.[1] By the end of the 18th century, an element was defined chemically as a simple substance that could not be broken down further into other substances. There were 14 elements that had been known since antiquity, including carbon, iron, copper, silver, gold, and mercury. Using alchemical (later chemical) methods, heat, and electricity, chemists had isolated 19 elements, including hydrogen, oxygen, sodium, and aluminium. Given that elements cannot be decomposed, a question to consider is: What does the existence of elements tell us about the nature of matter? An inference we can draw from this is that the fundamental components of one element (silver, for example) must be different from the fundamental components of another element (like gold).

In addition to the identification of elements, the law of conservation of mass was also a seminal development. Unlike the discovery of elements, however, the law of conservation of mass requires a sophisticated and precise system of measurement. With improvements in measurement, it was found that the mass of reactants and products in a chemical reaction would always be the same. The law of conservation of mass, in its original formulation, states that "We may lay it down as an incontestable axiom, that, in all operations of art and nature, nothing is created; an equal quantity of matter exists both before and after the experiment; the quantity of the elements remain precisely the same; and nothing takes place beyond changes and modifications of these elements."[2] While this law may or may not be familiar, a question that it inspires should be: What does the law of conservation of mass tell us about the nature of matter? Alone, the law of conservation of mass is a single principle, but it supports the inference that matter must consist of some sort of component that has mass and is persistent and unalterable.

Precise and accurate measurements provided the means for other important developments in atomic history. Consider, for example, carbon dioxide gas. When decomposed, carbon dioxide is always 27% carbon and 73% oxygen by mass. Similarly, when ammonia is broken down, it is always 75% hydrogen and 25% nitrogen by volume (or 18% hydrogen and 82% nitrogen by mass). These facts provided the empirical basis for the law of constant composition, also known as the law of definite proportions.[3] According to the law of constant composition, when broken down into its constituent elements, a pure substance will always give the same fraction of each element. We now call substances that abide by this rule compounds. Recall that the composition of mixtures, in contrast,

DOI: 10.1201/9781003487210-6

can show continuous variation in the relative amounts of each component (Chapter 2). What does the existence of the law of constant composition tell us about the nature of matter? We can infer that there are limited, fixed ways in which elements can combine to produce a compound.

Let us consider one last empirical result in the development of atomic theory: the law of multiple proportions.[4] For this we will consider sulfurous anhydride and sulfuric anhydride. An analysis of the compounds composition, by mass, shows that sulfurous anhydride has a sulfur-to-oxygen ratio of 1.00:1.00 and sulfuric anhydride has a sulfur-to-oxygen ratio of 1.00:1.50. The law of multiple proportions states that if we compare those oxygen-ratio values (1.00 and 1.50) they give a whole number ratio (2:3). From this ratio, the law states, we can infer that sulfurous anhydride (sulfur dioxide) has two oxygen particles for every one sulfur particle, whereas sulfuric anhydride (sulfur trioxide) has three oxygen particles for every one sulfur particle. These results are summarized in the law of multiple proportions: two elements (A and B) may combine in multiple different ways to create different compounds, and if the amount of A is fixed then the ratio of the mass of B in each compound will be a small, whole number ratio. With the law of multiple proportions, we can again ask what we can infer about the nature of matter. Given that the ratios are always whole numbers, we can infer that the fundamental particles of an element are discrete, countable units or particles.

Utilizing the evidence of elements, the law of conservation of mass, the law of constant proportions, and the law of multiple proportions we arrive at the inferences that:

(1) Each element is composed of extremely small, fundamental particles that are identical to each other, but the particles of one element are different from the atoms of another element.
(2) Matter consists of a particle that has mass; the particles of matter are persistent and unaltered during chemical reactions.
(3) Compounds form when two or more different types of particles combine.
(4) For a particular compound there are always the same kind and relative number of particles.

Following atomic theory, where these inferences are postulates, we term these small particles atoms. Atomic theory is the basis of modern chemistry.

THE PERIODIC TABLE

In the early 19th century, the relative atomic mass (A_r) values for elements began to be determined with reasonable accuracy. These values were and are frequently called atomic weight, though relative atomic mass is preferred. Chemists originally determined atomic mass values by defining 1/16th the atomic mass of oxygen as one atomic mass unit (1 amu). Today the atomic mass values are determined by defining $A_r(^{12}C)$ as *exactly* 12 unified atomic mass unit (12 u) or, equivalently, as *exactly* 12 daltons (12 Da).

Regardless of the scale, relative atomic mass values offer a way of ordering the elements from hydrogen, the lightest, to uranium, the heaviest naturally occurring. A list, by itself, does not provide any meaningful connections between elements, but scientists had noticed patterns in the properties of elements. For example, there are triads of similar elements, with the middle element showing a relative atomic mass that is roughly the average of the two end elements, e.g.: chlorine (35.5), bromine (78.4), and iodine (126.5).[5] There is also a periodic repetition of properties where every eighth element repeats in properties, which was termed the law of octaves.[6] This and other work showed that there was an underlying structure to how elements might be arranged. The first table of elements that gained widespread attention was put together between 1869 and 1871.[7,8] The elements were organized into groups based on similar properties and the similarities among the compounds they formed (Table 6.1). The focus on repeating patterns of properties is called the periodic law. The periodic table shows the common chemical formulae for each element in a group when it combines with hydrogen or with oxygen.

TABLE 6.1
An early periodic table from 1871

Row	Group I — R_2O	Group II — RO	Group III — R_2O_3	Group IV RH_4 RO_2	Group V RH_3 R_2O_5	Group VI RH_2 RO_3	Group VII RH R_2O_7	Group VIII — RO_4
1	H = 1							
2	Li = 7	Be = 9.4	B = 11	C = 12	N = 14	O = 16	F = 19	
3	Na = 23	Mg = 24	Al = 27.3	Si = 28	P = 31	S = 32	Cl = 35.5	
4	K = 39	Ca = 40	—=44	Ti = 48	V = 51	Cr = 52	Mn = 55	Fe = 56, Co = 59,
5	(Cu = 63)	Zn = 65	—=68	—=72	As = 75	So = 73	Br = 80	Ni = 59, Cu = 63
6	Rb = 85	Sr = 87	?Yt=88	Zr=90	Nb=94	Mo=96	—=100	Ru=104, Rh=104,
7	(Ag=108)	Cd=112	In=113	Sn=118	Sb=122	Te=128	I=127	Pd=106, Ag=108
8	Cs=133	Ba=137	?Di=136	?Ce=140	—	—	—	— — — —
9	(—)	—	—	—	—	—	—	
10	—	—	?Er=178	?La=180	Ta=182	W=184	—	Os=195, Ir=197,
11	(Au=199)	Hg=200	Tl=204	Pb=207	Bi=208	—	—	Pt=198, Au=199
12	—	—	—	Th=231	—	U=240	—	— — — —

The periodic law approach to a periodic table was successful for two reasons. First, it classified elements into groups with similar properties. Second, the table was assembled with the possibility that there may have been yet-undiscovered elements (shown as — in Table 6.1). A successful theory is one that can do two things. The first is to explain current information and the second is to provide testable hypotheses. For example, based on its position, eka-aluminium ("eka" meaning one below) was predicted, in 1871, to have a relative atomic mass of 68 and form an oxide with the formula R_2O_3. When gallium was discovered four years later, it was found to have a relative atomic mass of 69.723 and gallium oxide has the formula Ga_2O_3, which matched eka-aluminium. This finding gave significant support to the periodic table and further successful discoveries reinforced the first: scandium (discovered in 1879) matched the predicted properties of eka-boron, germanium (discovered in 1886) matched the predicted properties of eka-silicon, and technetium (discovered in 1937) matched the predicted properties of eka-manganese.

Together, the discoveries of predicted elements helped to underscore and support the periodic law. It should be noted that based on relative atomic mass there were some uncertainties of order (nickel and cobalt) and some anomalies like the ordering of tellurium and iodine. The refinement of the periodic table would take further work by many scientists: the discovery of noble gases,[9] the discovery that it was nuclear charge (now called atomic number) that was the organizing principle (not relative atomic mass),[10] and the reordering of the lanthanoids and actinoids as a separate block of the periodic table.[11] With these changes, the periodic table assumed its modern form (Figure 6.1) 70 years after the first version.

DISCOVERY OF ATOMIC STRUCTURE

Atomic theory postulated that atoms were the smallest, most fundamental unit of matter. The discovery of subatomic (smaller than atom) particles, and the inner structure of atoms, would radically

1	2	3	4	5	6	7	8	9	10	11	12	13	14	15	16	17	18
1 **H** 1.01																	2 **He** 4.00
3 **Li** 6.94	4 **Be** 9.01											5 **B** 10.8	6 **C** 12.0	7 **N** 14.0	8 **O** 16.0	9 **F** 19.0	10 **Ne** 20.2
11 **Na** 23.0	12 **Mg** 24.3											13 **Al** 27.0	14 **Si** 28.1	15 **P** 31.0	16 **S** 32.1	17 **Cl** 35.5	18 **Ar** 40.0
19 **K** 39.1	20 **Ca** 40.1	21 **Sc** 45.0	22 **Ti** 47.9	23 **V** 50.9	24 **Cr** 52.0	25 **Mn** 54.9	26 **Fe** 55.8	27 **Co** 58.9	28 **Ni** 58.7	29 **Cu** 63.5	30 **Zn** 65.4	31 **Ga** 69.7	32 **Ge** 72.6	33 **As** 74.9	34 **Se** 79.0	35 **Br** 79.9	36 **Kr** 83.8
37 **Rb** 85.5	38 **Sr** 87.6	39 **Y** 88.9	40 **Zr** 91.2	41 **Nb** 92.9	42 **Mo** 96.0	43 **Tc** –	44 **Ru** 101	45 **Rh** 103	46 **Pd** 106	47 **Ag** 108	48 **Cd** 112	49 **In** 115	50 **Sn** 119	51 **Sb** 122	52 **Te** 128	53 **I** 127	54 **Xe** 131
55 **Cs** 133	56 **Ba** 137	*	72 **Hf** 178	73 **Ta** 181	74 **W** 184	75 **Re** 186	76 **Os** 190.	77 **Ir** 192	78 **Pt** 195	79 **Au** 197	80 **Hg** 201	81 **Tl** 204	82 **Pb** 207	83 **Bi** 209	84 **Po** –	85 **At** –	86 **Rn** –
87 **Fr** –	88 **Ra** –	**	104 **Rf** –	105 **Db** –	106 **Sg** –	107 **Bh** –	108 **Hs** –	109 **Mt** –	110 **Ds** –	111 **Rg** –	112 **Cn** –	113 **Nh** –	114 **Fl** –	115 **Mc** –	116 **Lv** –	117 **Ts** –	118 **Og** –

*	57 **La** 139	58 **Ce** 140	59 **Pr** 141	60 **Nd** 144	61 **Pm** –	62 **Sm** 150.	63 **Eu** 152	64 **Gd** 157	65 **Tb** 159	66 **Dy** 163	67 **Ho** 165	68 **Er** 167	69 **Tm** 169	70 **Yb** 173	71 **Lu** 175
**	89 **Ac** –	90 **Th** 232	91 **Pa** 231	92 **U** 238	93 **Np** –	94 **Pu** –	95 **Am** –	96 **Cm** –	97 **Bk** –	98 **Cf** –	99 **Es** –	100 **Fm** –	101 **Md** –	102 **No** –	103 **Lr** –

FIGURE 6.1 Modern periodic table with the recommended relative atomic mass value of each element (Appendix 2) rounded to three significant figures.

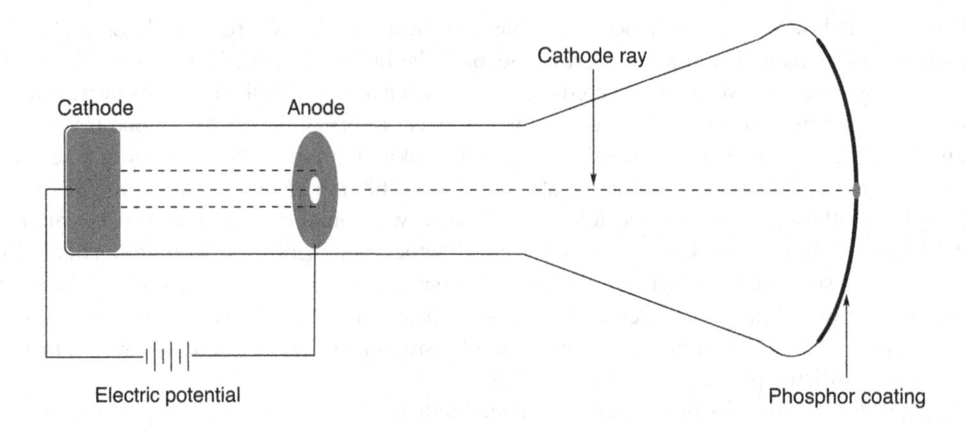

FIGURE 6.2 Diagram of a cathode ray tube, a partially evacuated glass tube with an electric potential between the cathode and the anode. This setup leads to the emission of invisible cathode rays (*dashed lines*), which cause the zinc sulfide (phosphor) coating to show a green spot where the cathode rays hit the coating.

reformulate atomic theory by showing that atoms were not fundamental particles. This revolution in atomic theory would ultimately provide for a deeper understanding of chemistry itself.

Cathode Rays

The first experimental evidence for subatomic particles came from studying the rays in a cathode ray tube.[12] A cathode ray tube (Figure 6.2) is a partially evacuated glass tube with an electric potential applied between a cathode (negative terminal) and an anode (positive terminal). A phosphor (zinc sulfide) coating then shows a green spot where the rays strike the end of the tube. It was found that the path of the rays could be deflected by electric and magnetic fields. The fact that the rays deflect means that they are charged, and they are repelled by negative and attracted to positive charge, indicating that the rays themselves carry a negative charge. The ratio of mass (measured in kilograms, kg) to charge (measured in coulombs, C)[13] for cathode rays was found to be -5.7×10^{-12} kg/C.[14]

The conclusion was that these negatively charged particles were a new subatomic (smaller than an atom) particle based upon the mass-to-charge ratio result and earlier results, which showed that cathode rays penetrated farther through various substances than do atoms.[15] This suggested that cathode rays were made up of negatively charged particles that are less massive than an atom.

The negatively charged particles that made up cathode rays are called electrons, the name for which had been proposed for the theoretical concept three years before their discovery.[16] With the discovery of electrons, several questions arise: What is the charge of the electron? What is the mass of the electron? What other subatomic particles exist? What is the inner structure of an atom?

Before continuing further with the discovery of atomic structure, it is worth making a note about some of the people cited in this chapter: Dalton (credited with developing modern atomic theory), Mendeleev (developer of the periodic law), J.J. Thomson (discoverer of the electron), Ernest Rutherford (discoverer of the nucleus and the proton), and James Chadwick (discoverer of the neutron). These individuals are honored for their contributions to the development of atomic science. While products of their time, and not necessarily above, reproach from a modern perspective, these individuals are acknowledged for their work to expand knowledge of the world. Also cited in this chapter are Philipp Lenard and Robert Millikan. Both conducted important work that earned them a Nobel Prize in Physics in 1905 (Lenard) and 1923 (Millikan). Despite their important work in the field of science, Lenard was a vocal supporter of the National Socialist German Workers' (Nazi) Party and an anti-Jewish "Deutsche Physik" (German Physics), and Millikan was an advocate of eugenics. These facts are addressed here as an important reminder that despite science's foundation on open inquiry and objective engagement with the world, prejudice has, does, and can negatively impact science, scientists, and society.

Problem 6.1. The mass-to-charge quotient of H^+ is 1.0×10^{-8} kg/C.

a. What is special about the element hydrogen?

b. Given your answer to part (a) and the mass-to-charge ratio magnitudes, what are the two possible inferences one could make about cathode rays?

Oil Drops

Since the mass-to-charge of the electron was known, determination of the charge or the mass would allow for determination of the other, unknown value. The first high-precision determination of the electron charge was reported in 1910 after studying the motion of ionized oil droplets in a mist of oil.[17] The oil droplets would fall, due to gravity, and rise, when an electric field was applied. Because the electric force experienced by the droplets is equal to the strength of the electric field multiplied by the charge, the value of the charge could be determined. It was found that the charge on the drops were multiples of one charge value 1.6021×10^{-19} C. This value is called the elementary charge. The electron charge is -1.6021×10^{-19} C.

With the value of the electron charge, the mass of the electron could be determined. The electron mass is 9.1094×10^{-31} kg (a relative mass of 0.000 55), which is smaller than the hydrogen ion by a factor of 1836. While this electron data answered some of the questions, chemists were left with other questions: What other subatomic particles exist? What is the structure of an atom? Given the small mass of the electron and negative charge, two facts were and are clear. First, the smallest atom is 1836 times larger than an electron and so most of an atom's mass must come from something other than electrons. Second, atoms are neutral and so another, positively charged particle must exist to balance the negatively charged electrons.

Alpha Particles and Hydrogen

After the discovery of electrons, an early atomic model hypothesized that the electrons are dispersed in a uniformly charged positive mass.[18] Given the resemblance to a plum pudding (Figure 6.3), this model is frequently referred to as the plum pudding model. To test this model, α particles (helium nuclei) were fired at metallic foil.[19] The expectation, if the plum pudding model were correct,

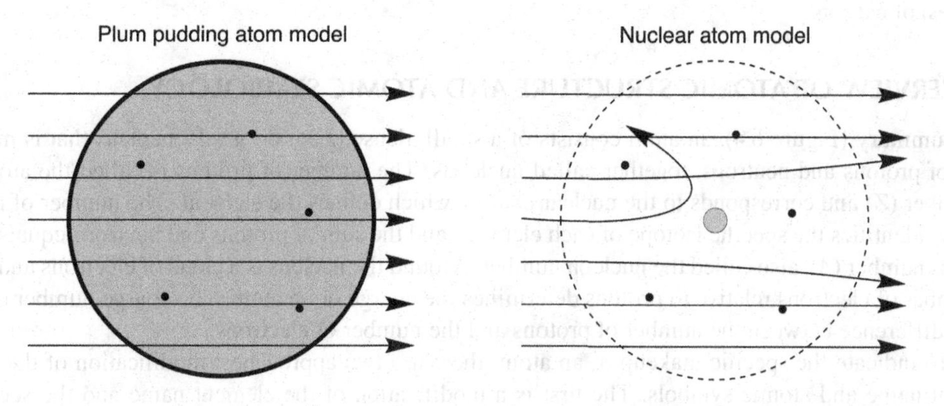

FIGURE 6.3 Plum pudding atom model (*left*) with electrons (*black*) dispersed in a positively charged mass (*light gray*). Nuclear atom model (*right*, not shown to scale) with electrons (*black*) dispersed around a small, positively charged nucleus (*light gray*). The arrows represent the path of α particles through the atom.

was that the positively charged α particles would move through the foil with minimal deflection (Figure 6.3).

While most particles did proceed through the foil without substantial deflection, some particles (roughly one in 8000) were deflected 90° or more by gold foil (Figure 6.3). To explain this result, a new atomic model was proposed: the nuclear atom. Here, the positive charge is concentrated in a very small nucleus that lies at the center of the atom. The nucleus constitutes most of the mass and the electrons are dispersed around the nucleus in a large cloud (Figure 6.3).[20,21]

The positively charged particle itself, the proton, was discovered through further experimentation with α particles. When α particles were fired into nitrogen gas, hydrogen-1 was produced.[22] This result first required repeated experimentation to ensure that the hydrogen-1 was not coming from other sources. After verifying no other source of hydrogen-1, the inference made was that the hydrogen-1 was dislodged from the nitrogen nucleus after colliding with an α particle. The hydrogen-1 nucleus is called a proton. The charge of a proton is equal and opposite that of an electron (1.6021×10^{-19} C). The proton resides in the nucleus, with a relative mass of 1.01. The number of protons is called the atomic number (*Z*) and corresponds to the nuclear charge. It is important to highlight that the number of protons determines the identity of an element. For example, any atom with one proton in the nucleus is hydrogen, any atom with two protons in the nucleus is helium, and so on.

Alpha Particles and "Unusually Penetrating Radiation"

Electrons and protons alone could not satisfactorily explain the structure of atoms. For example, helium has a relative atomic mass of 4.00 but an atomic number of 2. The nuclear charge number corresponds to two protons in the nucleus, but two protons would constitute a relative mass of only 2.02. In addition, the existence of isotopes, atoms with different relative atomic masses but the same atomic number, could also not be clearly explained with electrons and protons alone. The final subatomic particle, the neutron, was discovered when an unusually penetrating radiation resulted from α particles bombarding beryllium.[23] This unusually penetrating radiation was found to be a neutral particle, called a neutron. Neutrons are in the nucleus with protons, and so protons and neutrons are collectively called nucleons. The neutron has a relative mass of 1.01, and the number of neutrons identifies the specific isotope. For example, an atom with six protons in the nucleus is carbon. If the nucleus contains five neutrons that atoms is called carbon-11, 11 is the nucleon number (also called the mass number). If an atom contains six protons and six neutrons in its nucleus that atom is carbon-12. Six protons and seven neutrons constitute an atom of carbon-13, and carbon-14 contains six protons and eight neutrons. Carbon-11, carbon-12, carbon-13, and carbon-14 are isotopes (specific types) of carbon.

OVERVIEW OF ATOMIC STRUCTURE AND ATOMIC SYMBOLOGY

In summary (Figure 6.4), an atom consists of a small, dense (2×10^{14} g/mL) nucleus that is made up of protons and neutrons, together called nucleons. The number of protons is called the atomic number (*Z*) and corresponds to the nuclear charge, which defines the element. The number of neutrons identifies the specific isotope of each element, and the sum of protons and neutrons equals the mass number (*A*), also called the nucleon number. Around the nucleus is a cloud of electrons and the number of electrons relative to protons determines the charge of an atom. The charge number (*z*) is the difference between the number of protons and the number of electrons.

To indicate the specific makeup of an atom, there are two approaches: modification of the element name and atomic symbols. The first is a modification of the element name and the second is the annotation of atomic symbols. For example, if there were an element with nine protons, ten

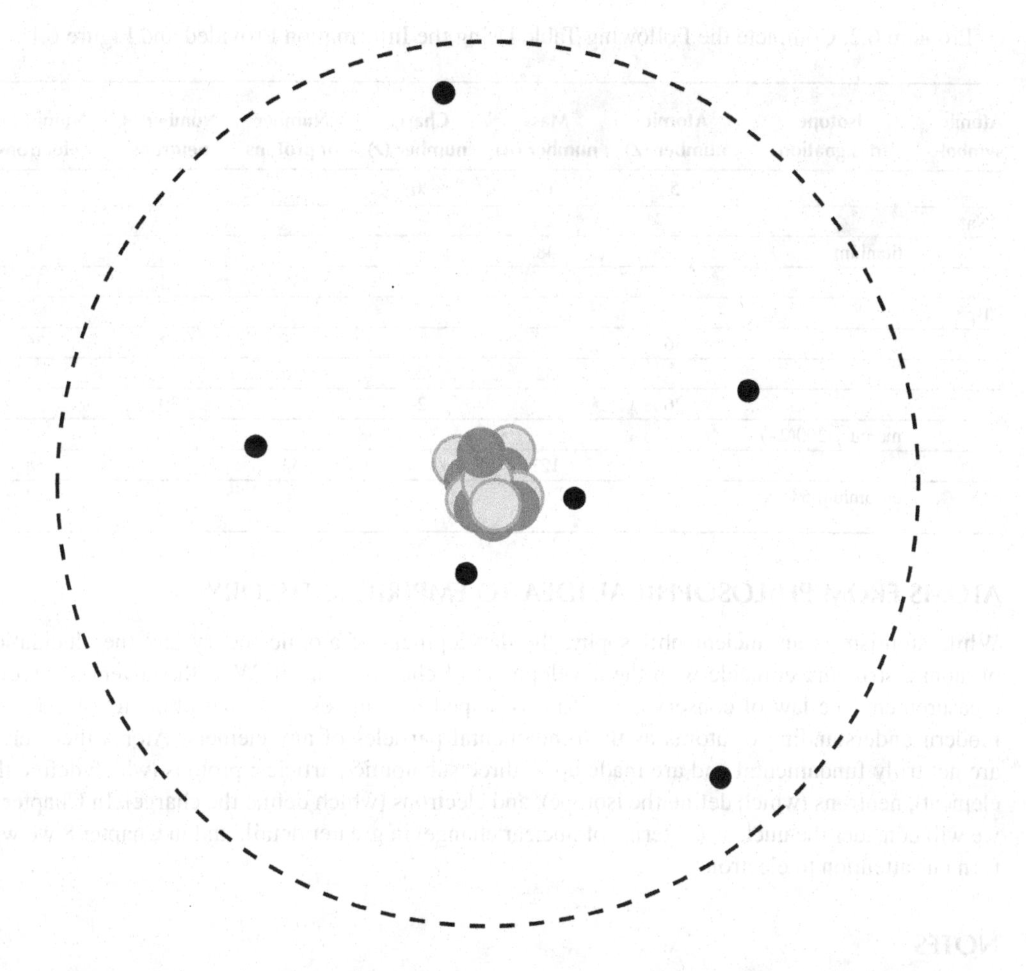

FIGURE 6.4 A cartoon representation of a carbon-12 atom showing a proton (*larger dark gray circles*) and neutron (*larger light gray circles*) in the nucleus and electrons (*small black circles*) in a surrounding cloud. The dashed circle represents the atomic size, circumference, but not to scale. If this drawing were to scale, the atomic radius should be 30 000 to 60 000 times larger than the nucleus.

neutrons, and ten electrons we would name that element fluoride-19(1–). Fluorine – here it is given the -ide suffix because it is a negatively charged ion (Chapter 10) – is the name of any atom with nine protons, and its mass number is 19. The charge number is 1– because there is one more electron than there are protons. We could also represent the specific isotope using the atomic symbol ${}^{19}_{9}F^{-}$. The general scheme for atomic symbols is:

$${}^{A}_{Z}E^{z}$$

E is the atomic symbol
Z is the atomic number, proton number
z is the charge number (the proton number less the electron number)
A is the mass number, nucleon number (the sum of protons and neutrons)

Note that the atomic number (*Z*) is often omitted from atomic symbols because it is redundant given a periodic table and the atomic symbol, and so ${}^{19}_{9}F^{-}$ is more commonly shown as just ${}^{19}F^{-}$.

Problem 6.2. Complete the Following Table Using the Information Provided and Figure 6.1

Atomic symbol	Isotope designation	Atomic number (Z)	Mass number (A)	Charge number (z)	Number of protons	Number of neutrons	Number of electrons
		5	10	0			
$^{23}Na^{+}$							
	titanium		48				18
					35	44	35
$^{79}Br^{-}$							
		36				47	36
			32	2–	16		
		26		2+		30	
	mercury-200(2+)						
			127		53		54
^{54}Cr	chromium-54						24

ATOMS FROM PHILOSOPHICAL IDEA TO EMPIRICAL THEORY

While atomism is an ancient philosophy, the development of atomic theory and the elucidation of atomic structure coincide with the development of chemistry itself. With the advent of precise measurement, the law of conservation of mass helped to spur research that ultimately led to our modern understanding of atoms as the fundamental particles of any element. Atoms themselves are not truly fundamental and are made up of three subatomic particles: protons (which define the element), neutrons (which define the isotope), and electrons (which define the charge). In Chapter 7, we will consider the nucleus (in terms of nuclear change) in greater detail, and in Chapter 8 we will turn our attention to electrons.

NOTES

1. Whitaker, R.D. An Historical Note on the Conservation of Mass. *J. Chem. Educ.*, **1975**, *52* (10), 658–659. DOI: 10.1021/ed052p658.
2. Lavoisier, A. *Elements of Chemistry, in a New Systematic Order, Containing All the Modern Discoveries*. Translated by Robert Kerr. William Creech: Edinburg, **1790**.
3. Proust, J. Recherches sur le Blue de Prusse. *Journal de Physique, de Chimie, d'Histoire Naturelle et des Arts*. **1794**, 334–341.
4. Dalton, J. *A New System of Chemical Philosophy, Part II*. S. Russell: Manchester, **1810**.
5. Döbereiner, J.W. Versuch zu einer Gruppirung der elementaren Stoffe nach ihrer Analogie. *Ann. Phys.*, **1829**, *91*, 301–307. DOI: 10.1002/andp.18290910217.
6. i) Newlands, J.A.R. On Relations Among the Equivalents. *Chemical News*, **1864**, *10*, 94–95.
   ii) Newlands, J.A.R. On the Law of Octaves. *Chemical News*, **1865**, *12*, 83.
7. Mendelejeff, D. Ueber die Beziehungen der Eigenschaften zu den Atomgewichten der Elemente. *Zeitschrift für Chemie*, **1869**, *12*, 405–406.
8. Mendeleev, D. *Osnovy Khimii* [The Principles of Chemistry]. St. Petersburg, **1869–71.**
9. i) Strutt, J.W.; Ramsay, W. VI. Argon, a New Constituent of the Atmosphere. *Philosophical Transactions of the Royal Society of London. (A.)*, **1895**, *186*, 187–241. DOI: 10.1098/rsta.1895.0006.
   ii) Ramsay, W. I. Helium, a Gaseous Constituent of Certain Minerals. Part I. *Proc. R. Soc. Lond.*, **1895**, *58* (347–352), 80–89. DOI: 10.1098/rspl.1895.0010.
   iii) Ramsay, W.; Travers, M.W. On the Companions of Argon. *Proc. R. Soc. Lond.*, **1898**, *63* (389–400), 437–440. DOI: 10.1098/rspl.1898.0057.
   iv) Ramsay, W.; Travers, M.W. On a New Constituent of Atmospheric Air. *Proc. R. Soc. Lond.*, **1898**, *63* (389–400), 405–408. DOI: 10.1098/rspl.1898.0051.

v) Ramsay, W.; Travers, M.W. On the Extraction from Air of the Companions of Argon and on Neon. *Report of the British Association for the Advancement of Science*, **1899**, 828–830. https://www.biodiversitylibrary.org/page/30221699 (Date Accessed: 04/15/23).

10. i) van den Broek, A. Das Mendelejeffsche „kubische“ periodische System der Elemente und die Einordnung der Radioelemente in dieses System. *Phys. Z.*, **1911**, *12*. 490–497.
   ii) van den Broek, A. Die Radioelemente, das periodische System und die Konstitution der Atome. *Phys. Z.*, **1913**, *14*, 32–41.
   iii) Moseley, H.G.J. XCIII. The High-frequency Spectra of the Elements. *Lond. Edinb. Dublin Philos. Mag. J. Sci. Series 6*, **1913**, *26* (156), 1024–1034. DOI: 10.1080/14786441308635052.
   iv) Moseley, H.G.J. LXXX. The High-frequency Spectra of the Elements. Part II. *Lond. Edinb. Dublin Philos. Mag. J. Sci. Series 6*, **1914**, *27* (160), 703–716. DOI: 10.1080/14786440408635141.
11. i) Seaborg, G.T. The Chemical and Radioactive Properties of the Heavy Elements. *Chem. Eng. News*, **1945**, *23* (23), 2190–2196. DOI: 10.1021/cen-v023n026.p2190.
   ii) Seaborg, G.T. The Transuranium Elements. *Science*, **1946**, *104* (2704), 379–386. DOI: 10.1126/science.104.2704.379.
12. Thomson, J.J. Cathode-rays. *Electrician*, **1897**, *39*, 104–109.
13. Thomson, J.J. XL. Cathode Rays. *Lond. Edinb. Dublin Philos. Mag. J. Sci. Series 5*, **1897**, *44* (269), 293–316. DOI: 10.1080/14786449708621070.
14. Committee on Data of the International Science Council (CODATA): Tiesinga, E.; Mohr, P.J.; Newell, D.B.; Taylor, B.N. CODATA Recommended Values of the Fundamental Physical Constants: 2018. *J. Phys. Chem. Ref. Data*. **2021**, *50* (3), 033105-1–033105-61. DOI: 10.1063/5.0064856.
15. Lenard, P. Ueber Kathodenstrahlen in Gasen von atmosphärischem Druck und im äussersten Vacuum. *Ann. Phys.*, **1894**, *287* (2), 225–267. DOI: 10.1002/andp.18942870202.
16. Stoney, G.J. XLIX. Of the “Electron,” or Atom of Electricity. *Lond. Edinb. Dublin Philos. Mag. J. Sci. Series 5*, **1894**, *38* (233), 418–420. DOI: 10.1080/14786449408620656.
17. Millikan, R.A. XXII. A New Modification of the Cloud Method of Determining the Elementary Electrical Charge and the Most Probable Value of That Charge. *Lond. Edinb. Dublin Philos. Mag. J. Sci. Series 6*, **1910**, *19* (110), 209–228. DOI: 10.1080/14786440208636795.
18. Thomson, J.J. XXIV. On the Structure of the Atom: An Investigation of the Stability and Periods of Oscillation of a Number of Corpuscles Arranged at Equal Intervals Around the Circumference of a Circle; with Application of the Results to the Theory of Atomic Structure. *Lond. Edinb. Dublin Philos. Mag. J. Sci. Series 6*, **1904**, *7* (39), 237–265. DOI: 10.1080/14786440409463107.
19. i) Geiger, H. On the Scattering of the Particles by Matter. *Proc. R. Soc. Land. A*, **1908**, *81*, 174–177. DOI: 10.1098/rspa.1908.0067.
   ii) Geiger, H.; Ernest, M. On a Diffusion Reflection of the Particles. *Proc. R. Soc. Land. A*, **1909**, *82*, 495–500. DOI: 10.1098/rspa.1909.0054.
   iii) Geiger, H. The Scattering of Particles by Matter. *Proc. R. Soc. Land. A*, **1910**, *83*, 492–504. DOI: 10.1098/rspa.1910.0038.
   iv) Geiger, H.; Marsden, E. LXI. The Laws of Deflexion of Particles Through Large Angles. *Lond. Edinb. Dublin Philos. Mag. J. Sci. Series 6*, **1913**, *25* (148), 604–626. DOI: 10.1080/14786440408634197.
20. Rutherford, E. LXXIX. The Scattering of and Particles by Matter and the Structure of the Atom. *Lond. Edinb. Dublin Philos. Mag. J. Sci. Series 6*, **1911**, *21* (125), 669–688. DOI: 10.1080/14786440508637080.
21. Nagaoka, H. LV. Kinetics of a System of Particles Illustrating the Line and the Band Spectrum and the Phenomena of Radioactivity. *Lond. Edinb. Dublin Philos. Mag. J. Sci. Series 6*, **1904**, *7* (41), 445–455. DOI: 10.1080/14786440409463141.
22. Rutherford, E. LIV. Collision of Particles with Light Atoms. IV. An Anomalous Effect in Nitrogen. *Lond. Edinb. Dublin Philos. Mag. J. Sci. Series 6*, **1919**, *37* (222), 581–587. DOI: 10.1080/14786440608635919.
23. i) Chadwick, J. Possible Existence of a Neutron. *Nature*, **1932**, *129*, 312. DOI: 10.1038/129312a0.
   ii) Chadwick, J. The Existence of a Neutron. *Proc. R. Soc. Lond. A*, **1932**, *136* (830), 692–708. DOI: 10.1098/rspa.1932.0112.

7 Nuclear Change

In Chapter 6, we discussed atomic theory and atomic structure. In this chapter, we will focus on the nucleus and nuclear change. According to the original formulation of atomic theory, atoms are persistent and unaltered by chemical reactions. The discovery of radioactivity required a revision of our understanding of atoms as we learned that atoms are not immutable and can transmute from one element into another element. Radioactivity also provided a new tool – in terms of α particles – that proved crucial in determining the structure of the nucleus (Chapter 6).

Radioactivity and radioactive decay are concepts with which you may be familiar, even if only from how they are portrayed in the media. When scientists talk about radioactivity, what we are discussing are isotopes that are unstable (radioactive). This is in contrast to isotopes that are stable (nonradioactive). When an isotope is radioactive, it will undergo radioactive decay.

Radioactive decay is when an atomic nucleus emits energy or particles to become a new, more stable nucleus. The emitted energy and particles from radioactive decay are invisible but can be seen in terms of how they affect other substances: photographic film will develop without light,[1] materials will weaken as chemical bonds are broken, and living organisms can become sick or develop mutations. There are three types of radioactive decay: α (alpha), β (beta), and γ (gamma). The first two types of decay, α and β, involve the emission of a particle from the nucleus along with the transmutation of the initial element into a new element.[2] The last type of decay, γ, involves the emission of high-energy electromagnetic radiation as an isotope goes from a higher energy state to a lower energy state. In γ decay, there is no transmutation of the isotope. In this chapter, we will consider all three types of radioactive decay before concluding with a discussion of decay chains.

α DECAY

We will start our discussion of radioactive decay by first considering α decay.[3] Let's first consider radon, a radioactive gas, for which homeowners must have their basements tested because the concentration can be dangerously high without proper ventilation. The isotope of radon with the longest half-life, the amount of time it takes a sample to decay by 50%, is radon-222. When radon-222 decays, it becomes polonium-218 (called a decay product). If we write an equation, with the information given, for this decay process, we have:

$$^{222}_{86}\text{Rn} \rightarrow {}^{218}_{84}\text{Po}$$

If we look at the mass numbers (212 ≠ 218) and the atomic numbers (86 ≠ 84), protons and neutrons are missing. Matter cannot be created or destroyed and so another particle must have been created. To balance the mass numbers, the missing particle must have a mass number of 4 (222 = 218 + **4**) and an atomic number of 2 (86 = 84 + **2**). An atomic number of two corresponds to helium, and a mass number of four means that a helium-4 particle is produced (specifically a helium-4(2+) particle):

$$^{222}_{86}\text{Rn} \rightarrow {}^{218}_{84}\text{Po} + {}^{4}_{2}\text{He}$$

In α decay, a helium-4(2+) particle, what we call an α particle, is always a product. This provides us with a tool to predict (provided we know a particular isotope undergoes α decay) what the decay product of a reaction is. For example, neptunium-237 undergoes α decay. To predict the decay product, we start by writing down the neptunium-237 starting material and the helium-4 α particle:

DOI: 10.1201/9781003487210-7

$$^{237}_{93}Np \rightarrow ^{4}_{2}He$$

To predict the decay product, we will consider the mass numbers and atomic numbers in turn. Currently, we have an unbalanced number of nucleons ($237 \neq 4$). To make this balanced, the decay product must have a mass number of 233 ($237 = 4 + 233$):

$$^{237}_{93}Np \rightarrow ^{4}_{2}He + ^{233}$$

Similarly, the atomic numbers show an unbalanced number of protons ($93 \neq 2$), and so the decay product must have an atomic number of 91 ($93 = 2 + 91$):

$$^{237}_{93}Np \rightarrow ^{4}_{2}He + ^{233}_{91}Pa$$

We then consult the periodic table to find that the element with atomic number 91 is protactinium (Pa), which allows us to complete the decay equation that shows that protactinium-233 is the decay product of neptunium-237.

$$^{237}_{93}Np \rightarrow ^{4}_{2}He + ^{233}_{91}Pa$$

It should be noted that because we are focused on the nuclei and not the whole atom, the charge number is often omitted. If we were to include the charge numbers, the balanced equation would be:

$$^{237}_{93}Np \rightarrow ^{4}_{2}He^{2+} + ^{233}_{91}Pa^{2-}$$

Radioactivity is dangerous to human health. Knowing that helium is produced by α decay, you may be surprised or perplexed, "wait, is helium dangerous?!" Helium-4 is not hazardous except that humans cannot breathe it, which presents a suffocation hazard. In contrast, helium-4(2+) is a hazardous chemical. It is important to note that the number of electrons an atom has (or in this case does not have) significantly changes its stability (and therefore its reactivity). The helium nucleus pulls strongly on electrons (Chapter 9) and will ionize (remove electrons) all chemicals it encounters to become a neutral atom. This makes α decay especially dangerous in living tissue.

Aside from the potential health hazard α decay represents, it is the only source of helium gas on Earth. That is, helium comes from the radioactive decay of heavy-metal ores in the Earth's crust. Despite its constant production, helium released into the atmosphere is lost to space due to its behavior in the atmosphere. As such, helium gas, as of 2024, is a byproduct of drilling for oil and natural gas, which makes helium a nonrenewable resource.

Now the question arises, why and how does α decay occur? More specifically, how can we predict whether α decay is going to occur? Isotopes above a certain size are highly destabilized by proton–proton repulsion in the nucleus. The largest isotope that is observationally stable has 82 protons and is lead-208. Bismuth-209 ($Z = 83$) and all isotopes with more than 82 protons undergo α decay. In these isotopes, the proton–proton repulsion provides enough destabilization to overcome the attractive nuclear force that holds nucleons together.

Problem 7.1. For each of the following, predict the α decay product isotopes or starting isotope.

a. $^{212}_{84}Po \rightarrow$

b. $^{239}_{94}Pu \rightarrow$

c. $\rightarrow ^{210}_{84}Po + ^{4}_{2}He$

d. $\rightarrow {}^{289}_{116}\text{Lv} + {}^{4}_{2}\text{He}$

e. ${}^{243}_{95}\text{Am} \rightarrow$

f. ${}^{252}_{101}\text{Md} \rightarrow$

g. $\rightarrow {}^{205}_{81}\text{Tl} + {}^{4}_{2}\text{He}$

h. $\rightarrow {}^{4}_{2}\text{He} + {}^{4}_{2}\text{He}$

β DECAY

We will now turn our attention to a second type of radioactive decay, β decay.[4,5] Only larger ($Z > 82$) nuclei tend to undergo α decay, but every element has isotopes that will undergo β decay, which occurs due to neutron–proton imbalances. There are two types of β decay: when an isotope is neutron-rich, it undergoes β^- decay and when an isotope is neutron-poor, it undergoes β^+ decay. Here, we will only consider the more common β^- decay.

Let's start by examining the simplest form of β^- decay, the decay of a neutron. Unlike free protons, free neutrons are unstable and have a half-life of only 848 seconds. When a neutron undergoes radioactive decay, a proton is the decay product:

$${}^{1}_{0}\text{n} \rightarrow {}^{1}_{1}\text{p}^{+}$$

As we saw in α decay, we have to consider the mass numbers and proton numbers. While there is no issue in terms of the mass numbers (1 = 1), there is an imbalance in terms of the charge ($0 \neq 1+$) and the atomic number ($0 \neq 1$). The missing particle would have a mass number of 0 and a negative charge, which corresponds to an electron being emitted.[6] The electron is assigned an atomic number of −1 to assist in balancing.

$${}^{1}_{0}\text{n} \rightarrow {}^{1}_{1}\text{p}^{+} + {}^{0}_{-1}\text{e}^{-}$$

As we saw for α decay, the charge number tends to be omitted in nuclear reactions. But the important takeaway about β^- decay is that an electron (a β^- particle) is always emitted (${}^{0}_{-1}\text{e}$) as a neutron transmutes into a proton, which results in a new isotope that has the same mass number (*A*) but an atomic number one larger than the initial isotope. To predict the outcome of a β^- decay, let's consider phosphorus-32. We start by writing down the starting isotope and the β^- particle (electron) product.

$${}^{32}_{15}\text{P} \rightarrow {}^{0}_{-1}\text{e}$$

To make this balanced, the decay product must have a mass number of 32 (32 = 0 + 32):

$${}^{32}_{15}\text{P} \rightarrow {}^{0}_{-1}\text{e} + {}^{32}$$

Similarly, the atomic numbers are not yet balanced ($15 \neq -1$), and so the decay product must have an atomic number of 16 (15 = −1 + 16):

$${}^{32}_{15}\text{P} \rightarrow {}^{0}_{-1}\text{e} + {}^{32}_{16}\text{S}$$

We would then consult the periodic table to find that element 16 is sulfur (S), which allows us to complete the decay equation and indicate that sulfur-32 is the β^- decay product of phosphorus-32.

$$^{32}_{15}P \rightarrow {}^{0}_{-1}e + {}^{32}_{16}S$$

As mentioned, β^- decay occurs when there is a neutron–proton imbalance and there are too many neutrons. There is no simple rule for predicting how many neutrons are too many. If asked to predict how an unstable nucleus with an atomic number less than 82 will decay, then β^- decay is the most likely decay pathway. In addition, if the mass number is substantially greater than the element's relative atomic mass ($A_r°$), that could be an indication that it is unstable. For example, carbon-14 is an unstable isotope of carbon. It has a mass number (14) that is almost two units greater than the relative atomic mass (12.011).

Problem 7.2. Predict the β^- decay product isotope or starting isotope.

a. $^{28}_{13}Al \rightarrow$

b. $^{98}_{43}Tc \rightarrow$

c. $\rightarrow {}^{3}_{2}He + {}^{0}_{-1}e$

d. $^{20}_{9}F \rightarrow$

e. $\rightarrow {}^{24}_{12}Mg + {}^{0}_{-1}e$

f. $\rightarrow {}^{206}_{82}Pb + {}^{0}_{-1}e$

g. $^{133}_{54}Xe \rightarrow$

h. $\rightarrow {}^{233}_{91}Pa + {}^{0}_{-1}e$

γ DECAY

Lastly, we will briefly consider γ decay.[7] Unlike α and β decay, there is no nuclear transmutation in γ decay. Instead, an excited state (indicated with an asterisk) nucleus emits γ rays, a form of electromagnetic radiation, as it relaxes to a ground state, for example:

$$^{237}Np^* \rightarrow {}^{237}Np + \gamma$$

After α and β decay, the resulting decay product nucleus can be left in an excited state. An excited state is where the nucleons are not in their lowest energy configuration. To return to ground state, γ rays are emitted. The topics of this chapter lie at the intersection of chemistry and physics, but γ decay is almost purely a physics phenomenon, and interested readers are encouraged to dig into the interesting world of nuclear physics.

ISOTOPE TRANSMUTATION SUMMARY AND DECAY CHAINS

The transmutation of isotopes through radioactive decay can be summarized graphically as shown in Figure 7.1.[8] We can use this graphical representation of decay to understand how an unstable isotope, like thorium-232, becomes a stable isotope (lead-208) through what is called a decay chain.

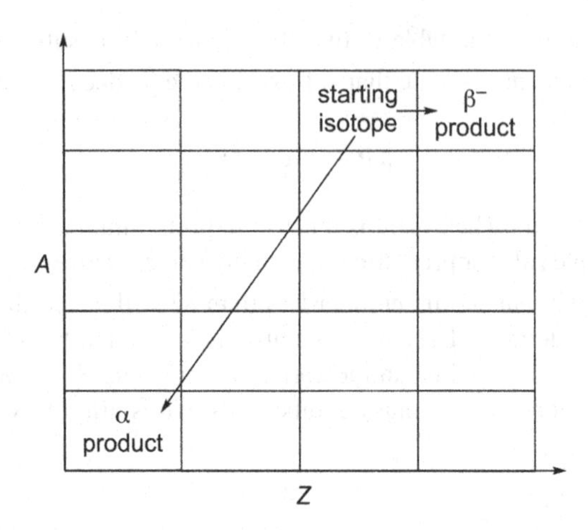

FIGURE 7.1 Isotope transmutation summary and the effect on mass number (*A*) and atomic number (*Z*).

Figure 7.2 shows the radioactive decay chain from thorium-232 to lead-208, which is often called the thorium series. Thorium-232 ($Z = 90$) is unstable and undergoes α decay to produce radium-228. But radium-228 is itself not stable. Not only are there more protons ($Z = 88$) than lead ($Z = 82$), which makes it likely to undergo α decay, but there is also a neutron–proton imbalance (and so it is also susceptible to β^- decay). In terms of order, β^- decay (correcting a neutron–proton imbalance) tends to occur before α decay (reducing proton–proton repulsion). And so, radium-228 undergoes two successive β^- decays to produce thorium-228, which then proceeds to undergo α decay. As you consider Figure 7.2, note that bismuth-212 can undergo both α decay (36% of the time) and β^- decay (64% of the time), which introduce a branch point into the decay chain. Ultimately, both branches after bismuth-212 arrive at lead-208, the stable decay product of the thorium decay chain. Decay chains, like those in Figure 7.2, represent the known pathways for naturally occurring radioisotopes to decay into stable isotopes (typically isotopes of lead).

Problem 7.3. Identify the decay products for each step of the uranium series. The uranium decay series starts with uranium-238 and undergoes the following decay steps to produce lead-206: α, β^-, β^-, α, α, α, α, α, β^-, β^-, α, β^-, β^-, α.

Aside from showing how unstable isotopes ultimately become stable, knowledge of decay chains provides us with a means of determining the age of rocks (radiometric dating). By comparing the ratio of lead-206 to uranium-238 (Problem 7.3) and knowing how long it takes this process to occur, we can determine the age of rocks that crystallized from one million years ago to over 4.5 billion years ago.

An important problem with radioactive materials, especially the waste from nuclear reactors, is that they present a storage problem. At its core, the problem is that not only is the original, purified material (^{235}U or ^{239}Pu) radioactive, but the byproducts are also radioactive (sometimes more dangerous than the original fuel). And so, the material must be stored such that human health is not impacted, and it must be safely stored for thousands of years. To understand the lifetime of radioactive materials, scientists measure the decay rates of isotopes and calculate the half-life. One half-life is the amount of time it takes for half of a radioactive sample to transmute, through either α decay or β^- decay, into another isotope.[9] After one half-life, 50% of the original isotope will have transmuted (into a new isotope). After two half-lives, 75% of the original isotope will have transmuted. After three half-lives, 87.5%, and 95.75% after four half-lives, and so forth. In the case of thorium-232, the half-life is 1.45×10^{10} years, so it takes 1.45×10^{11} years (10 half-lives, the amount recommended by nuclear safety guidelines) to decay to a safe level (Figure 4.12). The half-life of the uranium-238

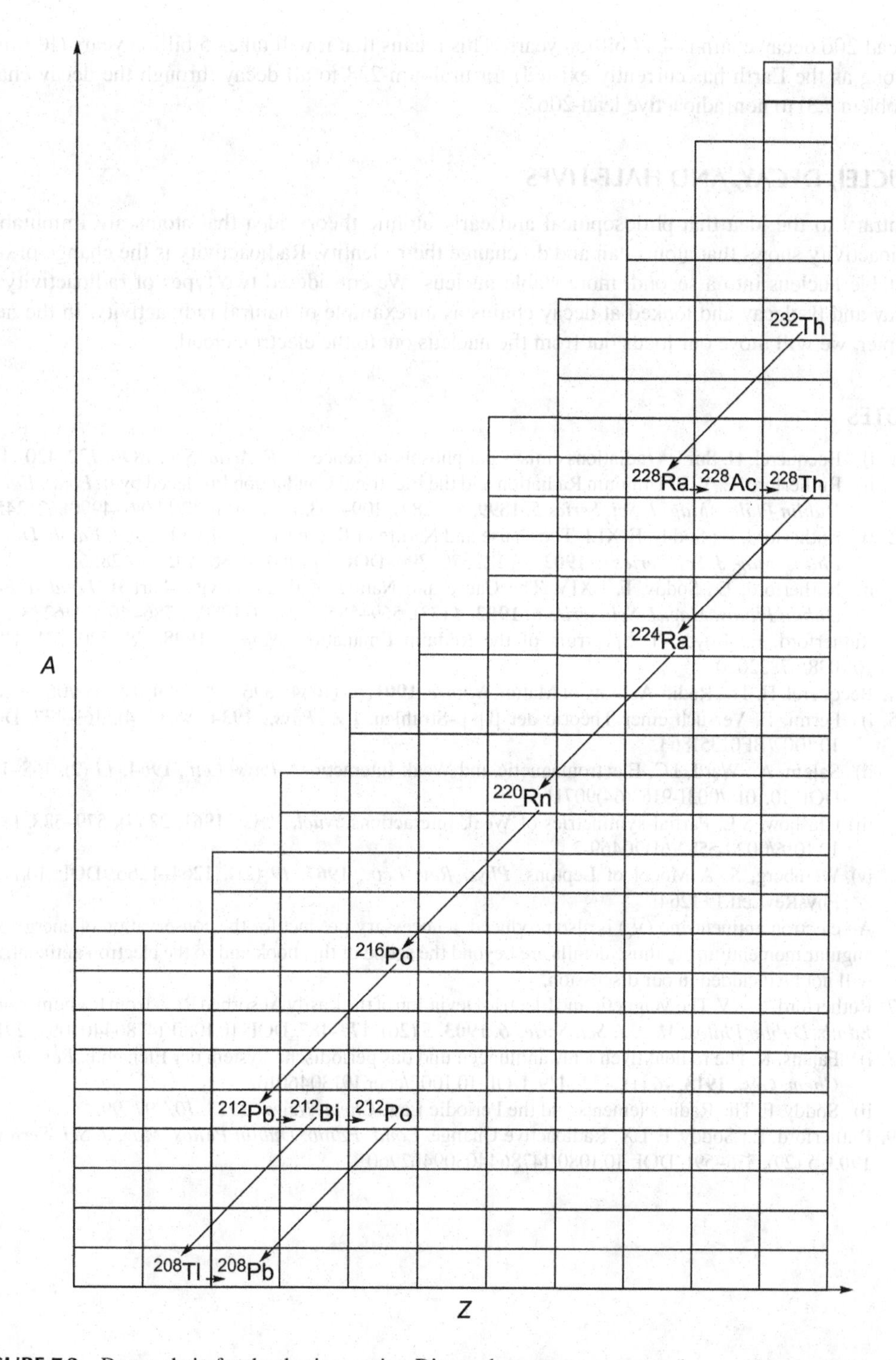

FIGURE 7.2 Decay chain for the thorium series. Diagonal arrows represent α decay and horizontal arrows represent β^- decay. The stable end isotope is lead-208. Empty cells represent other known isotopes, but they are omitted for clarity as they are not part of this decay series.

to lead-206 decay chain is 4.47 billion years. This means that it will take 45 billion years (10 times as long as the Earth has currently existed) for uranium-238 to all decay through the decay chain (Problem 7.3) to nonradioactive lead-206.

NUCLEI, DECAY, AND HALF-LIVES

Contrary to the idea that philosophical and early atomic theory idea that atoms are immutable, radioactivity shows that atoms can and do change their identity. Radioactivity is the change of one unstable nucleus into a second, more stable nucleus. We considered two types of radioactivity: α decay and β^- decay and looked at decay chains as an example of natural radioactivity. In the next chapter, we will move our focus out from the nucleus out to the electron cloud.

NOTES

1. i) Becquerel, H. Sur les radiations émises par phosphorescence. *C. R. Acad. Sci.*, **1896**, *122*, 420–421.
   ii) Rutherford, E. VIII. Uranium Radiation and the Electrical Conduction Produced by It. *Lond. Edinb. Dublin Philos. Mag. J. Sci. Series 5*, **1899**, *47* (284), 109–163. DOI: 10.1080/14786449908621245.
2. i) Rutherford, E.; Soddy, F. XLI. The Cause and Nature of Radioactivity.–Part I. *Lond. Edinb. Dublin Philos. Mag. J. Sci. Series 6*, **1902**, *4* (21), 370–396. DOI: 10.1080/14786440209462856.
   ii) Rutherford, E.; Soddy, F. LXIV. The Cause and Nature of Radioactivity.–Part II. *Lond. Edinb. Dublin Philos. Mag. J. Sci. Series 6*, **1902**, *4* (23), 569–585. DOI: 10.1080/14786440209462881.
3. Rutherford, E.; Royds, T. Spectrum of the Radium Emanation. *Nature*, **1908**, *78*, 220–221. DOI: 10.1038/078220c0.
4. Becquerel, H. The Radio-Activity of Matter. *Nature*, **1901**, *63* (1634), 396–398. DOI: 10.1038/063396d0.
5. i) Fermi, E. Versuch einer Theorie der β β β-Strahlen. I *Z. Phys.*, **1934**, *88* (3–4), 161–177. DOI: 10.1007/BF01351864.
   ii) Salam, A.; Ward, J.C. Electromagnetic and Weak Interactions. *Phys. Lett.*, **1964**, *13* (2), 168–171. DOI: 10.1016/0031-9163(64)90711-5.
   iii) Glashow, S.L. Partial-symmetries of Weak Interactions. *Nucl. Phys.*, **1961**, *22* (4), 579–588. DOI: 10.1016/0029-5582(61)90469-2.
   iv) Weinberg, S. A Model of Leptons. *Phys. Rev. Lett.*, **1967**, *19* (21), 1264–1266. DOI: 10.1103/PhysRevLett.19.1264.
6. An electron antineutrino ($\bar{\nu}_e$) is also produced, a necessary product for the conservation of energy and angular momentum, but those details are beyond the scope of this book and so the electron antineutrino will not be included in our discussion.
7. Rutherford, E. XV. The Magnetic and Electric Deviation of the Easily Absorbed Rays from Radium. *Lond. Edinb. Dublin Philos. Mag. J. Sci. Series 6*, **1903**, *5* (26), 177–187. DOI: 10.1080/14786440309462912.
8. i) Fajans, K. Die radioaktiven Umwandlungen und das periodische System der Elemente. *Ber. Dtsch. Chem. Ges.*, **1913**, *46* (1), 422–439. DOI: 10.1002/cber.19130460162.
   ii) Soddy, F. The Radio-elements and the Periodic Law. *Chem. News.*, **1913**, *107*, 97–99.
9. Rutherford, E.; Soddy, F. LX. Radioactive Change. *Lond. Edinb. Dublin Philos. Mag. J. Sci. Series 6*. **1903**, *5* (29), 576–591. DOI: 10.1080/14786440309462960.

8 Electron Configurations

While the nucleus is a small, self-contained, and extremely dense (2×10^{14} g/mL) package of protons and neutrons, electrons are dispersed throughout a cloud that is 30000 to 60000 times larger than the nucleus. In this chapter, we will discuss the arrangement and distribution of electrons in atoms.

WAVE–PARTICLE DUALITY

Before we start this chapter, we need to discuss the fact that electrons are not simple particles. Electrons are particles and they are also waves (this is referred to as wave–particle duality).[1] We do not need to focus on the particle–wave duality of electrons. The key idea we need to take from this is that electrons do not act like small moons orbiting the nucleus. Rather, they oscillate as complex and beautiful waves. If you pursue chemistry further, you may dig into the questions of "how" and "why." In this chapter, we will not dig into the underpinnings of electron configurations, which requires quantum mechanics; instead, we look to develop a sense for the description of and patterns in electron configurations.

ATOMIC EMISSION SPECTRA

When an atom is excited by the heat of a flame or with electricity, different color light is produced. If you have seen a fireworks display, you have seen the effect of exciting atoms to produce color. If the light produced by excited atoms is dispersed, with a prism, individual lines can be seen in what is called an atomic emission spectrum. In Figure 8.1, example atomic emission spectra are shown for hydrogen, helium, and lithium. Every element has its own unique pattern of lines in its atomic emission spectrum, like a fingerprint. As the periodic table expanded during the 19th century, atomic emission spectra were used to identify new elements.[2]

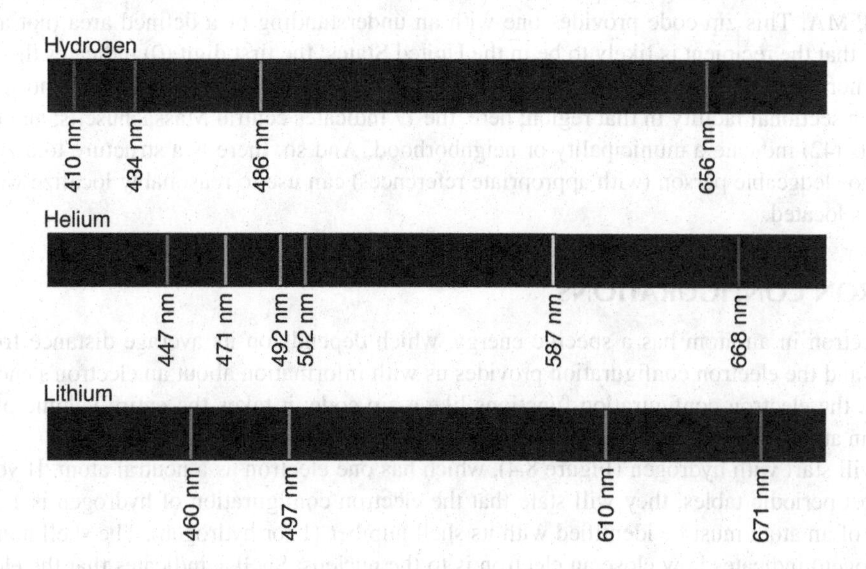

FIGURE 8.1 The atomic emission spectra of hydrogen (*top*), helium (*middle*), and lithium (*bottom*) showing the spectral lines in the visible range and their corresponding wavelengths.

DOI: 10.1201/9781003487210-8

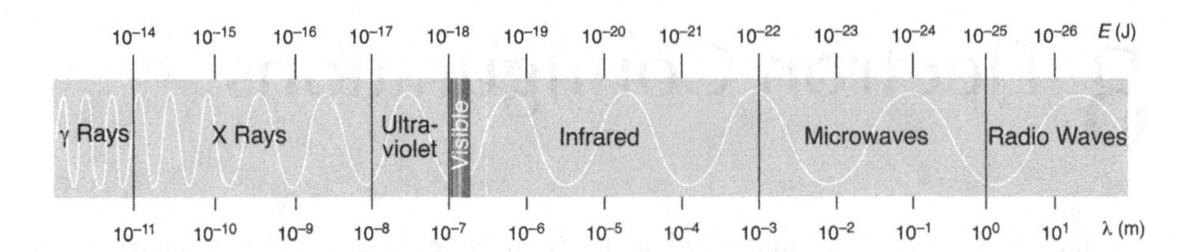

FIGURE 8.2 The electromagnetic spectrum showing the different regions – γ rays, X rays, ultraviolet (UV), visible, infrared (IR), microwaves, and radio waves – and their corresponding wavelength (λ) in meters (m) and photon energy (*E*) in joule (J). Figure is an original work based on that by Philip Ronan. https://commons.wikimedia.org/wiki/File:EM_spectrum.svg (Date accessed 22 May 2023).

Before we can fully understand what the lines in Figure 8.1 mean, we must first establish some basics about light. Visible light, the light that humans can perceive, is a very small subsection of the electromagnetic spectrum (Figure 8.2), which includes γ rays, X rays, ultraviolet light, infrared light, microwaves, and radio waves. Electromagnetic radiation has energy, which is carried in small packets called photons, and is inversely related to the wavelength of the electromagnetic radiation. Ultraviolet light, with wavelengths shorter than visible light, has more energy than blue light. The ultraviolet light in sunlight can cause skin (sunburn and melanoma) and eye (cataracts and photokeratitis) damage. In contrast, infrared light, with wavelengths longer than visible light, does not present a hazard to human health because the light has less energy than red light does.

Atomic emission spectra show specific discrete lines, which correspond to the energy associated with electrons changing energy levels. We can infer from the limited number of lines in the atomic emission spectra that there is only a limited number of energy changes that are possible because there is a limited number of energy levels an electron can have. In this chapter, our focus is on developing language to describe those energy levels.

ZIP CODES

To begin, let us first consider zip codes in the United States. Zip codes are a system that takes the entire area of the United States and breaks it down into a five-digit numerical code (Figure 8.3), which provides for a small geographic area where the residence is located. Let's consider 01742 for Concord, MA. This zip code provides one with an understanding of a defined area (not a single location) that the recipient is likely to be in the United States: the first digit (0) indicates the region, here the northeast, NJ, or Puerto Rico; the next two digits (17) indicate the county/counties served by a given sectional facility in that region; here, the 17 indicates central Massachusetts; and the last two digits (42) indicate a municipality or neighborhood. And so, there is a structure to a zip code that a knowledgeable person (with appropriate references) can use to reasonably localize where an address is located.

ELECTRON CONFIGURATIONS

Each electron in an atom has a specific energy, which depends on its average distance from the nucleus, and the electron configuration provides us with information about an electron's energy. In this way, the electron configuration functions like a zip code: it takes the entire volume of space around an atom and defines an average location, energy, for each electron.

We will start with hydrogen (Figure 8.4), which has one electron as a neutral atom. If you consider most periodic tables, they will state that the electron configuration of hydrogen is $1s^1$. Each electron of an atom must be identified with its shell number (1 for hydrogen). The shell number (1 for hydrogen) indicates how close an electron is to the nucleus. Shell 1 indicates that the electrons are closest to the nucleus and the lowest energy. For ground state atoms, shell numbers can be 1,

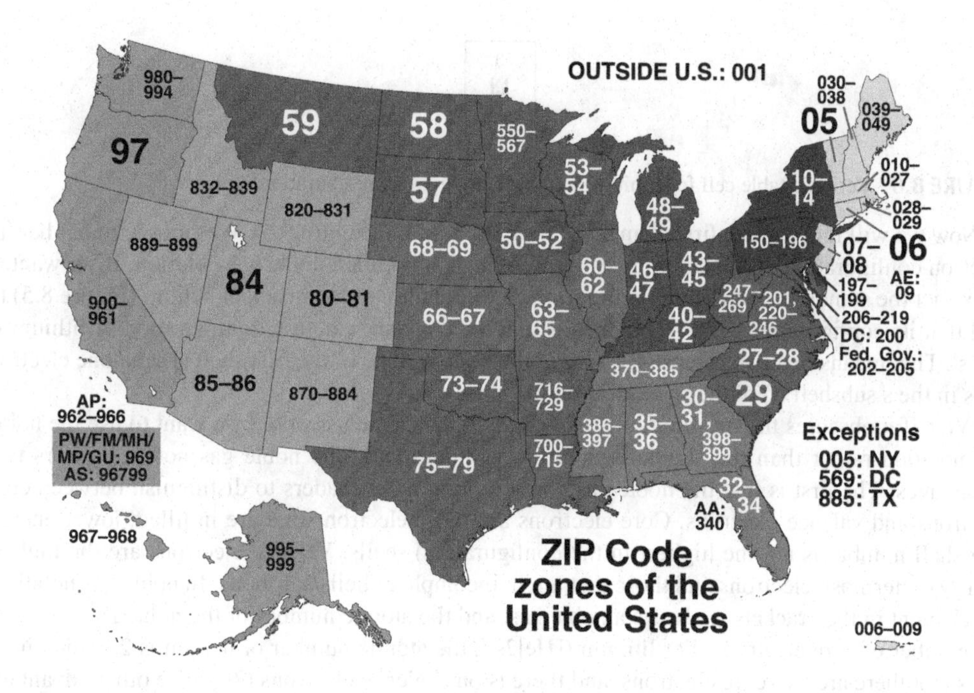

FIGURE 8.3 Zip code breakdown of the United States into regions/states. From "Zip Codes," Wikipedia.com https://en.wikipedia.org/wiki/ZIP_Code#/media/File:ZIP_Code_zones.svg released into public domain by user and creator: Denelson83 (Date accessed 15 November 2022).

1 H $1s^1$

FIGURE 8.4 Periodic table cell for hydrogen showing the electron configuration ($1s^1$).

2, 3, 4, 5, 6, and 7, with 1 being closest to the nucleus and lowest energy and 7 being farthest from the nucleus and highest energy. Note that these shell numbers correspond to the seven periods of the periodic table. Each electron of an atom must also be identified with its subshell identifier (s for hydrogen). The subshell identifier (s for hydrogen) indicates how an electron is vibrating within the shell. Since the waves have defined shapes, it can help to refine where electrons are (or are not) depending on the shape. The subshells are s, p, d, and f. Finally, the superscripted number after each subshell identifier indicates the number of electrons in that subshell. And so, the one electron (^{1}) of hydrogen is in the s subshell (s) of the first shell (1).

Moving onto helium on the periodic table (Figure 8.5), we can see that it has an electron configuration of $1s^2$. That is, its two electrons (^{2}) are both located in the s subshell (s) of the first shell (1). That completes the two elements of period 1. The highest shell number for hydrogen and helium is 1, and these highest energy electrons are called the valence electrons.

2 He $1s^2$

FIGURE 8.5 Periodic table cell for helium showing the electron configuration.

3
Li
[He]$2s^1$

FIGURE 8.6 Periodic table cell for lithium showing the electron configuration.

Now we will move to the first element of period 2 lithium (Figure 8.6). We can see that it has an electron configuration of [He]$2s^1$. This is written in what is called noble gas notation. If we wanted to extract the standard electron configuration, we would have to look back at helium (Figure 8.5) to find that its configuration is $1s^2$. This means that the standard electron configuration of lithium is $1s^22s^1$. That means that two electrons (^{2}) are in the s subshell (s) of the first shell (1) and one electron (^{1}) is in the s subshell (s) of the second shell (2).

Why, if it obscures the information about some of the elements, would we want to use the noble gas notation rather than the standard electron configuration? The noble gas notation offers two advantages. The first is that the noble gas configuration helps readers to distinguish between core electrons and valence electrons. Core electrons are those electrons that are in filled, lower energy (the shell number is not the highest in the configuration) shells. Valence electrons are the highest energy (outermost) electrons and those that are in incomplete shells/subshells. In noble gas notation, the element in the brackets is always a noble gas, and the atomic number of the noble gas indicates the number of core electrons. For lithium ([He]$2s^1$), the atomic number of helium is 2, which indicates that there are two core electrons, and there is one valence electrons ($2s^1$). The other advantage of the noble gas notation is that it can dramatically simplify writing out the electron configuration. Consider caesium whose configuration written in noble gas notation is [Xe]$6s^1$ and contrast the noble gas notation with the full configuration $1s^22s^22p^63s^23p^63d^{10}4s^24p^64d^{10}5s^25p^66s^1$. It is both significantly shorter to write [Xe]$6s^1$, and it is clearer that there is only one valence electron (and 54 core electrons) out of the 55 for caesium.

While the noble gas configuration is very useful for differentiating core and valence, in some instances there are electrons (in d and f subshells) that are not included in the noble gas notation that are also counted as core. The electrons in d and f subshells count as valence when the subshell is incomplete (the d subshell can hold 10 electrons and the f subshell can hold 14 electrons). If the configuration includes d^{10} or f^{14}, however, those electrons count as core and not towards the valence. For example, bromine, whose electron configuration is [Ar]$3d^{10}4s^24p^5$, has only seven valence electrons because the d subshell is full with 10 electrons. In contrast, iron has the configuration [Ar]$3d^64s^2$ and has eight valence electrons because the incomplete d subshell counts towards the total number of valence electrons.

Problem 8.1. Explain what each number and letter mean in the term: $3p^3$.

Problem 8.2. Let's consider phosphorus.

a. How many electrons does a neutral phosphorus atom have? Explain your answer.

b. Consider the expanded electron configuration for phosphorus: $1s^22s^22p^63s^23p^3$.
   i. How many electrons are in each shell?

   ii. For each shell, in what subshells are the electrons located? How many electrons in each subshell?

iii. Consider the noble gas notation electron configuration for phosphorus: $[Ne]3s^2 3p^3$
   1. Comparing the expanded electron configuration and the noble gas notation, what electrons are core electrons for phosphorus?

   2. Comparing the expanded electron configuration and the noble gas notation, what electrons are valence electrons for phosphorus?

Problem 8.3. How many valence electrons does each of the following have: N, P, As, Sb, and Bi? Do you notice a trend?

ELECTRON CONFIGURATIONS AND THE PERIODIC TABLE

The electron configurations of each element provide us with a means of identifying (by shell and subshell) the energy level, and, with quantum mechanics, a small region where each electron is likely located. The shell and subshell occupancies are reflected in the structure of the periodic table. Look at Figure 8.7, which shows the valence electron configuration for each element.[3]

1	2	3	4	5	6	7	8	9	10	11	12	13	14	15	16	17	18
1 H $1s^1$	2 He $1s^2$																
3 Li $2s^1$	4 Be $2s^2$											5 B $2s^2 2p^1$	6 C $2s^2 2p^2$	7 N $2s^2 2p^3$	8 O $2s^2 2p^4$	9 F $2s^2 2p^5$	10 Ne $2s^2 2p^6$
11 Na $3s^1$	12 Mg $3s^2$											13 Al $3s^2 3p^1$	14 Si $3s^2 3p^2$	15 P $3s^2 3p^3$	16 S $3s^2 3p^4$	17 Cl $3s^2 3p^5$	18 Ar $3s^2 3p^6$
19 K $4s^1$	20 Ca $4s^2$	21 Sc $3d^1 4s^2$	22 Ti $3d^2 4s^2$	23 V $3d^3 4s^2$	24 Cr $3d^5 4s^1$	25 Mn $3d^5 4s^2$	26 Fe $3d^6 4s^2$	27 Co $3d^7 4s^2$	28 Ni $3d^8 4s^2$	29 Cu $3d^{10} 4s^1$	30 Zn $3d^{10} 4s^2$	31 Ga $4s^2 4p^1$	32 Ge $4s^2 4p^2$	33 As $4s^2 4p^3$	34 Se $4s^2 4p^4$	35 Br $4s^2 4p^5$	36 Kr $4s^2 4p^6$
37 Rb $5s^1$	38 Sr $5s^2$	39 Y $4d^1 5s^2$	40 Zr $4d^2 5s^2$	41 Nb $4d^4 5s^1$	42 Mo $4d^5 5s^1$	43 Tc $4d^5 5s^2$	44 Ru $4d^7 5s^1$	45 Rh $4d^8 5s^1$	46 Pd $4d^{10}$	47 Ag $4d^{10} 5s^1$	48 Cd $4d^{10} 5s^2$	49 In $5s^2 5p^1$	50 Sn $5s^2 5p^2$	51 Sb $5s^2 5p^3$	52 Te $5s^2 5p^4$	53 I $5s^2 5p^5$	54 Xe $5s^2 5p^6$
55 Cs $6s^1$	56 Ba $6s^2$	*	72 Hf $5d^2 6s^2$	73 Ta $5d^3 6s^2$	74 W $5d^4 6s^2$	75 Re $5d^5 6s^2$	76 Os $5d^6 6s^2$	77 Ir $5d^7 6s^2$	78 Pt $5d^9 6s^1$	79 Au $5d^{10} 6s^1$	80 Hg $5d^{10} 6s^2$	81 Tl $6s^2 6p^1$	82 Pb $6s^2 6p^2$	83 Bi $6s^2 6p^3$	84 Po $6s^2 6p^4$	85 At $6s^2 6p^5$	86 Rn $6s^2 6p^6$
87 Fr $7s^1$	88 Ra $7s^2$	**	104 Rf $6d^2 7s^2$	105 Db $6d^3 7s^2$	106 Sg $6d^4 7s^2$	107 Bh $6d^5 7s^2$	108 Hs $6d^6 7s^2$	109 Mt $[6d^7 7s^2]$	110 Ds $[6d^8 7s^2]$	111 Rg $[6d^9 7s^2]$	112 Cn $[6d^{10} 7s^2]$	113 Nh $7s^2 7p^1$	114 Fl $[7s^2 7p^2]$	115 Mc $[7s^2 7p^3]$	116 Lv $[7s^2 7p^4]$	117 Ts $[7s^2 7p^5]$	118 Og $[7s^2 7p^6]$

*	57 La $5d^1 6s^2$	58 Ce $4f^1 5d^1 6s^2$	59 Pr $4f^3 6s^2$	60 Nd $4f^4 6s^2$	61 Pm $4f^5 6s^2$	62 Sm $4f^6 6s^2$	63 Eu $4f^7 6s^2$	64 Gd $4f^7 5d^1 6s^2$	65 Tb $4f^9 6s^2$	66 Dy $4f^{10} 6s^2$	67 Ho $4f^{11} 6s^2$	68 Er $4f^{12} 6s^2$	69 Tm $4f^{13} 6s^2$	70 Yb $4f^{14} 6s^2$	71 Lu $4f^{14} 5d^1 6s^2$
**	89 Ac $6d^1 7s^2$	90 Th $6d^2 7s^2$	91 Pa $5f^2 6d^1 7s^2$	92 U $5f^3 6d^1 7s^2$	93 Np $5f^4 6d^1 7s^2$	94 Pu $5f^6 7s^2$	95 Am $5f^7 7s^2$	96 Cm $5f^7 6d^1 7s^2$	97 Bk $5f^9 7s^2$	98 Cf $5f^{10} 7s^2$	99 Es $5f^{11} 7s^2$	100 Fm $5f^{12} 7s^2$	101 Md $5f^{13} 7s^2$	102 No $5f^{14} 7s^2$	103 Lr $5f^{14} 7s^2 7p^1$

FIGURE 8.7 Periodic table showing expected valence configuration for each element (note 1: core electrons are not noted; note 2: some elements vary slightly from the expected configuration due to subtle quantum mechanical effects). Appendix 4 shows the full electron configuration for each element.

If you look at Figure 8.7, you can see that elements in the same group have the same valence electron configurations, which is why elements in a group all show similar chemical properties. What should also become apparent is that the separate blocks of the periodic table are different based on where the valence electrons are. If you look at groups 1 and 2, the highest energy (valence) electrons are all in the s subshell, and so we call the group 1 and 2 elements the s block. In groups 13–18, the highest energy valence electrons are all in the p subshell and so we call those groups the p block. For groups 3–12, the highest energy valence electrons are all in the d subshell (the difference in shell number between s and d (or s and f) is due to a quantum mechanical effect that will not be considered here), and so groups 3–12 are called the d block. And finally, the rare earth metals are the f block because all the highest energy valence electrons are in the f subshell (Figure 8.8).

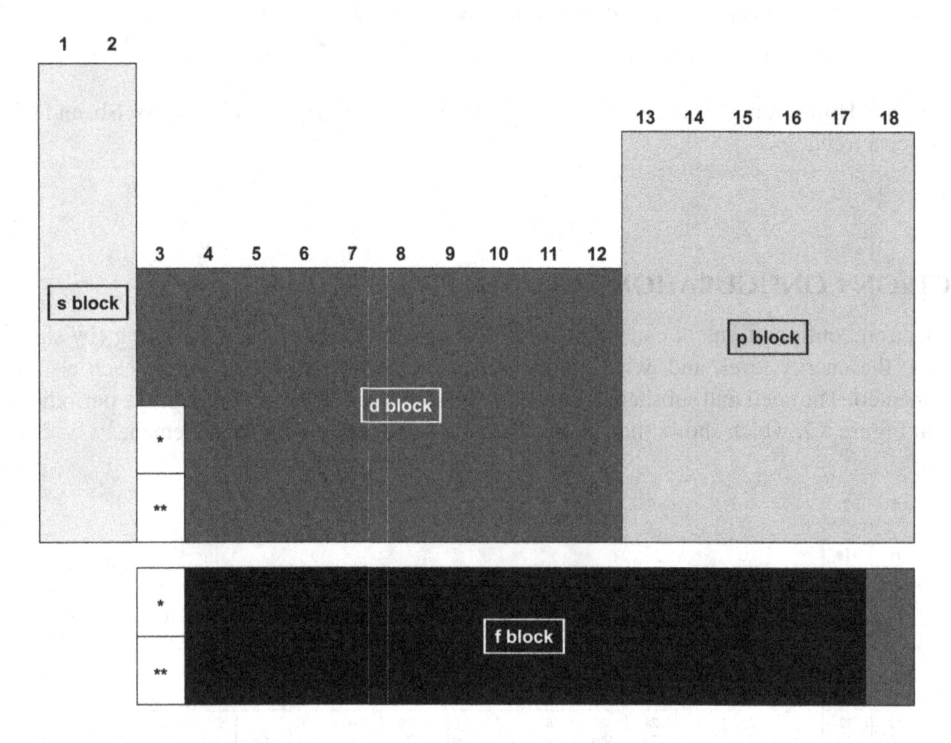

FIGURE 8.8 The different blocks of the periodic table, based on where the highest energy valence electrons are found.

The number of groups (columns) in each block is related to the number of electrons that can fit into each subshell. The s block contains only two groups because the s subshell can only hold two electrons. The p block contains six groups because the p subshell can hold six electrons. The d block contains ten groups because the d subshell can hold ten electrons. And the f block contains 14 groups because the f subshell can hold 14 electrons.

Problem 8.4. Using the periodic table, identify the element that each configuration corresponds to and the group that that element is in.

a. $1s^2 2s^2 2p^6$

b. $1s^2 2s^2 2p^4$

c. $[Ar]3d^{10} 4s^2 4p^5$

d. $[Ar]3d^2 4s^2$

e. $[Ne]3s^2$

f. $1s^2 2s^2 2p^6 3s^2 3p^6 3d^{10} 4s^2 4p^6 4d^{10} 5s^2 5p^6$

Problem 8.5. Identify the number of valence electrons in each element. Do you notice any trends?

H_____	He_____					
Li_____	Be_____	B_____	C_____	N_____	O_____	F_____
Na_____	Mg_____	Al_____	Si_____	P_____	S_____	Cl_____
K_____	Ca_____	Ga_____	Ge_____	As_____	Se_____	Br_____
Rb_____	Sr_____	In_____	Sn_____	Sb_____	Te_____	I_____
Cs_____	Ba_____	Tl_____	Pb_____	Bi_____	Po_____	At_____

Problem 8.6. Elements in the same group on the periodic table show similar chemical properties. Provide an explanation for this similarity in chemical properties (see Problem 8.5).

ELECTRON CONFIGURATION AND ELECTRONIC STABILITY

The last topic that we will consider in this chapter is the stability of electron configurations and how atoms will lose or gain electrons to achieve stability if they are not stable as neutral atoms. The only elements that have stable electron configurations are: He, Ne, Ar, Kr, Xe, Rn, and Og (the noble gases). This fact is why we use noble gases in the electron configuration shorthand. In general, however, noble gases are stable because they have their valence s and p subshells filled. Helium is unique in that it is stable and only has its s subshell filled (there is no p subshell in the first shell, this is for quantum mechanical reasons). Since it takes eight electrons (two for s and six for p) to fill the valence s and p subshells, this is often referred to as the octet rule.[4] It is important to be careful, however, as d-block and f-block elements can have eight or more valence electrons but are not stable. The eight electrons, implied by the term octet rule, must be in filled s and p subshells for an element to be stable.

All other, non-noble gas elements, therefore, have unstable electron configurations and they will lose or gain electrons to try to become more stable (lower in energy) and ideally achieve a noble gas configuration. In terms of losing or gaining electrons, atoms will proceed by whichever pathway is lower in energy. For example, an atom with three valence electrons (s^2p^1) will lose three electrons (lower energy) and not gain five electrons (higher energy).

If we consider the elements in group 1, we can see that they will tend to lose their one valence electron to achieve a noble gas configuration. After losing one, negatively charged, electron the elements become positively charged ions with a 1+ charge. Positively charged ions are called cations. For example, caesium atoms have $[Xe]6s^1$ configurations and will tend to lose the one valence electron to become Cs^+, which now has a configuration of [Xe].

In contrast, elements in group 17 (the halogens) all have one electron fewer than a noble gas, and so they will tend to gain an electron to achieve a noble gas configuration. After gaining one negatively charged electron, the elements become negatively charged ions with a 1– charge. Negatively charged ions are called anions. For example, fluorine has the configuration [He]$2s^22p^5$. When it gains an electron (to become F^-), the configuration of the fluoride anion is [He]$2s^22p^6$, which is the same as that for neon. With this information in hand, then, we can see how the preferred charges (Figure 8.9) that each element adopts when it loses or gains electrons comes about.

1	2	3	4	5	6	7	8	9	10	11	12	13	14	15	16	17	18
1 H 1+/1−																	2 He 0
3 Li 1+	4 Be 2+											5 B 3+	6 C variable	7 N 3−	8 O 2−	9 F 1−	10 Ne 0
11 Na 1+	12 Mg 2+											13 Al 3+	14 Si variable	15 P 3−	16 S 2−	17 Cl 1−	18 Ar 0
19 K 1+	20 Ca 2+	21 Sc 3+	22 Ti variable	23 V variable	24 Cr variable	25 Mn variable	26 Fe variable	27 Co variable	28 Ni variable	29 Cu variable	30 Zn 2+	31 Ga 3+	32 Ge variable	33 As variable	34 Se 2−	35 Br 1−	36 Kr 0
37 Rb 1+	38 Sr 2+	39 Y 3+	40 Zr 4+	41 Nb variable	42 Mo variable	43 Tc variable	44 Ru variable	45 Rh variable	46 Pd variable	47 Ag 1+	48 Cd 2+	49 In variable	50 Sn variable	51 Sb variable	52 Te variable	53 I 1−	54 Xe 0
55 Cs 1+	56 Ba 2+	*	72 Hf 4+	73 Ta variable	74 W variable	75 Re variable	76 Os variable	77 Ir variable	78 Pt variable	79 Au variable	80 Hg variable	81 Tl variable	82 Pb variable	83 Bi variable	84 Po variable	85 At variable	86 Rn 0
87 Fr 1+	88 Ra 2+	**	104 Rf unknown	105 Db unknown	106 Sg unknown	107 Bh unknown	108 Hs unknown	109 Mt unknown	110 Ds unknown	111 Rg unknown	112 Cn unknown	113 Nh unknown	114 Fl unknown	115 Mc unknown	116 Lv unknown	117 Ts unknown	118 Og unknown

*	57 La 3+	58 Ce 3+	59 Pr 3+	60 Nd 3+	61 Pm 3+	62 Sm 3+	63 Eu 3+	64 Gd 3+	65 Tb 3+	66 Dy 3+	67 Ho 3+	68 Er 3+	69 Tm 3+	70 Yb 3+	71 Lu 3+
**	89 Ac 3+	90 Th variable	91 Pa variable	92 U variable	93 Np variable	94 Pu variable	95 Am 3+	96 Cm 3+	97 Bk 3+	98 Cf 3+	99 Es 3+	100 Fm 3+	101 Md 3+	102 No variable	103 Lr 3+

FIGURE 8.9 Periodic table showing the common charge each element tends to take on in ionic compounds (if known). 0 means that monatomic ions of this element do not naturally form.

Problem 8.7. Why are the noble gases electronically stable? What rule do we use as a shorthand to explain their stability?

Problem 8.8. For the following electron configurations, is the configuration stable? If not, would the atom with this electronic configuration tend to lose electrons or gain electrons? Explain your answer.

a. $1s^22s^22p^6$

b. $1s^22s^22p^4$

c. [Ar]$3d^{10}4s^24p^5$

d. [Ar]$3d^24s^2$

e. $[Ne]3s^2$

f. $1s^22s^22p^63s^23p^63d^{10}4s^24p^64d^{10}5s^25p^6$

Problem 8.9. The common charges the elements take when they make ionic compounds are as follows: 1+ for group 1, 2+ for group 2, 3+ for group 13, 3– for group 15, 2– for group 16, and 1– for group 17. Explain why these groups of atoms take on these charges (make sure to consider their electron configurations).

Problem 8.10. Group 14 elements can readily take on charges between 4+ and 4–. In terms of their electron configurations, why would these elements show this ambivalent behavior?

SHELLS, SUBSHELLS, AND PERIODICITY

Electrons can occupy specific energy levels: shells and subshells. This is a quantum mechanical result, the details of which are beyond the scope of this book. What is important to note is that trends in chemical properties – ion(s) formed and whether an atom gains or loses electrons – is dictated by the number of highest energy (valence) electrons. Only the noble gases have a truly stable electron configuration, and other elements react in order to, ideally, achieve a noble gas configuration. The repeating patterns in shell and subshell occupancies also lead to patterns in periodic properties (Chapter 9) and the types of reactions elements engage in (Chapter 16).

NOTES

1. This is a publication of de Broglie's thesis in full: de Broglie, L. Recherches Sur la Théorie des Quanta. *Ann. Phys.* (Paris), **1925**, *10* (3), 22–128. DOI: 10.1051/anphys/192510030022.
2. The use of emission spectra to identify elements was first reported by Charles Wheatstone (Wheatstone, C. On the Prismatic Decomposition of Electrical Light. *Br. Assoc. Adv. Sci., Report*, **1835**, 11–12.) and Angstrom first demonstrated the spectra lines of gases (Angstrom, A.J. Otiska undersökningar. *Kungl. Svenska Vetenskapsakad. Handl.*, **1852**, *40*, 339–366.)
3. All reported configurations conform with those reported by NIST Atomic Spectra Database: Kramida, A.; Ralchenko, Y.; Reader, J.; NIST ASD Team. *NIST Atomic Spectra Database* (ver. 5.10) [Online]. Available: https://physics.nist.gov/asd. [2023, April 28]. National Institute of Standards and Technology, Gaithersburg, MD. DOI: https://doi.org/10.18434/T4W30F.
4. Langmuir, I. The Arrangement of Electrons in Atoms and Molecules. *J. Am. Chem. Soc.*, **1919**, *41* (6), 868–934. DOI: 10.1021/ja02227a002.

9 Periodic Trends

Electron configurations (Chapter 8) show repeating (periodic) patterns in the valence shell configuration of atoms. This periodicity in electron configurations – and their chemical properties – is why atoms are grouped together. In this chapter, we turn our attention to some of the specific properties of atoms: atomic radii (r), ionization energy (E_i), electron affinity (E_{ea}), and electronegativity (χ). We will use electron configurations and electrostatic attraction (the force between the positively charged nucleus and negatively charged electrons) to explain the trends in these properties.

ELECTROSTATIC ATTRACTION: COULOMB'S LAW

Before looking at any periodic trend, we must first establish a framework for considering the interactions of charged particles. If there are two positively charged protons near each other, they will repel each other. Similarly, if there are two negatively charged electrons near each other, they will repel each other. But a positively charged proton and a negatively charged electron will attract one another. The repulsion of like charges (proton–proton or electron–electron) and the attraction of opposite charges (proton–electron) are basic examples of what is called electrostatics. Mathematically, the electrostatic repulsion or attraction between charges is described by Coulomb's law (Equation 9.1).[1] The electrostatic force in units of newtons (N) between two charged particles is proportional to the product of their charges (Q_1 and Q_2) measured in units of coulombs (C). The electrostatic force is inversely proportional to the square of the distance (r^2) between the charged particles in units of meters (m). In Coulomb's law, the force is repulsive when $F_{\text{electrostatic}}$ is positive and is attractive when $F_{\text{electrostatic}}$ is negative.

$$F_{\text{electrostatic}} = k_e \frac{Q_1 Q_2}{r^2} \quad (9.1)$$

$F_{\text{electrostatic}}$ is the electrostatic force between two particles (N)
k_e is the Coulomb constant ($8.988 \times 10^9 \frac{\text{N m}^2}{\text{C}^2}$)
Q_1 is the charge of particle one (C)
Q_2 is the charge of particle two (C)
r is the interparticle distance (m)

Coulomb's law works for describing the behavior of two particles – one proton and one electron, for example – but in an atom, there can be many protons in the nucleus and many electrons in the electron cloud. We can apply a modified version of Coulomb's law (Equation 9.2) to model the attraction of one valence electron (one particle) to the nucleus (the second particle).

$$F_{\text{electrostatic}} = -k_e e^2 \frac{Z_{\text{eff}}}{r^2} \quad (9.2)$$

$F_{\text{electrostatic}}$ is the electrostatic force of attraction between the nucleus and a valence electron (N)
k_e is the Coulomb constant ($8.988 \times 10^9 \frac{\text{N m}^2}{\text{C}^2}$)
e is the elementary charge of an electron and a proton (1.602×10^{-19} C)
Z_{eff} is the effective charge of the nucleus (unitless)
r is the distance between the nucleus and an electron (m)

DOI: 10.1201/9781003487210-9

There are several things to notice about Equation 9.2. First, the attraction between a valence electron (negatively charged) and the nucleus (positively charged) is always attractive, and so the sign has been fixed as negative. A proton and electron both have the same charge magnitude (*e*) and so e^2 appears next to the Coulomb constant to account for the charges, though we must take the number of protons in the nucleus into account. The difference from one atom to the next is the number of protons in the nucleus (*Z*). Equation 9.2, however, has the effective nuclear charge (Z_{eff}) not the actual nuclear charge (*Z*). To explain why, let's consider hydrogen. For hydrogen, the nuclear charge that the valence electron feels is the full nuclear charge (1.00). But for lithium, however, the valence electron feels a nuclear charge of 1.26 (Z_{eff}) not the full 3.00 nuclear charge (*Z*) of the lithium nucleus. Why? Unlike in hydrogen (where there is nothing between the valence electron and the nucleus), in lithium the valence $2s^1$ electron is shielded (or screened) from the nucleus by the core $1s^2$ electrons. As an analogy, if the nucleus were performers at a concert, the $1s^2$ core electrons would be standing right in front of the performers (experiencing the full force of the concert). In contrast, the $2s^1$ valence electron would be standing behind the $1s^2$ electrons and so its view (and experience) would be blocked (shielded, screened) by the core $1s^2$ electrons. Standing behind the $1s^2$ electrons, the $2s^1$ only experiences part of the performance and so it only experiences part of the concert. If we keep looking down group 1, we can see that after lithium sodium has an effective nuclear charge of 1.84 (despite the nuclear charge being 11.00). This is because the valence $3s^1$ electron is now being shielded by a larger $1s^22s^22p^6$ core. And so, while *Z* increases substantially from lithium to sodium (3.00–11.00), Z_{eff} increases by a very small amount (1.26–1.84).

The effective nuclear charge (Z_{eff}) can be calculated as the nuclear charge less some empirical shielding constant. The effective nuclear charge can also be measured experimentally. Figure 9.1 shows the effective nuclear charge measured from X-ray photoelectron spectroscopy.[2]

Aside from the effective nuclear charge (Z_{eff}), the other important term of Equation 9.2 is the distance between a valence electron and the nucleus (*r*). The distance between an electron and the nucleus is dependent upon the shell number. As the shell number increases, so does the shell radius.

1	2	3	4	5	6	7	8	9	10	11	12	13	14	15	16	17	18
1 H 1.00																	2 He
3 Li 1.26	4 Be 1.66											5 B 1.56	6 C 1.82	7 N 2.07	8 O 2.00	9 F 2.26	10 Ne
11 Na 1.84	12 Mg 2.25											13 Al 1.99	14 Si 2.32	15 P 2.63	16 S 2.62	17 Cl 2.93	18 Ar
19 K 2.26	20 Ca 2.68	21 Sc 2.77	22 Ti 2.83	23 V 2.82	24 Cr 2.82	25 Mn 2.96	26 Fe 3.04	27 Co 3.04	28 Ni 3.00	29 Cu 3.01	30 Zn 3.32	31 Ga 2.66	32 Ge 3.05	33 As 3.40	34 Se 3.39	35 Br 3.73	36 Kr
37 Rb 2.77	38 Sr 3.23	39 Y 3.45	40 Zr 2.55	41 Nb 3.56	42 Mo 3.61	43 Tc 3.66	44 Ru 3.68	45 Rh 3.70	46 Pd 3.13	47 Ag 3.73	48 Cd 4.06	49 In 3.26	50 Sn 3.67	51 Sb 3.98	52 Te 4.07	53 I 4.38	54 Xe
55 Cs 3.21	56 Ba 3.71	*	72 Hf 4.46	73 Ta 4.57	74 W 4.60	75 Re 4.57	76 Os 4.74	77 Ir 4.91	78 Pt 4.88	79 Au 4.94	80 Hg 5.25	81 Tl 4.02	82 Pb 4.43	83 Bi 4.39	84 Po 4.72	85 At 4.96	86 Rn
87 Fr	88 Ra	**	104 Rf	105 Db	106 Sg	107 Bh	108 Hs	109 Mt	110 Ds	111 Rg	112 Cn	113 Nh	114 Fl	115 Mc	116 Lv	117 Ts	118 Og

*	57 La 3.84	58 Ce 3.87	59 Pr 3.79	60 Nd 3.81	61 Pm 3.83	62 Sm 3.86	63 Eu 3.88	64 Gd 4.04	65 Tb 3.93	66 Dy 3.96	67 Ho 3.99	68 Er 4.02	69 Tm 4.04	70 Yb 4.46	71 Lu
**	89 Ac	90 Th 4.65	91 Pa	92 U 4.65	93 Np	94 Pu	95 Am	96 Cm	97 Bk	98 Cf	99 Es	100 Fm	101 Md	102 No	103 Lr

FIGURE 9.1 Effective nuclear charge (Z_{eff}) of each element from X-ray spectroscopic data.

TABLE 9.1
Shell Number and Shell Radius (pm) for a Rubidium Atom Calculated Using the Electron Localization Function (ELF)

Shell number	Shell radius in a rubidium atom (pm)
1	3.38
2	13.60
3	50.27
4	196.85

And so, a valence electron that is in shell 1 is much closer (r is much smaller) than an electron in shell 2. Table 9.1 shows the radius in picometers (1 pm is 1×10^{-12} m) for shells 1, 2, 3, and 4 of rubidium. [3]

Problem 9.1. Looking at Equation 9.2, what will make the magnitude of the attractive force ($F_{electrostatic}$) between a valence electron and the nucleus stronger (in terms of Z_{eff} and r)?

Problem 9.2. Looking at Equation 9.2, what will make the magnitude of the attractive force ($F_{electrostatic}$) between a valence electron and the nucleus weaker (in terms of Z_{eff} and r)?

Problem 9.3. Looking at a period, sodium to chlorine, what is changing in the atoms to cause Z_{eff} to increase?

Problem 9.4. Looking at a group, why does Z_{eff} for silicon (2.32) show only a 27% increase compared to carbon (1.82) while Z shows a 133% increase?

Problem 9.5. What happens to the magnitude of the attractive force ($F_{electrostatic}$) between the nucleus and a valence electron as the shell number increases (Table 9.1)?

Problem 9.6. Thinking about valence electron shell numbers, what would you predict for the trend in atomic size for an atom in period 1 compared to an atom in period 2 compared to an atom in period 3? Explain.

ATOMIC RADIUS (*R*)

Atoms can be treated as nanoscopic spheres, for which there is a measurable atomic radius and atomic volume.[4] There are two common approaches for determining the radius of an atom: the covalent radius (r_{cov}) and the van der Waals radius (r_{vdW}). The covalent atomic radius represents the atomic radius when an atom is bonded to another atom.[5] The van der Waals radius represents

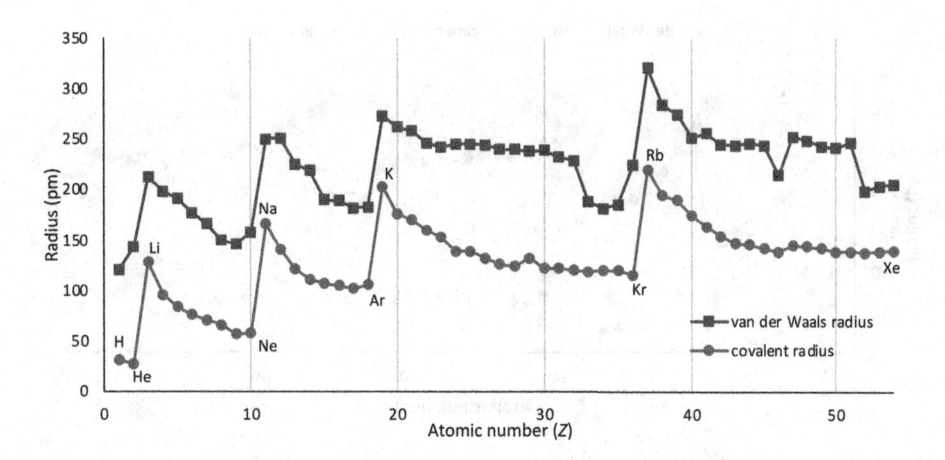

FIGURE 9.2 The covalent (r_{cov}) and van der Waals (r_{vdW}) atomic radii of elements 1–54.

the size of a free atom and is the balance point between intermolecular attraction (Chapter 13) and repulsion between two atoms. In other words, the van der Waals radius is the closest another atom can approach.[6] Figure 9.2 shows the covalent and van der Waals radii of elements 1–54.

One of the first things to notice in Figure 9.2 is the difference between the two radii. The van der Waals radius (r_{vdW}(H) = 120 pm) is significantly larger than the covalent radius (r_{cov}(H) = 31 pm) for the same atom. This is because for atoms to bond, they must get close enough for their electron clouds to overlap (Chapter 11). The second thing to notice in Figure 9.2 is a regular (periodic) pattern in atomic size, particularly with the covalent radii. There is an increase in atomic size within a group as one moves down a group from period 1 (r_{cov}(He) = 28 pm) to period 2 (r_{cov}(Ne) = 58 pm) to period 3 (r_{cov}(Ar) = 106 pm) and so on. There is a decrease in atomic size across a period as one moves across a period from group 1 (r_{cov}(Li) = 128 pm) to group 18 (r_{cov}(Ne) = 58 pm).

Problem 9.7. Consider the elements in group 18 (Figure 9.2). Use the electron configuration of each element and Equation 9.2 to provide an explanation for the trend r_{cov}(He) < r_{cov}(Ne) < r_{cov}(Ar) < r_{cov}(Kr) < r_{cov}(Xe).

Problem 9.8. Consider the elements in period 3, Na to Ar (Figure 9.2). Using the electron configuration of each element and Equation 9.2, provide an explanation for the general decrease in covalent radius from sodium to argon.

Figure 9.2 only includes the radii of neutral atoms. The radius of an atom can and does change upon becoming an ion (Figure 9.3).[7] When a neutral atom becomes a cation, the radius of an atom (r_{vdW}(Li) = 212 pm) decreases significantly when it loses electrons to become a cation (r_{cation}(Li^+) = 74 pm). In contrast, there is no significant change when a neutral atom (r_{vdW}(Cl) = 182 pm) gains electrons to become an anion (r_{anion}(Cl^-) = 181 pm). Consider Figure 9.3 and answer the questions that follow.

Problem 9.9. Provide an explanation for why cations are smaller than their corresponding neutral atoms. Think about how the electron configuration changes, for example, when sodium becomes sodium(1+).

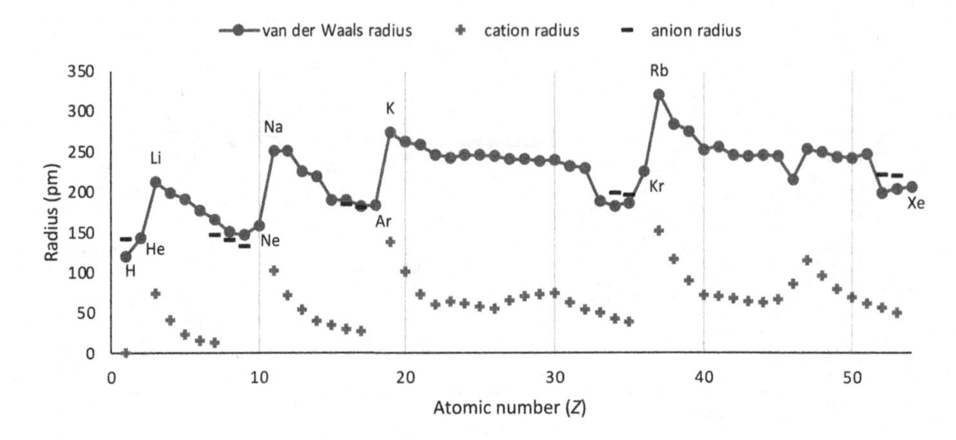

FIGURE 9.3 Atomic (van der Waals) radii of elements 1–54 and their cationic and anionic radii.

Problem 9.10. Provide an explanation for why anions are similar in size to their corresponding neutral atoms. Think about how the electron configuration changes, for example, when chlorine becomes chloride (1–).

Problem 9.11. Order the following ions in order of radius size (from smallest to largest): F^-, Na^+, O^{2-}, Mg^{2+}, N^{3-}, and Al^{3+}. What is the determining factor (Z_{eff} or shell number) in this order?

ENERGY CHANGE ASSOCIATED WITH FORMING IONS

Ionization Energy (E_I)

When an atom ionizes (forms a cation), there is a significant change in atomic size (Figure 9.3). Ionizing an atom does not happen spontaneously and requires energy to be added (an endothermic process). Ionization energy (E_i) is the term for the energy needed to overcome the electrostatic attraction ($F_{electrostatic}$) and remove one electron from an atom to form a cation (Equation 9.4). The first ionization energy of elements 1–54 is shown in Figure 9.4.[8]

$$X + E_i \rightarrow X^+ + 1\, e^- \tag{9.4}$$

For an atom with multiple electrons, successive ionization energy values can be measured where the first ionization energy is the energy it takes to form a cation from a neutral atom. The second ionization energy is the energy it takes to form a dication from a cation, and so forth.

Electron Affinity (E_{EA})

When an electron is added to an atom, energy is released, which, as an exothermic process (Chapter 5), would be reported as a negative value. Electron affinity (E_{ea}) is the term for the energy released when an electron is added to an atom to form an anion (Equation 9.5).

$$X + 1\, e^- \rightarrow X^- + E_{ea} \tag{9.5}$$

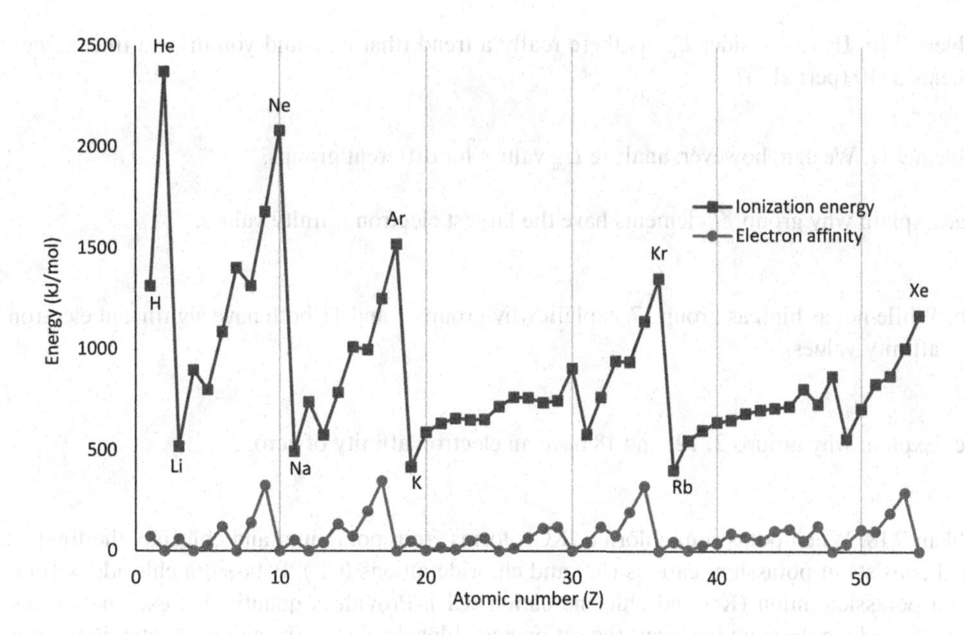

FIGURE 9.4 First ionization energy (E_i) and electron affinity (E_{ea}) for elements 1–54.

To measure electron affinity, it is easier to measure the amount of energy that must be added (endothermic process) to remove an electron from an anion (Equation 9.6). Therefore, the values for electron affinity are reported as positive values. While the values are positive, a better sense of electron affinity comes from thinking of it in the context of Equation 9.5, where electron affinity measures how much more stable the atom becomes when an electron is added.

$$X^- + E_{ea} \rightarrow X + 1\ e^- \tag{9.6}$$

Considering Figure 9.4, which shows both the ionization energy and electron affinity for elements 1–54. A larger ionization energy means that it is harder to remove an electron from the atom. A larger electron affinity means that the atom is more stable when an electron is added. An atom with an electron affinity of zero means that the anion was unstable and so the energy required to remove the electron could not be measured.

Problem 9.12. Notice that the E_i(Li) < E_i(Be). Provide an explanation for this.

Problem 9.13. Provide an explanation for the trend in ionization: E_i(He) > E_i(Ne) > E_i(Ar) > E_i(Kr) > E_i(Xe).

Problem 9.14. Moving from sodium to argon we cover the entire third period of elements. Can you make a generalization for the overall trend for E_i within a period?

Problem 9.15. Does the generalized trend you proposed in Problem 9.14 hold up for the other periods in Figure 9.4?

Problem 9.16. If we consider E_{ea}, is there really a trend (that is, could you draw a linear line) for elements 3–10 (period 2)?

Problem 9.17. We can, however, analyze E_{ea} values for different groups.

a. Explain why group 17 elements have the largest electron affinity values.

b. While not as high as group 17, explain why groups 1 and 11 both have significant electron affinity values.

c. Explain why groups 2, 12, and 18 have an electron affinity of zero.

Problem 9.18. When potassium chloride (KCl) forms from potassium and chlorine, the final compound consists of potassium cations (K^+) and chloride anions (Cl^-). Potassium chloride will *never* form a potasside anion (K^-) and chlorine cation (Cl^+). Provide a quantitative explanation (using Figure 9.4) why potassium is always the cation and chloride always the anion in potassium chloride in terms of:

a. The ionization energy values (E_i) of potassium and chlorine.

b. The electron affinity values (E_{ea}) of potassium and chlorine.

c. The overall energy of forming K^+ and Cl^- versus K^- and Cl^+ (a and b).

ELECTRONEGATIVITY (χ)

Understanding the factors that affect an atom's ionization energy and electron affinity is useful for understanding the properties of an atom in a vacuum. As we move forward in our study of chemistry, however, we tend to focus more on atoms within molecules. In compounds, electronegativity (χ) is a more useful concept when talking about atoms because it is the power of an atom to attract electrons to itself.[9] Linus Pauling first proposed the concept of electronegativity, and the Pauling electronegativity scale (χ_P) remains one of the most widely used (Figure 9.5).[10]

To determine the Pauling electronegativity (χ_P) values (Figure 9.5), the strength of a bond between two atoms (H–F, for example) is measured. This bond strength is then compared to the average strength of a bond between each atom with itself, H–H and F–F. The change in bond strength for the heteroatomic (here HF) bond compared to the average homoatomic (HH and FF) bonds is ascribed to a difference in how much one atom pulls on the electrons of the other, that is, to the electronegativity difference between two atoms. To set the scale, hydrogen was arbitrarily set to 2.20. Electronegativity values greater than that of hydrogen (most nonmetals) were chosen for elements that have a greater tendency to pull electrons towards themselves than hydrogen does. Electronegativity values less than hydrogen (most metals and metalloids) indicate that those elements have a weaker tendency to pull electrons towards themselves than hydrogen does.

1	2	3	4	5	6	7	8	9	10	11	12	13	14	15	16	17	18
1 H 2.20																	2 He
3 Li 0.98	4 Be 1.57											5 B 2.04	6 C 2.55	7 N 3.04	8 O 3.44	9 F 3.98	10 Ne
11 Na 0.93	12 Mg 1.31											13 Al 1.61	14 Si 1.90	15 P 2.19	16 S 2.58	17 Cl 3.16	18 Ar
19 K 0.82	20 Ca 1.00	21 Sc 1.36	22 Ti 1.54	23 V 1.63	24 Cr 1.66	25 Mn 1.55	26 Fe 1.83	27 Co 1.88	28 Ni 1.91	29 Cu 1.90	30 Zn 1.65	31 Ga 1.81	32 Ge 2.01	33 As 2.18	34 Se 2.55	35 Br 2.96	36 Kr 3.0
37 Rb 0.82	38 Sr 0.95	39 Y 1.22	40 Zr 1.33	41 Nb 1.6	42 Mo 2.16	43 Tc 1.9	44 Ru 2.2	45 Rh 2.28	46 Pd 2.20	47 Ag 1.93	48 Cd 1.69	49 In 1.78	50 Sn 1.96	51 Sb 2.05	52 Te 2.30	53 I 2.66	54 Xe 2.6
55 Cs 0.79	56 Ba 0.89	*	72 Hf 1.3	73 Ta 1.5	74 W 2.36	75 Re 1.9	76 Os 2.0	77 Ir 2.20	78 Pt 2.28	79 Au 2.54	80 Hg 2.00	81 Tl 2.04	82 Pb 2.33	83 Bi 2.02	84 Po 2.0	85 At 2.2	86 Rn
87 Fr 0.7	88 Ra 0.9	**	104 Rf	105 Db	106 Sg	107 Bh	108 Hs	109 Mt	110 Ds	111 Rg	112 Cn	113 Nh	114 Fl	115 Mc	116 Lv	117 Ts	118 Og

*	57 La 1.1	58 Ce 1.12	59 Pr 1.13	60 Nd 1.14	61 Pm 1.2	62 Sm 1.17	63 Eu 1.1	64 Gd 1.20	65 Tb 1.2	66 Dy 1.22	67 Ho 1.23	68 Er 1.24	69 Tm 1.25	70 Yb 1.1	71 Lu 1.27
**	89 Ac 1.1	90 Th 1.0	91 Pa 1.5	92 U 1.38	93 Np 1.36	94 Pu 1.28	95 Am 1.30	96 Cm	97 Bk	98 Cf	99 Es	100 Fm	101 Md	102 No	103 Lr

FIGURE 9.5 Pauling electronegativity (χ_P) values.

It is not immediately clear from the formal definition of the Pauling scale how electronegativity relates to periodic trends, the focus of this chapter. To make this connection, let us consider a second electronegativity scale. The Allred–Rochow electronegativity (χ_{AR}) scale defines electronegativity as the attraction between the nucleus and an electron on the surface of an atom (a valence electron).[11] This is expressed mathematically by Equation 9.5. If you compare Equation 9.5 with Equation 9.2, you can see that the definition of the Allred–Rochow electronegativity is the modified form of Coulomb's law with the Coulomb constant and the elementary charge terms removed.

$$\chi_{AR} = \frac{Z_{eff}}{r^2} \tag{9.5}$$

Z_{eff} is the effective nuclear charge
r is the covalent radius of the atom (pm)

Given the familiarity and widespread use of Pauling's scale, the χ_{AR} values are linearly transformed to be comparable to Pauling's values. Allred–Rochow values calculated from Z_{eff} (Figure 9.1) and r_{cov} (Figure 9.2) are shown in Figure 9.6.

Now, instead of switching between Figure 9.5 and Figure 9.6 to compare the χ_P and χ_{AR} values, we can plot the two data sets together (Figure 9.7). The first point to notice is that the χ_P and χ_{AR} values are almost identical. Starting from two very different points for χ_P, bond strengths, and χ_{AR}, nucleus–valence electron attraction, ends with the same result. This demonstrates that electronegativity is an intrinsic property that can be derived from different properties of atoms, whether in molecules or free. The second point to notice is that electronegativity relates to electrostatic attraction (one of the bases of periodic trends) and relates to bonding. That is, electronegativity relates to and can help explain a variety of trends, as you shall see throughout your study of chemistry.

1	2	3	4	5	6	7	8	9	10	11	12	13	14	15	16	17	18
1 H																	2 He
3 Li 1.18	4 Be 1.80											5 B 2.04	6 C 2.60	7 N 3.17	8 O 3.46	9 F 3.87	10 Ne
11 Na 1.12	12 Mg 1.40											13 Al 1.53	14 Si 1.84	15 P 2.09	16 S 2.14	17 Cl 2.40	18 Ar
19 K 1.05	20 Ca 1.24	21 Sc 1.29	22 Ti 1.38	23 V 1.44	24 Cr 1.59	25 Mn 1.63	26 Fe 1.76	27 Co 1.86	28 Ni 1.88	29 Cu 1.75	30 Zn 2.05	31 Ga 1.79	32 Ge 1.98	33 As 2.15	34 Se 2.12	35 Br 2.27	36 Kr
37 Rb 1.06	38 Sr 1.23	39 Y 1.29	40 Zr 1.22	41 Nb 1.51	42 Mo 1.63	43 Tc 1.73	44 Ru 1.75	45 Rh 1.82	46 Pd 1.69	47 Ag 1.78	48 Cd 1.89	49 In 1.69	50 Sn 1.85	51 Sb 1.95	52 Te 2.00	53 I 2.07	54 Xe
55 Cs 1.04	56 Ba 1.20	*	72 Hf 1.59	73 Ta 1.66	74 W 1.77	75 Re 1.92	76 Os 2.08	77 Ir 2.19	78 Pt 2.29	79 Au 2.31	80 Hg 2.52	81 Tl 1.86	82 Pb 1.96	83 Bi 1.92	84 Po 2.16	85 At 2.04	86 Rn
87 Fr	88 Ra	**	104 Rf	105 Db	106 Sg	107 Bh	108 Hs	109 Mt	110 Ds	111 Rg	112 Cn	113 Nh	114 Fl	115 Mc	116 Lv	117 Ts	118 Og

*	57 La 1.26	58 Ce 1.28	59 Pr 1.27	60 Nd 1.28	61 Pm 1.30	62 Sm 1.31	63 Eu 1.31	64 Gd 1.35	65 Tb 1.34	66 Dy 1.36	67 Ho 1.37	68 Er 1.39	69 Tm 1.39	70 Yb 1.48	71 Lu
**	89 Ac	90 Th	91 Pa	92 U	93 Np	94 Pu	95 Am	96 Cm	97 Bk	98 Cf	99 Es	100 Fm	101 Md	102 No	103 Lr

FIGURE 9.6 Allred–Rochow electronegativity (χ_{AR}) values transformed into values comparable to χ_P values.

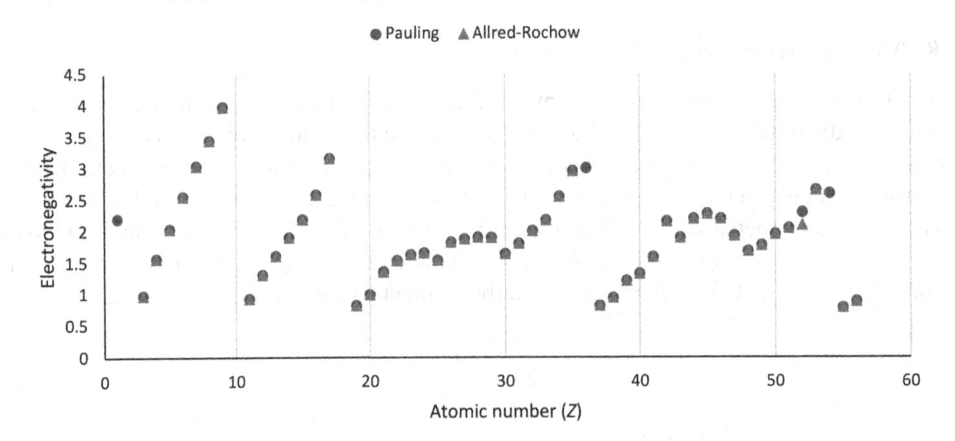

FIGURE 9.7 Comparison of Pauling and Allred-Rochow electronegativity values for elements 1–56. Note that all values for the Allred-Rochow electronegativity scale have been transformed into values comparable to χ_P values.

Problem 9.19. Consider the electrostatic potential maps of sodium hydride, dihydrogen, and hydrogen chloride.

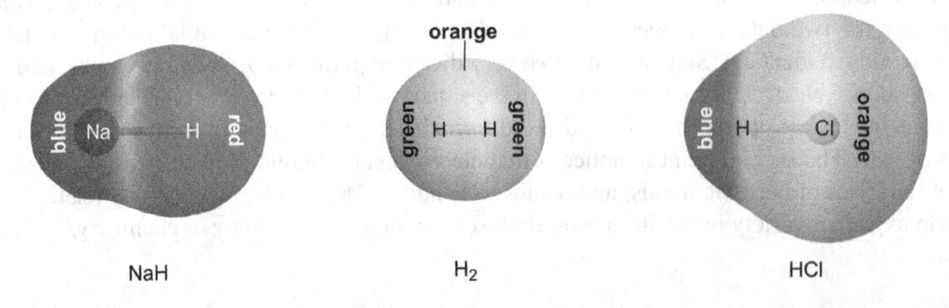

Red indicates areas of high electron density, blue indicates areas of low electron density, and green indicates a middling amount of electron density.

Using the Pauling electronegativity values in Figure 9.5, can you explain the differences in color for H in NaH, H_2, and HCl?

Problem 9.20. Look at the electronegativity values in Figure 9.5.

a. LiCl and CaF_2 are examples of ionic compounds. What is the difference in electronegativity ($\Delta\chi_P$) for Li and Cl? For Ca and F?

b. Now consider NH_3 and CCl_4, example molecular compounds. What is the difference in electronegativity ($\Delta\chi_P$) for N and H? For C and Cl?

c. Provide a generalization about the difference in electronegativity ($\Delta\chi_P$) for ionic compounds (compounds composed of a metal and nonmetal) compared to the difference in electronegativity for molecular compounds (compounds composed of nonmetals only).

d. For each pair of elements, determine the electronegativity difference ($\Delta\chi_P$), indicate whether the compound formed is most likely ionic or molecular (based on your answer to part c.), and which atom will pull on the electrons more.

i. N and Cl

$\Delta\chi_P$:
Type of compound (ionic or molecular):
Atom that pulls on the electrons more:

ii. H and C

$\Delta\chi_P$:
Type of compound (ionic or molecular):
Atom that pulls on the electrons more:

iii. H and Si

$\Delta\chi_P$:
Type of compound (ionic or molecular):
Atom that pulls on the electrons more:

iv. Br and Na

$\Delta\chi_P$:
Type of compound (ionic or molecular):
Atom that pulls on the electrons more:

v. Al and P

$\Delta\chi_P$:
Type of compound (ionic or molecular):
Atom that pulls on the electrons more:

THE EFFECTS OF NUCLEAR CHARGE, SHELL NUMBER, AND ELECTRON CONFIGURATIONS

Due to the repeating, periodic patterns in electron configurations (Chapter 8) and Coulomb's law, there are repeating patterns in atomic size (radius), ionization energy (E_i), electron affinity (E_{ea}), and electronegativity (χ). In this chapter, we investigated how the shell number and nuclear charge affected these properties and led to differences between metallic elements (which tend to become cations) and nonmetallic elements (which tend to become anions). In the next chapter, we will begin to investigate the differences in how different elements combine to produce ionic (metal and nonmetal) and molecular (nonmetal and nonmetal) compounds.

NOTES

1. Coulomb, C.A. Premier mémoire sur l'éctricité et le magnétisme. *Histoire de l'Academie Royale des Sc.*, **1785**, 569–577.
2. Husain, M.; Batra, A. Electronegativity Scale from X-ray Photoelectron Spectroscopic Data. *Polyhedron*, **1989**, *8* (9), 1233–1237. DOI: 10.1016/S0277-5387(00)81146-8.
3. i) Shell radii can be calculated using the Electron Localization Function (ELF): Becke, A.D.; Edgecombe, K.E. A Simple Measure of Electron Localization in Atomic and Molecular Systems. *J. Chem. Phys.*, **1990**, *92* (9), 5397–5403. DOI: 10.1063/1.458517.
   ii) The ELF results for iron were for rubidium in its ^2S ground state: Kahout, M.; Savin, A. Atomic Shell Structure and Electron Numbers. *Int. J. Quantum Chem.*, **1996**, *60* (4), 875–882. DOI: 10.1002/(SICI)1097-461X(1996)60:4<875::AID-QUA10>3.0.CO;2-4.
4. First developed from X-ray diffraction of crystals. Bragg, W.L. XVIII. The Arrangement of Atoms in Crystals. *Lond. Edinb. Dublin Philos. Mag. J. Sci.*, **1920**, *40* (236), 169–189. DOI: 10.1080/14786440808636111.
5. i) Slater, J.C. Atomic Radii in Crystals. *J. Chem. Phys.*, **1964**, *41* (10), 3199–3204. DOI: 10.1063/1.1725697.
   ii) Cordero, B.; Gómez, V.; Platero-Prats, A.E.; Revés, M.; Echeverría, J.; Cremades, E.; Barragán, F.; Alvarez, S. Covalent Radii Revisited†‡. *Dalton Trans.*, **2008**, *21*, 2832–2838. DOI: 10.1039/B801115J.
   iii) Bańkowski, Z. Covalent Radii of Noble Gas Atoms and Position of Hydrogen in the Periodic System of Elements. *Nature*, **1966**, *209*, 71–72. DOI: 10.1038/209071a0.
6. i) Bondi, A. van der Waals Volumes and Radii. *J. Phys. Chem.*, **1964**, *68* (3), 441–451. DOI: 10.1021/j100785a001.
   ii) Alvarez, S. A Cartography of the van der Waals territories†‡. *Dalton Trans.*, **2013**, *42*, 8617–8636. DOI: 10.1039/c3dt50599e.
7. All values are for a coordination number of 6. For transition metals, the radii are for low-spin states and are either for the most common oxidation number or for the oxidation state of the element as it appears in a naturally occurring oxide.
   i) Shannon, R.D.; Prewitt, C.T. Effective Ionic Radii in Oxides and Fluorides. *Acta Cryst.*, **1969**, B*25*, 925–946. DOI: 10.1107/S0567740869003220.
   ii) Shannon, R.D. Revised Effective Ionic Radii and Systemic Studies of Interatomic Distances in Halides and Chalcogenides. *Acta Cryst.*, **1976**, *A32*, 751–767. DOI: 10.1107/S0567739476001551.
   iii) Marcus, Y. Ionic Radii in Aqueous Solutions. *Chem. Rev.*, **1988**, *88* (8), 1475–1498. DOI: 10.1021/cr00090a003.
   iv) Libowitz, G.G. *The Solid-State Chemistry of Binary Metal Hydrides.* W.A. Benjamin Inc.: New York, **1965**.
8. Sansonetti, J.E.; Martin, W.C.; Young, S.L. *Handbook of Basic Atomic Spectroscopic Data.* NIST Standard Reference Database 108. DOI: 10.18434/T4FW23.
9. i) Pauling, L. The Nature of the Chemical Bond. IV. The Energy of Single Bonds and the Relative Electronegativity of Atoms. *J. Am. Chem. Soc.*, **1932**, *54* (9), 3570–3582. DOI: 10.1021/ja01348a011.
   ii) "Electronegativity." IUPAC. *Compendium of Chemical Terminology*, 2nd ed. (the "Gold Book"). Compiled by A.D. McNaught, A. Wilkinson. Blackwell Scientific Publications: Oxford, 1997. Online version (2019-) created by S. J. Chalk. DOI: 10.1351/goldbook.E01990.

10. i) The data most widely cited today are the revised Pauling values given by Allred: Allred, A.L. Electronegativity Values from Thermochemical Data. *J. Inorg. Nucl.*, **1961**, *17* (3–4), 215–221. DOI: 10.1016/0022-1902(61)80142-5.
ii) Tellurium value from: Huggins, M.L. Bond Energies and Polarities[1]. *J. Am. Chem. Soc.*, **1953**, *75* (17), 4123–4126. DOI: 10.1021/ja01113a001.
iii) Noble gas values from: Allen, L.C.; Huheey, J.E. The Definition of Electronegativity and the Chemistry of the Noble Gases. *J. Inorg. Nucl. Chem.*, **1980**, *42* (10), 1523–1524. DOI: 10.1016/0022-1902(80)80132-1.
iv) All other values with only one decimal place are from: Gordy, W.; Thomas, W.J.O. Electronegativity of the Elements. *J. Chem. Phys.*, **1956**, *24* (2), 439–444. DOI: 10.1063/1.1742493.
11. Allred, A.L.; Rochow, E.G. A Scale of Electronegativity Based on Electrostatic Force. *J. Inorg. Nucl.*, **1958**, *5* (4), 264–268. DOI: 10.1016/0022-1902(58)80003-2. This system was first proposed by Gordy: Gordy, W. A New Method of Determining Electronegativity from Other Atomic Properties. *Phys. Rev.*, **1946**, *69* (11–12), 604–607. DOI: 10.1103/PhysRev.69.604.

10 Chemical Formulae and Nomenclature

In this chapter, we will move our attention out from the scale of a singular atom to consider atoms in compounds. Following the law of definite proportions (Chapter 6), a compound is a pure substance that is made up of two or more different elements that are in a fixed ratio. Atoms with unstable electron configurations form compounds to achieve greater stability, and the resulting compounds have no charge. We will investigate how chemical formulae are determined and what they mean before turning our attention to the basics of compositional nomenclature.

EMPIRICAL FORMULAE

To begin our discussion of chemical formulae, we will start with empirical formulae. An empirical formula is formed from atomic symbols with appropriate subscripts to give the simplest formula expressing the composition of a compound.[1] For example, H_2O shows that the composition of water is two parts hydrogen for every one part oxygen. But where did this formula come from? Empirical formulae are based on data from experimental observations (empirical data). Let us look at how, for example, the empirical formula of water is determined.

To determine the empirical formula of a compound, it must first be broken down (or decomposed) into its constituent elements. This can be done with heat (thermal decomposition), with fire (pyrolysis), or with electricity (electrolysis). Water can be decomposed through electrolysis into hydrogen gas, H_2(g), and oxygen gas, O_2(g). For our purposes, let's consider an electrolysis experiment that produces 5.2 mL of hydrogen gas and 2.6 mL of oxygen gas, both at STP. We can use this information to determine the empirical formula.

First, the amount (mol) of each element must be determined from the experimental data. In 5.2 mL of hydrogen gas, there are 4.6×10^{-4} mol hydrogen:

5.2 ~~mL~~ hydrogen gas	1×10^{-3} ~~L~~	1 ~~mol~~	$= 4.6 \times 10^{-4}$ mol H
	1 ~~mL~~	22.7 ~~L~~	

And in 2.6 mL of oxygen gas, there are 2.3×10^{-4} mol oxygen:

2.6 ~~mL~~ oxygen gas	1×10^{-3} ~~L~~	1 ~~mol~~	$= 2.3 \times 10^{-4}$ mol O
	1 ~~mL~~	22.7 ~~L~~	

This tells us that the ratio of hydrogen to oxygen is 4.6×10^{-4} mol H:2.3×10^{-4} mol O. To find a whole number mole-to-mole ratio, we divide both sides by the smallest amount (2.3×10^{-4}), which gives us a ratio of 2 mol H:1 mol O. So, when we write H_2O, we are indicating this ratio of 2 mol H:1 mol O with the atomic symbols and subscripts. And remember that an equivalent amount (mol) of two chemicals means equivalent numbers of atoms, which means that 2 mol H:1 mol O corresponds to a ratio of 2 H atoms:1 O atom. The atom-to-atom ratio comes from the mole-to-mole ratio.

A good check on our work is to compare our empirical formula's molar mass to the molar mass of the compound, which can be determined independently through a number of techniques including mass spectrometry. Using the recommended relative atomic mass values (Appendix 2), we find that our empirical formula's molar mass is 18.015 g/mol (2(1.0080 g/mol) + 1(15.999 g/mol)).

DOI: 10.1201/9781003487210-10

The molar mass from mass spectrometry is 18.02 g/mol. Given their identical values, when rounded to two decimal places, we can have confidence that the empirical formula we have determined is a reasonable formula for the compound.

Problem 10.1. Nitrogen trichloride decomposes to produce 0.47 L nitrogen gas, $N_2(g)$, and 1.41 L chlorine gas, $Cl_2(g)$, at STP. Using this information, determine the empirical formula of the compound. Mass spectrometry gives a molar mass value of 120.36 g/mol.

Not all compounds decompose to give pure gaseous elements. More frequently, elemental analysis gives a percent composition for each element in the compound. For example, cobalt(II) bromide is found to be 26.94% cobalt by mass and 73.06% bromine by mass. We will now consider how we deal with these data for determining empirical formulae. The first thing to notice is that 26.94% cobalt by mass is another example (like amount concentration) of a hidden conversion factor. 26.94% cobalt by mass is a way of representing the conversion factor 100.0 g compound = 26.94 g Co, and 73.06% bromine by mass is a way of representing the conversion factor 100.0 g compound = 73.06 g Br. We are not typically given exactly how much of the compound is analyzed, and so we will assume a 100.0 g sample, which means that there are 26.94 g cobalt and 73.06 g bromine. Now, using these masses of the elements and the relative atomic masses (Appendix 2), we can determine the amount of each element:

26.94 ~~g~~ Co	1 ~~mol~~	= 0.4571 mol Co
	58.933 ~~g~~	

73.06 ~~g~~ Br	1 ~~mol~~	= 0.9143 mol Br
	79.904 ~~g~~	

This tells us that the ratio of cobalt to oxygen is 0.4571 mol Co:0.9143 mol Br or simplified it reduces to a ratio of 1 mol Co:2 mol Br. This means our empirical formula is $CoBr_2$. Comparing the experimentally determined molar mass (218.74 g/mol) to our empirical formula's molar mass (218.741 g/mol), we can have confidence that the empirical formula we have determined is a reasonable formula for the compound.

Problem 10.2. Determine the empirical formula for each compound.

a. Cisplatin, the common name for a platinum compound used in medicine, has the following percent composition values: 65.02% Pt, 9.34% N, 2.02% H, 23.63% Cl. Determine the empirical formula for cisplatin. The mass spectrometry molar mass is 300.1 g/mol.

b. Calculate the empirical formula for calcium given the percent composition values: 29.44% Ca, 23.55% S, and 47.01% O. The experimental molar mass is 136.13 g/mol.

MOLECULAR FORMULAE

So far, we have only considered empirical formulae of compounds where the empirical formula molar mass corresponds with the molar mass of the compound. For molecular compounds, those involving only nonmetal elements, the empirical formula molar mass can diverge from the molar mass of the compound. To find the molecular formula, the formula that is in accord with the compound molar mass,[2] we utilize the empirical formula approach that we have seen already, and then there is one additional step.

Let's consider glucose, a carbohydrate, with a mass spectrum molar mass of 180.16 g/mol. Glucose has a chemical composition of 40.00% C, 6.71% H, and 53.29% O. Since no amount of chemical is given, we will assume an initial 100.0 g of compound, which means that we have 40.00 g C, 6.71 g H, and 53.29 g O. We must first determine the amount (mol) of each element:

40.00 ~~g~~ C	1 ~~mol~~	= 3.330 mol C
	12.011 ~~g~~	

6.71 ~~g~~ H	1 ~~mol~~	= 6.66 mol H
	1.0080 ~~g~~	

53.29 ~~g~~ O	1 ~~mol~~	= 3.331 mol O
	15.999 ~~g~~	

Based on the amount (mol) ratios 3.330 mol C:6.66 mol H:3.331 mol O, the empirical formula for glucose is CH_2O. Note that this empirical formula is why carbohydrates are called carbohydrates, that is they are carbon (C) hydrated with water (H_2O). Now when we compare our empirical formula molar mass (30.026 g/mol) with the molar mass of the compound (180.16 g/mol), we can see that they do not equal each other. If we look at the ratio of the two, we can see that the empirical formula molar mass is one-sixth of the compound molar mass $\left(\frac{30.026 \text{ g/mol}}{180.16 \text{ g/mol}} = \frac{1}{6}\right)$. That tells us that the empirical formula is one-sixth of the molecular formula and that the molecular formula of glucose is found by multiplying each subscript in the empirical formula by six: $C_6H_{12}O_6$. The empirical formula only provides the lowest whole-number ratio of elements. To find the chemical formula for a molecule, the empirical formula is multiplied through by a whole-number value as we saw for glucose.

Problem 10.3. Determine the empirical formula and molecular formula for each compound.

a. Ribose, with a molar mass of 150.13 g/mol, has a chemical composition of 40.00% C, 6.72% H, and 53.28% O.

b. Ethylene glycol (antifreeze), with a molar mass of 62.07 g/mol, has a chemical composition of 38.70% C, 9.74% H, and 51.55% O.

IUPAC Nomenclature[3]

Nomenclature is a framework for naming chemical formulae. In an ideal nomenclature system, each chemical formula would have a unique name, and each name corresponds to a specific chemical formula. In chemistry, nomenclature follows the rules and guidelines from the International Union of Pure and Applied Chemistry (IUPAC). Given the large variety of types of chemical compounds and the complexity of coherently naming them, there is not a singular set of rules for naming all chemical compounds. The basic nomenclature system presented here is the compositional nomenclature of binary compounds. If you continue to study chemistry, there are different nomenclature systems that you may encounter: substitutive nomenclature in organic chemistry (the chemistry of carbon and carbon-rich compounds), additive nomenclature in inorganic chemistry (the chemistry of the other 117 elements other than carbon), and biochemical nomenclature in biochemistry (the chemistry in living systems).

Naming Atoms

There are very few elements that naturally exist as free monatomic, single-atom, species. Only in the noble gases (He, Ne, Ar, Kr, Xe, and Rn) and mercury (Hg) do the atoms exist as monatomic species. For all other elements, they naturally occur as diatomic (O_2) (two-atom) or polyatomic (P_4) (many-atom) species. When discussing monatomic atoms of an element, it is best practice to clearly state this. For example, saying oxygen gas implies the chemical formula O_2 because elemental oxygen is naturally a diatomic molecule; however, saying atomic oxygen or oxygen atoms clearly indicates that one is speaking about the monatomic species O.

For atoms and compounds, we can indicate the phase of a substance in chemical formulae with abbreviations: (s) for solid, (l) for liquid, and (g) for gas. Another common notation is aqueous (aq), which is used to indicate that a substance is dissolved in water to give a homogenous mixture. This is not formally part of nomenclature, but a way of noting the phase in the formula. For example, liquid mercury would be Hg(l).

Homonuclear Molecular Compounds

Most pure elements exist as diatomic (two-atom) or polyatomic (many-atom) species. Most metals and metalloids form three-dimensional networks where all the atoms in the solid are bonded to one or more atoms in the solid. For example, when you grab a piece of aluminium foil all the aluminium atoms are bonded to one or more other aluminium atoms. As a chemical formula, we write only Al for aluminium, which is the empirical formula.

Some elements naturally exist, when pure, as homonuclear diatomic and polyatomic molecules (Table 10.1). The proper IUPAC name uses a prefix to indicate the number of atoms in the chemical formula (Table 10.2). We will use these prefixes and slightly modified versions throughout chemical nomenclature. While the IUPAC guidelines provide a clear, unambiguous approach (Br_2 is

TABLE 10.1
Elements That, When Pure, Naturally Exist as Homonuclear Diatomic and Polyatomic Molecules

Element	Chemical formula	Common name (IUPAC)	Element	Chemical formula	Common name (IUPAC)
bromine	Br_2	bromine (dibromine)	nitrogen	N_2	nitrogen (dinitrogen)
chlorine	Cl_2	chlorine (dichlorine)	oxygen	O_2	oxygen (dioxygen)
fluorine	F_2	fluorine (difluorine)	oxygen	O_3	ozone (trioxygen)
hydrogen	H_2	hydrogen (dihydrogen)	phosphorus	P_4	phosphorus (tetraphosphorus)
iodine	I_2	iodine (diiodine)	sulfur	S_8	sulfur (octasulfur)

TABLE 10.2
The First Ten IUPAC Multiplicative Prefixes Used to Indicate the Number of an Element in a Chemical Formula

IUPAC prefix	Corresponding number of atoms	IUPAC prefix	Corresponding number of atoms
mono	one	hexa	six
di	two	hepta	seven
tri	three	octa	eight
tetra	four	nona	nine
penta	five	deca	ten

dibromine), these guidelines are not always followed. And so, it is common for chemists to say bromine and assume that the reader or listener understands that they mean dibromine (from contexts).

A useful mnemonic for remembering the seven elements that are naturally diatomic is their seven atomic symbols listed in the order HOFBrINCl read as "hoff-BRINK-uhl." As mentioned, it is always best to assume that if someone is discussing one of these elements – not in a compound but pure – then they are implying the diatomic form of that element even if the prefix di is not included in the name.

The last term to be familiar with for a pure element is allotrope. Oxygen can exist as two different forms (allotropes) that are bonded together differently: dioxygen (O_2) and trioxygen (ozone, O_3). At sea level, dioxygen is the predominant form, and ozone is only persistent in the upper atmosphere where it provides a protective shield against ultraviolet (UV) radiation. Other examples of allotropes are graphite and diamond (allotropes of carbon) and red phosphorus (a network solid) and white phosphorus (tetraphosphorus, P_4).

Naming Ions

Ions are charged particles, which have either more or fewer electrons than the number of protons. If there are more electrons than the number of protons, the ion will have a negative charge and is called an anion. If there are fewer electrons than the number of protons, the ion will have a positive charge and is called a cation. As we saw in Chapter 6, ions have the charge number written after the element name (Table 10.3).

The name of the element in cations, positively charged ions, are not changed at all. To highlight the difference in charge of anions, negatively charged ions, the element name is changed. The

TABLE 10.3
Selected Main-Group, Monatomic Ions, Their IUPAC Name, and Their Atomic Symbol

Number of protons	Number of electrons	IUPAC name	Common name	Atomic symbol
1	0	hydron or hydrogen(1+)	hydrogen	H^+
3	2	lithium(1+)	lithium	Li^+
4	2	beryllium(2+)	beryllium	Be^{2+}
7	10	nitride(3–)	nitride	N^{3-}
8	10	oxide(2–)	oxide	O^{2-}
9	10	fluoride(1–)	fluoride	F^-
11	10	sodium(1+)	sodium	Na^+
12	10	magnesium(2+)	magnesium	Mg^{2+}
15	18	phosphide(3–)	phosphide	P^{3-}
16	18	sulfide(2–)	sulfide	S^{2-}
17	18	chloride(1–)	chloride	Cl^-

TABLE 10.4
Examples of Element Name Change (Natural Ending to -ide Suffix) for Ions

Element	Common anion	IUPAC anion name	Common anion name
hydrogen	H^-	hydride(1–)	hydride
nitrogen	N^{3-}	nitride(3–)	nitride
oxygen	O^{2-}	oxide(2–)	oxide
fluorine	F^-	fluoride(1–)	fluoride
phosphorus	P^{3-}	phosphide(3–)	phosphide
sulfur	S^{2-}	sulfide(2–)	sulfide
chlorine	Cl^-	chloride(1–)	chloride
selenium	Se^{2-}	selenide(2–)	selenide
bromine	Br^-	bromide(1–)	bromide
tellurium	Te^{2-}	telluride(2–)	telluride
iodine	I^-	iodide(1–)	iodide

existing element name ending (-ine, -ium, -ogen, -on, -ur, -us, -ygen) is replaced with -ide suffix (Table 10.4).

The charges shown in Tables 10.3 and 10.4 come from the common charges (Figure 10.1) for each element. Because these elements have a single common charge, however, it is common for the charge number to be omitted (Table 10.4) in the common anion name because the charge number can be considered assumed knowledge by practicing chemists. For cations with a set charge, the common cation name also tends to omit the charge number, for example, the term "lithium ion" implies the 1+ charge number, lithium(1+), of the ion. In contrast, the term copper ion could equally well correspond to the copper(1+) or the copper(2+) ion. For ions without a set charge, the charge must be indicated.

1	2	3	4	5	6	7	8	9	10	11	12	13	14	15	16	17	18
1 H 1+/1–																	2 He 0
3 Li 1+	4 Be 2+											5 B 3+	6 C variable	7 N 3–	8 O 2–	9 F 1–	10 Ne 0
11 Na 1+	12 Mg 2+											13 Al 3+	14 Si variable	15 P 3–	16 S 2–	17 Cl 1–	18 Ar 0
19 K 1+	20 Ca 2+	21 Sc 3+	22 Ti variable	23 V variable	24 Cr variable	25 Mn variable	26 Fe variable	27 Co variable	28 Ni variable	29 Cu variable	30 Zn 2+	31 Ga 3+	32 Ge variable	33 As variable	34 Se 2–	35 Br 1–	36 Kr 0
37 Rb 1+	38 Sr 2+	39 Y 3+	40 Zr 4+	41 Nb variable	42 Mo variable	43 Tc variable	44 Ru variable	45 Rh variable	46 Pd variable	47 Ag 1+	48 Cd 2+	49 In variable	50 Sn variable	51 Sb variable	52 Te variable	53 I 1–	54 Xe 0
55 Cs 1+	56 Ba 2+	*	72 Hf 4+	73 Ta variable	74 W variable	75 Re variable	76 Os variable	77 Ir variable	78 Pt variable	79 Au variable	80 Hg variable	81 Tl variable	82 Pb variable	83 Bi variable	84 Po variable	85 At variable	86 Rn 0
87 Fr 1+	88 Ra 2+	**	104 Rf unknown	105 Db unknown	106 Sg unknown	107 Bh unknown	108 Hs unknown	109 Mt unknown	110 Ds unknown	111 Rg unknown	112 Cn unknown	113 Nh unknown	114 Fl unknown	115 Mc unknown	116 Lv unknown	117 Ts unknown	118 Og unknown

*	57 La 3+	58 Ce 3+	59 Pr 3+	60 Nd 3+	61 Pm 3+	62 Sm 3+	63 Eu 3+	64 Gd 3+	65 Tb 3+	66 Dy 3+	67 Ho 3+	68 Er 3+	69 Tm 3+	70 Yb 3+	71 Lu 3+
**	89 Ac 3+	90 Th variable	91 Pa variable	92 U variable	93 Np variable	94 Pu variable	95 Am 3+	96 Cm 3+	97 Bk 3+	98 Cf 3+	99 Es 3+	100 Fm 3+	101 Md 3+	102 No variable	103 Lr 3+

FIGURE 10.1 Periodic table showing the common charge each element tends to take on in ionic compounds (if known). 0 means that monatomic ions of this element do not naturally form.

TABLE 10.5
Example Chemical Formulae and Names for Binary Ionic Compounds of Main Group Elements with a Single Common Charge

Chemical formula	IUPAC name
NaBr	sodium bromide
AlP	aluminium phosphide
CaF_2	calcium fluoride
HCl	hydrogen chloride
K_2O	potassium oxide
Mg_3N_2	magnesium nitride

Naming Compounds Composed of Main-Group Metals and Nonmetals

Compounds composed of hydrogen, main-group metals (groups 1, 2, and 13–15), and metalloids (groups 1–3) with nonmetals (groups 15–17) are typically simple binary compounds. Those involving metals/metalloids and nonmetals are termed ionic compounds or salts. Our first discussion of nomenclature will look at the compositional nomenclature of main-group binary ionic compounds. Consider some example compounds shown in Table 10.5.

There are some points to note about the chemical formulae and the IUPAC names. First, the less electronegative element (metal or hydrogen cation) is listed first in the formula and first in the name. Second, the more electronegative element (nonmetal anion) is listed second in the formula and second in the name. Since the nonmetal is an anion in these compounds, the nonmetal element name also gets the -ide suffix (Table 10.4).

Let's consider some practical examples of going from chemical formula to name and name to chemical formula. If we have the chemical formula SrS, we first need to identify the two elements present in the compound. The first element (Sr) is strontium, which will appear in the compositional name without alteration because it is the metal cation. The second element is sulfur, which will appear in the compositional name with the -ide suffix because it is the nonmetal anion. Altogether, strontium sulfide is the complete compositional name for SrS.

To go from a compositional name to a formula, we will need both the atomic symbols and the common charges of each ion. We will derive the formula for calcium phosphide. To begin, we need to write down the atomic symbol and charge for each part of the name. Calcium in ionic compounds has a 2+ charge (Figure 10.1) and the atomic symbol is Ca^{2+}. In ionic compounds, phosphide has the charge of 3– (Figure 10.1) and the atomic symbol is P^{3-}. In the chemical formula, the metal (Ca) is listed first, and the nonmetal (P) is listed second. To determine the appropriate subscripts to give a formula with zero overall charge, we can use the approach of crossing over the charge-number magnitudes (Figure 10.2).

$$Ca^{2+} P^{3-} \longrightarrow Ca_3P_2$$

FIGURE 10.2 Example of charge-number magnitude crossover to arrive at the chemical formula for a neutral ionic compound.

Ionic compound chemical formulae are always empirical formulae, and so, if possible, we reduce the subscripts to the lowest, whole-number ratios. In this case, the 3:2 ratio is irreducible, and Ca_3P_2 is the chemical formula for calcium phosphide.

Problem 10.4. Using the rules of IUPAC nomenclature, write a chemical formula for each compound.

calcium chloride	aluminium oxide
hydrogen selenide	caesium nitride
magnesium sulfide	hydrogen chloride
lithium phosphide	potassium fluoride
rubidium selenide	barium iodide

Problem 10.5. Name the following compounds according to the rules for IUPAC nomenclature.

$SrCl_2$	BaO
Na_2S	Rb_3N
LiI	MgF_2
H_3P	HF
Mg_3P_2	CaSe

Before we move onto more rules of compositional nomenclature, a brief mention needs to be given to a nonsystematic (trivial) nomenclature that does not follow IUPAC recommendations of compositional nomenclature but is still very prevalent: binary acid nomenclature. Acid nomenclature is unique in that it describes a chemical property (Chapter 16), not the composition of the compound. Examples of trivial, nonsystematic, acid names that are commonly encountered include hydrofluoric acid (HF(aq)), hydrochloric acid (HCl(aq)), hydrobromic acid (HBr(aq)), and hydroiodic acid (HI(aq)). The state of matter or phase – here (aq) meaning aqueous or dissolved in water – does not have a role in compositional nomenclature, which means that the preceding examples should more properly be named hydrogen fluoride, hydrogen chloride, hydrogen bromide, and hydrogen iodide following the conventions we saw for binary nomenclature of main group elements. If noting the phase is important then HF(aq), for example, would be called aqueous hydrogen fluoride. The term "aqueous" is an adjective for the name hydrogen fluoride. This book will follow the IUPAC conventions in these cases, but hydrochloric acid is an especially common example of a binary acid name.

Naming Compounds Composed of Transition Metals and Nonmetals or Only Nonmetals

So far, we considered only those compounds that consisted of elements that adopt a single, set charge. In this section, we will consider how to expand the basic nomenclature system already introduced to account for compounds containing metals with a variable charge and for those compounds that contain only nonmetals (molecular compounds), where charge does not play a role because those compounds do not involve any ions. For the nomenclature that we have seen so far, the formula could be determined because there are known set charges for each element. To accommodate transition-metal ionic compounds and molecular compounds in our existing structure, one approach is to incorporate charge numbers and oxidation numbers (discussed below) into the chemical nomenclature (Table 10.6).

For transition metals, the charge number is being added as we saw in Table 10.3. And since iron, for example, does not have a fixed common charge, the charge number cannot be omitted as it often is for main-group elements (Tables 10.3 and 10.4). Another approach, which is more commonly seen than charge numbers, is to state the oxidation number (Chapter 16), as Roman numerals,[4] next to the less electronegative atom. For transition metals, the charge is typically the oxidation number

TABLE 10.6
Example Chemical Formulae and Names for Compounds Where the Electropositive Element Does Not Have a Single Common Charge and for Molecular (Only Nonmetal Elements) Compounds

Chemical formula	IUPAC name (charge number)	IUPAC name (oxidation number)
FeO	iron(2+) oxide	iron(II) oxide
Fe_2O_3	iron(3+) oxide	iron(III) oxide
PbS	lead(2+) sulfide	lead(II) sulfide
PbS_2	lead(4+) sulfide	lead(IV) sulfide
VCl_3	vanadium(3+) chloride	vanadium(III) chloride
CO	–	carbon(II) oxide
CO_2	–	carbon(IV) oxide
NF_3	–	nitrogen(III) fluoride
SF_4	–	sulfur(IV) fluoride
SF_6	–	sulfur(VI) fluoride

and so you can see that there is no difference in the value in parentheses between column two and column three (Table 10.6). For molecular compounds, none of the atoms exist as ions and so there is no charge number to write. What oxidation numbers do is that they assume, this is a simplifying assumption only, all compounds are perfectly ionic. In this hypothetical situation, we further assume that the more electronegative atom has its common charge (Figure 10.1). For the compound CO, then, the oxygen would be oxide(2–) and therefore the carbon must be carbon(2+). But we do not write carbon 2+ because it is not actually ionic, but we indicate that its oxidation number is (II). Aside from the introduction of charge numbers and oxidation numbers, the nomenclature for transition metals (where charge numbers and oxidation numbers can both be used) and molecular compounds (only oxidation numbers can be used) functions identically to the basic system already introduced. Note that for molecular compounds, the subscripts are always left unreduced.

Problem 10.6. Using the rules of IUPAC nomenclature, write a chemical formula for each compound.

cobalt(3+) oxide	iron(III) chloride
copper(II) bromide	lithium phosphide
chromium(III) sulfide	phosphorus(V) chloride
sulfur(II) fluoride	osmium(4+) nitride
potassium nitride	iron(II) phosphide
sodium selenide	silver(II) fluoride
tin(2+) oxide	gold(III) chloride

Problem 10.7. Name the following compounds according to the rules for IUPAC nomenclature.

CrO_3	Ni_3P_2
PbO_2	Cs_2O
$ScCl_3$	MgI_2

PCl_3	SnF_4
$PbCl_2$	CuSe
RuO_3	LiBr
GaN	CoN

Compositional Nomenclature with Multiplicative Prefixes

Another approach (Table 10.7) for treating transition metal compounds and molecular compounds is to use the multiplicative prefixes in Table 10.2. Here we do not worry about charges at all. Instead, a prefix is added in front of the element name to indicate the subscript number for that element. The prefix mono can be left off the name of the first element, but it can never be left off the name of the second element. For some compounds prefixes are the only way to ensure the right subscripts (see N_2Cl_2 in Table 10.7).

TABLE 10.7
Example Chemical Formulae and Names (with Charge Numbers, Oxidation Numbers, and Prefixes) for Compounds Where the Electropositive Element Does Not Have a Single Common Charge and For Molecular (Only Nonmetal Elements) Compounds

Chemical formula	IUPAC name (charge number)	IUPAC name (oxidation number)	IUPAC name (prefixes)
FeO	iron(2+) oxide	iron(II) oxide	iron monoxide
Fe_2O_3	iron(3+) oxide	iron(III) oxide	diiron trioxide
PbS	lead(2+) sulfide	lead(II) sulfide	lead monosulfide
PbS_2	lead(4+) sulfide	lead(IV) sulfide	lead disulfide
VCl_3	vanadium(3+) chloride	vanadium(III) chloride	vanadium trichloride
CO	–	carbon(II) oxide	carbon monoxide
CO_2	–	carbon(IV) oxide	carbon dioxide
NF_3	–	nitrogen(III) fluoride	nitrogen trifluoride
SF_4	–	sulfur(IV) fluoride	sulfur tetrafluoride
SF_6	–	sulfur(VI) fluoride	sulfur hexafluoride
N_2F_2	–	–	dinitrogen difluoride

Problem 10.8. Using the rules of IUPAC nomenclature, write a chemical formula for each compound.

tin(IV) oxide	chlorine dioxide
phosphorus triiodide	calcium arsenide(3–)
boron trifluoride	dioxygen difluoride
carbon dioxide	sulfur hexafluoride
iron trichloride	osmium tetroxide
osmium(VIII) oxide	bromine pentafluoride

Problem 10.9. Name the following compounds according to the rules for IUPAC nomenclature.

N_2F_2 OF_2
SBr_2 BCl_3
XeF_4 $BiCl_3$
P_4O_3 CS_2
Na_2O ClF_3
Co_2O_3 HBr

Compositional Nomenclature with Polyatomic Ions

Finally, the last compositional nomenclature topic that we will consider is compounds like NaOCl. You may be familiar with this compound already as it is the active chemical in bleach. What will likely stand out as you look at this compound now, however, is that this does not look like a binary (two) compound because there are three elements present. In this compound, there is the monoatomic ion sodium(1+) (Na^+) and the polyatomic (many atom) ion hypochlorite (OCl^-). Both polyatomic cations (Table 10.8) and anions (Table 10.9) are possible.

TABLE 10.8
Common and IUPAC Nomenclature of Polyatomic Cations

Common name (IUPAC)	Chemical formula	Common name (IUPAC)	Chemical formula
ammonium (azanium)	NH_4^+	mercury(I) (dimercury(2+))	Hg_2^{2+}

Polyatomic ions have their own unique names, and the -ide, -ate, and -ite suffixes denote that they are negatively charged, which means that polyatomic ion suffixes do not get replaced with the -ide suffix that is added to monatomic anions. The only significant change from the chemical formulas that we have seen so far is the use of parentheses. Because polyatomic ions are groups, parentheses are used (Table 10.10) when there is more than one polyatomic ion in the formula. This helps to show that the subscript applies to the group of atoms and not a singular atom.

In addition, in naming some polyatomic ions (highlighted with an asterisk in Table 10.9) we cannot use the multiplicative prefixes in Table 10.2. Consider the polyatomic ion phosphate (PO_4^{3-}) as it appears in the compound $Ca_3(PO_4)_2$, we could name this following the rules of compositional nomenclature for metals (calcium) and nonmetals (phosphate) as calcium phosphate. If, instead, we used multiplicative prefixes and named the compound tricalcium diphosphate, that would be incorrect. Diphosphate (Table 10.9) is a different polyatomic ion: $P_2O_7^{2-}$. For phosphate and the other polyatomic ions indicated by an asterisk in Table 10.9, we must use the modified multiplicative prefixes (Table 10.11) and use parentheses for added clarity. Together the correct name with multiplicative prefixes for $Ca_3(PO_4)_2$ is tricalcium tris(phosphate).

Problem 10.10. Using the rules of IUPAC nomenclature, write a chemical formula for each compound.

iron triacetate aluminium hydroxide
strontium sulfate lead(II) sulfate
hydrogen telluride(2–) cobalt(2+) nitrate
potassium permanganate magnesium carbonate
copper(I) sulfide ammonium carbonate
potassium cyanide dimercury(2+) chloride

TABLE 10.9
Common and IUPAC Nomenclature of Polyatomic Anions

Common name (IUPAC)	Chemical formula	Common name (IUPAC)	Chemical formula
acetate	$CH_3CO_2^-$	hydroxide (oxidanide)	OH^-
amide (azanide)	NH_2^-	hypobromite	BrO^-
azide (trinitride(1–))	N_3^-	hypochlorite	ClO^-
bromate	BrO_3^-	hypoiodite	IO^-
bromite	BrO_2^-	iodate	IO_3^-
bicarbonate (hydrogencarbonate)*	HCO_3^-	iodite	IO_2^-
bifluoride (difluoridohydrogenate(1–))*	HF_2^-	nitrate	NO_3^-
bisulfate (hydrogensulfate)*	HSO_4^-	nitrite	NO_2^-
bisulfide (sulfanide)*	HS^-	perbromate	BrO_4^-
bisulfite (hydrogensulfite)*	HSO_3^-	perchlorate	ClO_4^-
carbonate*	CO_3^{2-}	periodate	IO_4^-
chromate*	CrO_4^{2-}	permanganate	MnO_4^-
chlorate	ClO_3^-	peroxide (dioxide(2–))*	O_2^{2-}
chlorite	ClO_2^-	phosphate*	PO_4^{3-}
cyanate	OCN^-	pyrophosphate (diphosphate)*	$P_2O_7^{4-}$
cyanide	CN^-	sulfate	SO_4^{2-}
dichromate*	$Cr_2O_7^{2-}$	sulfite	SO_3^{2-}
dihydrogenphosphate*	$H_2PO_4^-$	tetrafluoridoborate*	BF_4^-
formate	HCO_2^-	thiocyanate	SCN^-
hydrogenphosphate*	HPO_4^{2-}	triiodide	I_3^-

Note that for those polyatomic ions whose names have been highlighted with an asterisk, the multiplicative prefixes already seen (di, tri, tetra, etc.) cannot be used and modified prefixes must be used instead (bis, tris, tetrakis, pentakis, etc.).

TABLE 10.10
Example Chemical Formulae and IUPAC Names of Compounds Containing Polyatomic Ions

Chemical formula	IUPAC name (charge number)	IUPAC name (oxidation number)	IUPAC name (prefixes)
$NaCN$	sodium cyanide	sodium cyanide	sodium monocyanide
$Mg(CN)_2$	magnesium cyanide	magnesium cyanide	magnesium dicyanide
$Co(OH)_3$	cobalt(3+) hydroxide	cobalt(III) hydroxide	cobalt trihydroxide
CaS	calcium sulfide	calcium sulfide	calcium monosulfide
$Al_2(SO_4)_3$	aluminium sulfate	aluminium sulfate	dialuminium trisulfate
$MnCO_3$	manganese(2+) carbonate	manganese(II) carbonate	manganese monocarbonate
$Au(CH_3CO_2)_3$	gold(3+) acetate	gold(III) acetate	gold triacetate
$Ca(H_2PO_4)_2$	calcium dihydrogenphosphate	calcium dihydrogenphosphate	calcium bis(dihydrogenphosphate)
$(NH_4)_2SO_4$	ammonium sulfate	ammonium sulfate	diammonium monosulfate
Hg_2Cl_2	dimercury(2+) chloride	mercury(I) chloride	dimercury dichloride
$Fe_2(CrO_4)_3$	iron(3+) chromate	iron(III) chromate	diiron tris(chromate)

TABLE 10.11
The First Ten IUPAC Multiplicative Prefixes Used to Indicate the Number of Groups in a Chemical Formula

IUPAC prefix	Corresponding number of atoms	IUPAC prefix	Corresponding number of atoms
mono	one	hexakis	six
bis	two	heptakis	seven
tris	three	octakis	eight
tetrakis	four	nonakis	nine
pentakis	five	decakis	ten

Problem 10.11. Name the following compounds according to the rules for IUPAC nomenclature.

$Cu(OH)_2$	$Sn(CH_3CO_2)_4$
$Zn_3(PO_4)_2$	$Pb(NO_3)_2$
H_2S	Ni_2O_3
NH_4F	$MgSO_4$
$NaHCO_3$	Li_2O_2
$LiBrO_4$	$AgNO_3$

Oxyacid Nomenclature

Lastly, we will consider the nomenclature of oxyacids, which are nonsystematic but are still accepted by IUPAC. Consider Table 10.11, which provides a list of common oxyacid formulae and their name.

TABLE 10.11
Example Chemical Formulae and Names for Binary Ionic Compounds of Main Group Elements with a Single Common Charge

Chemical formula	Name
CH_3CO_2H	acetic acid
H_2CO_3	carbonic acid
$HClO_3$	chloric acid
$HClO_2$	chlorous acid
$HClO$	hypochlorous acid
HNO_3	nitric acid
HNO_2	nitrous acid
$HMnO_4$	permanganic acid
H_3PO_4	phosphoric acid
H_2SO_4	sulfuric acid
H_2SO_3	sulfurous acid

An acid, here, is defined as any oxyanion with a hydrogen(1+) cation. The names of each are derived from the polyatomic anion name: an -ate suffix becomes -ic acid and an -ite suffix becomes -ous acid. For example, nitrate (NO_3^-) with hydrogen(1+) gives the formula HNO_3 and the nitr<u>ate</u> anion becomes nitr<u>ic acid</u>.

Problem 10.12. Using the rules of IUPAC nomenclature, write a chemical formula for each compound.

acetic acid	phosphoric acid
sodium hypobromite	sodium acetate
silver fluoride	bromous acid
perbromic acid	bromic acid
hypobromous acid	tin(II) bromide
dichromic acid	tin tetrabromide

Problem 10.13. Name the following compounds according to the rules for IUPAC nomenclature.

$HMnO_4$	H_2CrO_4
HNO_2	$BaCl_2$
H_2CO_3	$Fe(NO_3)_3$
HBr	$Ca(CN)_2$
Na_2CO_3	HNO_3
$LiMnO_4$	CS_2

CHEMICAL FORMULAE AND NOMENCLATURE

In this chapter we looked at the meaning of chemical formulae as a means of representing the amount ratio – equivalently either mole-to-mole or atom-to-atom – of each element in a compound. We then turned our attention to the IUPAC compositional nomenclature for binary compounds. In the following chapter, we will take these formulae and look at how the atoms are connected through ionic or covalent bonding.

NOTES

1. "Empirical formula." IUPAC. Compendium of Chemical Terminology, 2nd ed. (the "Gold Book"). Compiled by A.D. McNaught, A. Wilkinson. Blackwell Scientific Publications: Oxford, **1997**. Online version (2019-) created by S.J. Chalk. DOI: 10.1351/goldbook.E02063.
2. "Molecular formula." IUPAC. *Compendium of Chemical Terminology*, 2nd ed. (the "Gold Book"). Compiled by A.D. McNaught, A. Wilkinson. Blackwell Scientific Publications: Oxford, **1997**. Online version (2019-) created by S.J. Chalk. DOI: 10.1351/goldbook.M03987.
3. The rules and guidelines here follow those from:
   i) *Principles of Chemical Nomenclature: A Guide to IUPAC Recommendations*, 2011 ed.; Ed. G.J. Leigh. RSC Publishing: Cambridge, **2011**.
   ii) IUPAC. *Nomenclature of Inorganic Chemistry, IUPAC Recommendations 2005*. (the "Red Book"). Prepared for publication by N.G. Connelly, T. Damhus, R.M. Hartshorn, A.T. Hutton. RSC Publishing: Cambridge, **2005**.
4. Oxidation numbers, also called oxidation states, were first proposed by Lavoisier: Lavoisier, A.L. *Elements of Chemistry*. Creech: Edinburgh, **1790**, pp. 159–172; Latimer, W.M. *The Oxidation States of the Elements and Their Potentials in Aqueous Solution*. Prentice-Hall: New York, NY, **1938**.
   ii) Usage of Roman numerals, was formerly called the Stock number: Stock, A. Einige Nomenklaturfragen der anorganischen Chemie. **1919**, *32* (98), 373–374. DOI: 10.1002/ange.19190329802.

11 Bonding

We have considered nomenclature for compounds, but how are the atoms in a compound held together? In this chapter, we will explore frameworks for representing and understanding what holds atoms together, that is, chemical bonds. Note that a thorough explanation of bonding requires the use of quantum mechanics, which is beyond the scope of this book.

LEWIS DOT SYMBOLS

In Chapter 8, we discussed the electron configuration of an atom, which is represented with a string of alphanumeric characters that identify the number of electrons and their energy. For example, the full electron configuration of nitrogen is $1s^22s^22p^3$, which can be abbreviated with the noble gas notation to be $[He]2s^22p^3$. Each of these notations, while rooted in quantum mechanical theory, is cumbersome. It is for this reason that Lewis dot symbols[1] are widely employed by chemists. Consider the Lewis dot symbol for nitrogen (Figure 11.1).

FIGURE 11.1 The Lewis dot symbol for nitrogen showing the valence configuration.

From the Lewis dot symbol, it is immediately apparent that nitrogen has five valence electrons. If we consider a periodic table of Lewis dot symbols (Figure 11.2), the patterns and group trends in electron configurations are directly visualizable.

1	2	3	4	5	6	7	8	9	10	11	12	13	14	15	16	17	18
H·	He:																
Li·	Be:											·B:	·C:	·N:	:O:	:F:	:Ne:
Na·	Mg:											·Al:	·Si:	·P:	:S:	:Cl:	:Ar:
K·	Ca:	Sc:	Ti:	V:	·Cr·	·Mn:	·Fe:	·Co:	·Ni:	:Cu·	:Zn:	·Ga:	·Ge:	·As:	:Se:	:Br:	:Kr:
Rb·	Sr:	Y:	Zr:	Nb·	·Mo·	·Tc:	·Ru·	·Rh·	:Pd	:Ag·	:Cd:	·In:	·Sn:	·Sb:	:Te:	:I:	:Xe:
Cs·	Ba:	*	Hf:	Ta:	W:	·Re:	·Os:	·Ir:	·Pt·	:Au·	:Hg:	·Tl:	·Pb:	·Bi:	:Po:	:At:	:Rn:
Fr·	Ra:	**	Rf:	Db:	Sg:	·Bh:	·Hs:	·Mt:	·Ds·	·Rg:	:Cn:	·Nh:	·Fl:	·Mc:	:Lv:	:Ts:	:Og:

*	La:	·Ce:	·Pr:	·Nd:·	·Pm:·	·Sm:·	·Eu:·	·Gd:·	·:Tb:·	·:Dy:·	·:Ho:·	::Er:·	::Tm:·	::Yb::	Lu:
**	Ac:	·Th:	·Pa:	·U:·	·Np:·	·Pu:·	·Am:·	·Cm:·	·:Bk:·	·:Cf:·	·:Es:·	::Fm:·	::Md:·	::No::	Lr:

FIGURE 11.2 Lewis dot symbols for the elements.

DOI: 10.1201/9781003487210-11

Recall from Chapter 8 that atoms are at their most stable (lowest energy) when the valence shell is filled. For hydrogen and helium, there is no 1p subshell and so they only need a duet of electrons to fill the 1s subshell and be stable. Main group elements, aside from hydrogen and helium, follow the octet rule,[2] which states that a stable atom has eight valence electrons. Specifically, those eight valence electrons correspond to filled s and p subshells. Transition metals require 12 electrons to achieve stability, which corresponds to filling their valence s and d subshells.[3]

Looking at the main group elements in Figure 11.2, we can see that only group 18 elements have a stable configuration, which is why the noble gas elements naturally exist as monatomic gases. For all other elements, they can only achieve electronic stability through forming diatomic or polyatomic molecules or through forming compounds. Atoms in molecules and compounds are held together by bonds. Bonds can be thought of as the energy, or force, holding atoms together. More specifically, when ions and atoms are brought together, forming ionic bonds or covalent bonds, respectively, they release energy (an exothermic process). To break ions and atoms in a compound apart, energy must be added to break the bond (an endothermic process).

IONIC BONDS

Ions, and hence ionic bonds, are formed in cases where the interacting atoms have a large electronegativity difference. The electronegativity difference is so large that electrons are not shared but transferred from one to the other. The atom that loses electrons, called a cation, empties its valence shell, leaving a stable, noble-gas core. The atom that gains electrons fills its valence shell to achieve a noble-gas configuration and an octet. This exchange of electrons to form cations and anions is a redox reaction (Chapter 16). An ionic bond refers to the electrostatic attraction experienced between a positively charged ion (cation) and a negatively charged ion (anion).[4] The electrostatic attraction between a cation and an anion is described mathematically by Coulomb's law (Chapter 9).[5]

It is common practice to roughly divide covalent bonds as those where $\Delta \chi_p < 1.7$ and ionic bonds as those where $\Delta \chi_p > 1.7$, where a difference of 1.7 corresponds to 50% ionic character.[6] Ionic compounds form crystal lattices (Figure 11.3). Crystal lattices are regular repeating unit cells of ions. This is in contrast to covalent compounds that exist as discrete sets of bonded atoms. Ionic compounds tend to have high melting points (hundreds to thousands of degrees Celsius) and to be solids at normal temperature and pressure (20 °C and 0.987 atm).

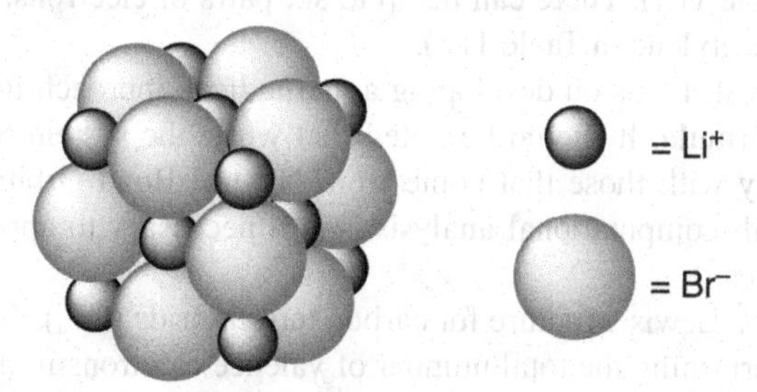

FIGURE 11.3 An example crystal lattice of lithium bromide. Based on a lithium bromide unit cell from "Lithium Bromide," Wikipedia.com https://en.wikipedia.org/wiki/Lithium_bromide#/media/File:Lithium-bromide-3D-ionic.png by user: Benjah-bmm27. Date accessed 15 January 2024.

COVALENT BONDS

A covalent bond forms when two atoms share a pair of electrons.[7] Covalent bonds are formed in cases where the interacting atoms have little to no electronegativity difference. A small electronegativity difference means that neither atom can pull the electrons away from the other and so they are shared between the two. Covalent bonds come in two types: purely covalent and polar covalent.

Purely covalent bonds occur when the electronegativity difference is very low ($\Delta \chi_p < 0.4$) and the electrons are equally shared between the two atoms. Polar covalent bonds are formed when the electronegativity difference is between 0.4 and 1.7. In polar covalent bonds, the electrons are shared but unequally, which renders one atom electron poor (partially positive) and the other electron rich (partially negative). This leads to electrostatic attraction – partial ionic bonding – which makes polar covalent bonds stronger.

Covalent compounds tend to have low melting points, and they can be solids, liquids, or gases at normal temperatures and pressure (20.0 °C and 1.00 atm). Covalent compounds consist of discretely bonded atoms with a well-defined structure. These structures, molecules, can be represented with Lewis structures.

Lewis Structures[8]

The most common model for depicting molecules was developed by G.N. Lewis in 1916. Lewis structures are the medium that chemists use most often to think about, communicate about, and represent the structure of molecules. Table 11.1 shows three example Lewis structures.

TABLE 11.1
Name, Chemical Formula, and Lewis Structure of Three Example Molecules

Name	Acetylene (ethyne)	Ammonia	Dioxygen
Chemical formula	HCCH	NH_3	O_2
Lewis structure	H–C≡C–H	H–N̈–H (with H below N)	:Ö=Ö: (lone pairs above and below)

The assumption in a Lewis structure is that a pair of electrons can be localized on a single atom, which is called a lone pair and is drawn as two dots next to the atomic symbol (Table 11.1); alternatively, a pair of electrons can be shared between two atoms and is drawn as a line connecting the two atomic symbols (Table 11.1). There can be up to six pairs of electrons, three lines, connecting two main group atoms (acetylene in Table 11.1).

The following pages will focus on developing a methodical approach for drawing Lewis structures from a chemical formula. It should be noted that while the system presented here will give results that match closely with those that come from Natural Bond Orbital[9] analysis (a quantum mechanical theory), a full computational analysis is still necessary to appreciate the subtleties of bonding in most structures.

To start, we will draw a Lewis structure for carbon tetrafluoride (CF_4). To draw any Lewis structure, the first step is determining the total number of valence electrons in the system. In this work, Figure 11.2 will be useful.

The carbon atom has four valence electrons:	$1C(4\ e^-) = \ \ 4\ e^-$
Each fluorine atom has seven valence electrons:	$+4F(7\ e^-) = 28\ e^-$
The total number of valence electrons. is:	$32\ e^-$ total

The second step is to determine the number of electrons needed to satisfy the valence shell of each atom. Each hydrogen atom needs two electrons total, each main group atom needs eight electrons total, and each transition metal needs 12 electrons.[10]

The carbon atom needs eight electrons: $1C(8\ e^-) = 8\ e^-$
Each fluorine atom needs eight electrons: $+4F(8\ e^-) = 32\ e^-$
The total number of electrons required is: $40\ e^-$ needed

There are not enough valence electrons (32 is less than 40) for each atom to have a full valence shell alone, and so the atoms must share electrons to complete their valence shells. The number of shared electrons is the difference between the number of electrons needed and the total number of electrons.

$$\begin{array}{r} 40\ e^-\ \text{needed} \\ -32\ e^-\ \text{total} \\ \hline 8\ e^-\ \text{shared} \end{array}$$

A bond, which is drawn as a line in Lewis structures, represents a shared pair of electrons. And so, the eight shared electrons in carbon tetrafluoride represent four bonds.

$$8e^-\left(\frac{1\ \text{bond}}{2\ e^-}\right) = 4\ \text{bonds}$$

Finally, by subtracting the number of shared electrons from the total number of valence electrons., we can find the number unshared. In the case of carbon tetrafluoride, this gives us 24 unshared electrons., which will correspond to 12 lone pairs.

$$\begin{array}{r} 32\ e^-\ \text{total} \\ -8\ e^-\ \text{shared} \\ \hline 24\ e^-\ \text{unshared} \end{array}$$

Now, we use the calculation results (4 lines and 24 dots) to draw a Lewis structure. First, the least electronegative, non-hydrogen atom is placed in the middle, and the other atoms are arranged around that central atom. Here that is carbon, which goes in the middle, and the fluorine atoms are arranged around the central carbon atom (Figure 11.4).

```
  F
F C F
  F
```

FIGURE 11.4 Arrangement of atoms in carbon tetrafluoride.

Lines, bonds, are then drawn to connect the peripheral atoms to the central atom (Figure 11.5). After which we take stock of how many electrons are around each atom.

```
  F
  |
F–C–F
  |
  F
```

FIGURE 11.5 Partial Lewis structure of carbon tetrafluoride with bonds drawn between connected atoms.

With just the lines drawn, each fluorine atom is connected by one bond to the central carbon atom. Each bond represents two electrons, which means that the valence shell of the carbon atom is now full. For each fluorine atom, it is connected by one line, which means it only has two electrons around it. To satisfy the valence shells of any atoms left unfilled, we add unshared electrons as lone pairs, and we start at the most electronegative atoms first. Here, three lone pairs are drawn on each of the fluorine atoms, which will satisfy each fluorine atom's valence shell: one line (two electrons) and three lone pairs (six electrons) give a full octet. This completes the Lewis structure of carbon tetrafluoride (Figure 11.6).

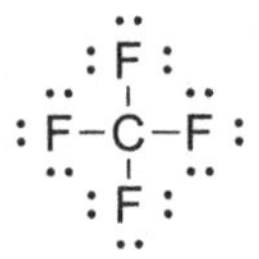

```
      ..
     :F:
  ..  |  ..
 :F - C - F:
  ..  |  ..
     :F:
      ..
```

FIGURE 11.6 Complete Lewis structure of carbon tetrafluoride with bonds and lone pair.

For polyatomic ions, the same approach holds, but the charge must be considered when calculating the valence electron total. For each negative charge, one additional electron is added to the total number of valence electrons. For example, amide (H_2N^-) has eight valence electrons: one electron from each hydrogen atom (two total), five electrons from the nitrogen atom, and one electron represented by the negative charge. For each positive charge, one electron is subtracted from the total number of valence electrons. For example, ammonium (H_4N^+) has eight valence electrons: one from each hydrogen atom (four total), five from the nitrogen atom, and one electron is subtracted because of the 1+ charge. Otherwise, mathematical steps proceed as seen in the example of carbon tetrafluoride to give the structures of amide and ammonium as seen in Figure 11.7. Note the brackets with the charge to the right. This is a way of denoting that the entire structure has a charge. We will see a way of formally designating where this charge is, which will render the bracket notation unnecessary.

```
     ..  ]-             H   ]+
   H-N:                 |
     |                H-N-H
     H                  |
                        H
```

FIGURE 11.7 The Lewis structures of amide (*left*) and ammonium (*right*).

A final type of molecule that will be considered in this introduction to Lewis structures are relatively rare, odd-electron species, which are referred to as radicals. Notice all the examples given so far have even numbers of electrons. If we consider CH_3, the calculation of the number of bonds and unshared electrons finds that CH_3 should have seven shared electrons (3.5 bonds). What this means is that there are three bonds and the 0.5 bond is a single, unshared electron (also called a radical electron). A Lewis structure for CH_3 is shown in Figure 11.8. Odd-electron (radical) species like CH_3 are typically highly reactive and do not persist for very long because at least one atom, here the carbon atom, does not have a full valence shell.

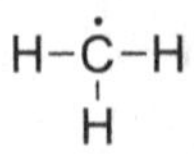

```
    .
  H-C-H
    |
    H
```

FIGURE 11.8 Lewis structure of the methyl radical (CH_3).

Problem 11.1. Draw Lewis structures for each of the following chemical formulae: CH_2O, nitrogen trifluoride, hydrogen iodide, SH_3^+, NF_4^+, $AlCl_4^-$, hydroxide, and OH.

Formal Charge[11]

In Figure 11.7, the amide anion (H_2N^-) is drawn using a bracket notation to show that there is an overall negative charge. It is useful to formally identify the specific atom(s) that carry the charge(s). There are several systems to keep track of electrons in molecules. In Lewis structures, formal charge is most used. Formal charge is a bookkeeping tool that chemists use to gauge how many electrons are around a given atom compared to the neutral atom by itself. The word "formal" in formal charge is short for formalism, which is an agreed upon convention for determining a bookkeeping charge value. To calculate charge, the formal charge assumption is that all bonds are purely covalent. The formula for formal charge (*FC*) is shown in Equation 11.1. It is important to note that the formal charge must be calculated for each atom in a molecule and that all nonzero formal charges must be included next to the atom. The sum of all individual formal charges must equal the overall charge. Note that formal charge should not be confused with actual charge, which can be best approximated computationally.

$$FC = e_v - e_{lp} - b \tag{11.1}$$

e_v is the number of valence electrons
e_{lp} is the number of unshared electrons
b is the number of bonds

With this information in hand, the formal charge for the hydrogen atoms in amide would be 0 (1 – 0 – 1) and the nitrogen atom would be 1– (5 – 2 – 4). Typically, when the formal charge is 0, then no formal charge is noted. For a formal charge of 1+, this is denoted by a plus sign inscribed in a circle, and a charge of 1– is denoted by a negative sign inscribed in a circle. A complete Lewis structure always includes all nonzero formal charges, and the complete Lewis structures for amide and ammonium are shown in Figure 11.9.

FIGURE 11.9 The Lewis structures of amide (*left*) and ammonium (*right*) showing formal charges.

Problem 11.2. For each of the following molecules, calculate the formal charge on each atom and indicate any formal charges by appropriately drawing in the formal charge(s).

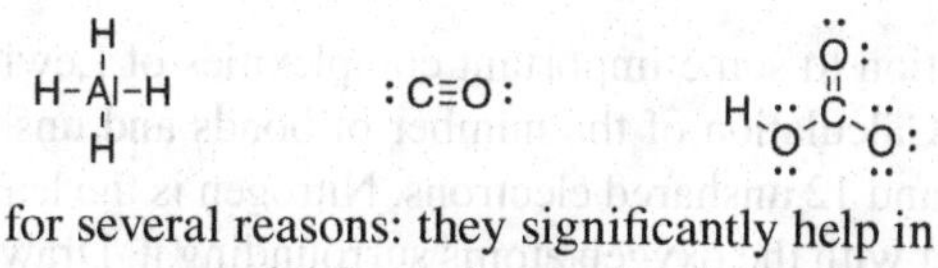

Formal charges are useful for several reasons: they significantly help in predicting the reactivity of a molecule, and formal charges are useful for trying to decide among possible Lewis structures for a given molecular formula. Consider the formula N_2H_4. One could potentially draw either of the two Lewis structures in Figure 11.10.

FIGURE 11.10 Possible Lewis structures of hydrazine (N_2H_4).

The best Lewis structure is typically the structure with the fewest number of formal charges (Figure 11.11), which allows us to identify the structure on the right as the better structure.

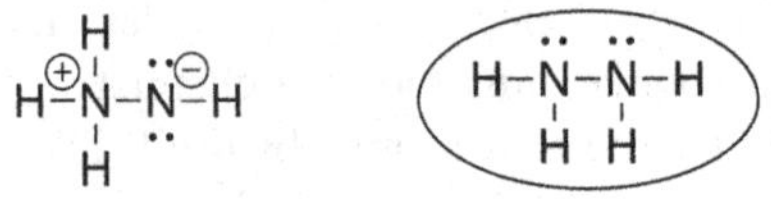

FIGURE 11.11 Possible Lewis structures of hydrazine with formal charges indicated and the best structure circled.

Looking at the Lewis structures with formal charges (Figure 11.11), the second Lewis structure is the correct and best structure for hydrazine. An important takeaway is that the Lewis structure with fewer formal charges tends to be, but is not always, the better structure. Consider two possible Lewis structures of thionyl chloride ($SOCl_2$) (Figure 11.12).

FIGURE 11.12 Possible Lewis structures of thionyl chloride with formal charges indicated.

On the basis of formal charges, the linear structure on the left might be thought to be the best structure; however, the actual structure of thionyl chloride is the radial structure on the right (Figure 11.12). Radial structures (with a central atom and the rest around the periphery) are typically more stable than linear structures, which must be considered along with formal charges when determining the most likely Lewis structure for a molecule. Note that following the approach to drawing Lewis structures presented here (putting the least electronegative atom in the middle and all others around that central atom) will ensure that less stable, linear structures are avoided.

Problem 11.3. Draw a Lewis structure for each of the following. Include all nonzero formal charges: sulfate, HPO_3^{2-}, hydrogen cyanide, HCO_2^-.

Contributing Structures

We will now turn our attention to some important complexities of Lewis structures. Let's start by considering nitrite (NO_2^-). Calculation of the number of bonds and unshared electrons shows that nitrite should have 3 bonds and 12 unshared electrons. Nitrogen is the least electronegative element, and so it will go in the center with the oxygen atoms surrounding it. Drawing in the bonds, lone-pair electrons, and formal charges, we could draw nitrite as shown in Figure 11.13.

The Lewis structure in Figure 11.13 provides us with a model of the molecular structure for nitrite. The model indicates that the two oxygen atoms are inequivalent. One oxygen is negatively

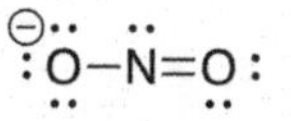

⊖ :O–N=O:

FIGURE 11.13 A Lewis structure for nitrite (NO_2^-).

charged, the other is neutral. One oxygen is making a double bond to nitrogen, and one is making a single bond. We would expect to see these differences reflected in their bond lengths because double bonds are shorter than single bonds. When nitrite is analyzed by X-ray crystallography – a method for determining the relationships between atoms by using X-rays – the two oxygen atoms are equivalent: the N–O bond lengths are both 123.3 pm.[12] What does this result tell us about the validity of the Lewis structure of nitrite? The immediate takeaway is that the Lewis structure in Figure 11.13 is not a good model for nitrite. This tells us that the Lewis structure system has limitations. Lewis structures assume that all electrons are localized – that is they are either on a single atom as a lone pair or they are between two atoms as a bond. This assumption is invalid. If we consider the Lewis structure that was drawn in Figure 11.13, it was arbitrarily decided, by the author, that the double bond would be drawn between the nitrogen atom and the right oxygen atom. However, it would be equally acceptable to draw either of the Lewis structures shown in Figure 11.14.

⊖ :O–N=O: :O=N–O: ⊖

FIGURE 11.14 Valid Lewis structures of nitrite.

Both Lewis structures in Figure 11.14 are equally correct to draw (in that they all follow the rules of drawing good Lewis structures). Each Lewis structure differs from the other only in terms of where we drew the lone-pair electrons and the double bond. However, each of these Lewis structures is, when considered alone, a bad depiction of nitrite. In these situations, where multiple Lewis structures are acceptable to draw for a molecule, to get an accurate depiction of the molecule we must consider all of them. We call all the valid Lewis structures for a single molecule contributing structures.[13,14] To show their relationship as contributing structures, we draw a double-headed arrow in between each (Figure 11.15).

⊖ :O–N=O: ⟷ :O=N–O: ⊖

FIGURE 11.15 Two contributing structures of nitrite.

Contributing structures indicate that the electrons in a molecule are spread among multiple atoms, that is that the electrons are delocalized. It is vital to note that nitrite, and any molecule for which there are contributing structures, does not oscillate back and forth between different forms. Each Lewis structure is itself a poor approximation by chemists to describe a more complex reality. The actual structure of nitrite is a weighted average of all significant contributing structures. Contributing structures must be included whenever an arbitrary choice is made in a Lewis structure drawing, here the choice of whether to draw the double bond between the right oxygen or the left oxygen. If you make an arbitrary choice, then the other option must be drawn and considered.

Problem 11.4. For each of the following chemical formulae, draw all contributing structures.
Disulfur monoxide

Orthoborate (BO_3^{3-})

Thiocyanate (CSN^-)

Carbonyl fluoride (COF_2)

Hypercoordinate Molecules[15]

When in the center of a molecule, the smaller atoms of the second period (B, C, N, O, and F) can have at most four areas of electron density (four bonded atoms and/or lone pairs). Atoms in period three and beyond can have more than four areas of electron density around the central atom, e.g., PF_5, SF_6, and IF_7. These compounds are hypercoordinate: that is they have more than four areas of electron density. Let us start by drawing a Lewis structure for xenon difluoride (XeF_2). Calculation of the number of bonds and unshared electrons finds that xenon difluoride should have 1 bond and 20 unshared electrons. Xenon is the least electronegative atom, and so it will go in the center with the fluorine atoms surrounding it. We then draw in the bond to the central xenon, complete the remaining octets by adding in lone-pair electrons, and add in any nonzero formal charges (Figure 11.16). We can see that xenon is hypercoordinate because there are five areas of electron density around the xenon (three lone pairs and two bonded atoms).

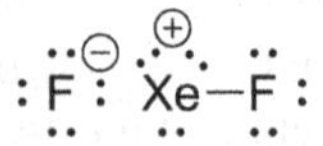

FIGURE 11.16 Xenon difluoride Lewis structure.

In Figure 11.16 the fluorine on the left side was arbitrarily chosen not to have a line drawn. To fully account for the structure of XeF_2, all the contributing structures (Figure 11.17) need to be drawn.

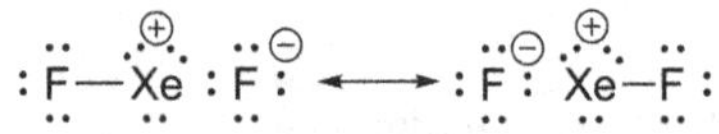

FIGURE 11.17 Xenon difluoride contributing structures.

What we can gather from the contributing structures of XeF_2 is that the two fluorine atoms are bonded to xenon through a mixture of covalent and ionic bonds. The exact nature of the bonding in these compounds has been debated for nearly a century. On one side is the view that the octet rule generally holds true for most main group elements and that the bonding is best represented as shown in Figure 11.17.[16] On the other side, it is proposed that these hypercoordinate compounds are also hypervalent (Figure 11.18) and that these compounds involve the central atom having an expanded octet.[17]

FIGURE 11.18 Xenon difluoride hypervalent structure.

Both computational and theoretical evidence support the octet-rule structures (Figure 11.17) over expanded octet (hypervalent) structures.[18] Hypervalent structures are commonly shown – because they avoid formal charges[19] – in many books and online.

Problem 11.5. Draw the Lewis structures for each of the following chemical formulae: chlorine trifluoride, SO_2Cl_2, xenon trioxide, SeO_4^{2-}, ClO_3^-, and Br_3^-. Check the Lewis structures you have drawn in this problem against the results found online, where hypervalent structures are almost exclusively presented.

BONDING: STRIVING FOR STABILITY

As we saw in Chapter 8, only the noble gases have truly stable electron configurations. All other elements tend to combine – exchanging electrons (ionic compounds) or sharing electrons (molecular compounds) – to achieve stable electron configurations. In this chapter, we considered Lewis dot symbols and Lewis structures as tools to help understand and show bonding in compounds. In the following chapters, we will look at how striving for stability (lower energy) can help to explain the three-dimensional shape of molecules (Chapter 12) and the interactions between molecules (Chapter 13).

NOTES

1. Lewis, G.N. The Atom and the Molecule. *J. Am. Chem. Soc.*, **1916**, *38* (4), 762–785. DOI: 10.1021/ja02261a002.
2. Langmuir, I. The Arrangement of Electrons in Atoms and Molecules. *J. Am. Chem. Soc.*, **1919**, *41* (6), 868–934. DOI: 10.1021/ja02227a002.
3. This corresponds to filling the s and d subshells. The p subshell is not substantially involved in transition-metal bonding: Landis, C.R.; Weinhold, F. Valence and Extra-Valence Orbitals in Main Group and Transition Metal Bonding. *J. Comput. Chem.*, **2007**, *28* (1), 198–203. DOI: 10.1002/jcc.20492.
4. "Ionic bond." IUPAC. *Compendium of Chemical Terminology*, 2nd ed. (the "Gold Book"). Compiled by A.D. McNaught and A. Wilkinson. Blackwell Scientific Publications: Oxford, **1997**. Online version (2019-) created by S.J. Chalk. DOI: 10.1351/goldbook.IT07058.
5. Coulomb, C.A. Premier mémoire sur l'électricité et le magnétisme. *Histoire de l'Academie Royale des Sc.*, **1785**, 569–577.
6. Pauling, L. *The Nature of the Chemical Bond and the Structure of Molecules and Crystals. An Introduction to Modern Structural Chemistry* (3rd ed.). Cornell University Press: Ithaca, NY, **1960**, p. 100.
7. A covalent bond is a region of relatively high electron density between nuclei which arises at least partly from sharing of electrons and gives rise to an attractive force. "Covalent bond." IUPAC. *Compendium of Chemical Terminology* (2nd ed.). (the "Gold Book"). Compiled by A.D. McNaught and A. Wilkinson. Blackwell Scientific Publications: Oxford, **1997**. Online version (2019-) created by S.J. Chalk. DOI: 10.1351/goldbook.C01384.
8. i) Lewis, G.N. The Atom and the Molecule. *J. Am. Chem. Soc.*, **1916**, *38* (4), 762–785. DOI: 10.1021/ja02261a002.
   ii) Lewis, G.N. The Magnetochemical Theory. *Chem. Rev.*, **1924**, *1* (2), 231–248. DOI: 10.1021/cr60002a003.
   iii) Lewis, G.N. The Chemical Bond. *J. Chem. Phys.*, **1933**, *1* (1), 17–28. DOI: 10.1063/1.1749214.
9. Glendening, E.D.; Badenhoop, J.K.; Reed, A.E.; Carpenter, J.E.; Bohmann, J.A.; Morales, C.M.; Landis, C.R.; Weinhold, F.A. *NBO 6.0.* Theoretical Chemistry Institute: University of Wisconsin: Madison, **2013**.

10. This corresponds to filling the s and d subshells. The p subshell is not substantially involved in transition-metal bonding: Landis, C.R.; Weinhold, F. Valence and Extra-Valence Orbitals in Main Group and Transition Metal Bonding. *J. Comput. Chem.*, **2007**, *28* (1), 198–203. DOI: 10.1002/jcc.20492.
11. Formal charge comes from Langmuir's concept of residual charge: Langmuir, I. Types of Valence. *Science*, **1921**, *54*, 59–67. DOI: 10.1126/science.54.1386.510.
12. Carpenter, G.B. The Crystal Structure of Sodium Nitrite. *Acta Cryst.*, **1952**, *5* (1), 132–135. DOI: 10.1107/S0365110X52000265.
13. This concept is often termed "resonance" following the name given by Pauling:
    i) Pauling, L. The Nature of the Chemical Bond. III. The Transition from One Extreme Bond Type to Another. *J. Am. Chem. Soc.*, **1932**, *54* (3), 988–1003. DOI: 10.1021/ja01342a022.
    ii) Pauling, L.; Sherman, J. The Nature of the Chemical Bond. VI. The Calculation from Thermochemical Data of the Energy of Resonance of Molecules Among Several Electronic Structures. *J. Chem. Phys.*, **1933**, *1* (8), 606–617. DOI: 10.1063/1.1749335.
14. Ingold's development of similar ideas, along with his development of electron-pushing arrows, led him to term this phenomenon mesomerism: Ingold, C.K. Principles of an Electronic Theory of Organic Reactions. *Chem. Rev.*, **1934**, *15* (2), 225–274. DOI: 10.1021/cr60051a003.
15. Hypercoordinate simply means that there are more than four areas of electron density at the central atom. Following the recommendation of Paul von Ragué Schleyer (Hypervalent compounds in LETTERS *Chem. Eng. New.*, **1984**, *62* (22), 4), "hypercoordinate" is the term that shall be used in this text, but the terms hypervalent and expanded octet are other terms that are also used for these molecules.
16. Langmuir, I. Types of Valence. *Science*, **1921**, *54*, 59–67. DOI: 10.1126/science.54.1386.510.
17. Lewis, G.N. *Valence and the Structure of Atoms and Molecules*. The New York Catalog Co.: New York. **1923**.
18. i) Suidan, L.; Badenhoop, J.K.; Glendening, E.D.; Weinhold, F. Common Textbook and Teaching Misrepresentations of Lewis Structures. *J. Chem. Educ.*, **1995**, *72* (7), 583–586. DOI: 10.1021/ed072p583.
    ii) Collins, G.A.D.; Cruickshank, D.W.J.; Breeze, A. Bonding in Krypton Difluoride. *J. Chem. Soc., Faraday Trans. 2*, **1974**, *70*, 393–397. DOI: 10.1039/F29747000393.
    iii) Kutzelnigg, W. Chemical Bonding in Higher Main Group Elements. *Angew. Chem. Int. Ed.*, **1984**, *23* (4), 272–295. DOI: 10.1002/anie.198402721.
    iv) Reed, A.E.; Weinhold, F. On the Role of d Orbitals in SF_6. *J. Am. Chem. Soc.*, **1986**, *108* (13), 3586–3593. DOI: 10.1021/ja00273a006.
19. While IUPAC recommendations and conventions are closely followed in this book, the recommended depiction of hypervalent structures (Brecher, J. Graphical Representation Standards for Chemical Structure Diagrams. *Pure Appl. Chem.*, **2008**, *80* (2), 277–410. DOI: 10.1351/pac200880020277.) will not be observed.

12 Three-Dimensional Structure

Lewis structures provide a means of representing the bonding, using a combination of lines and dots (Chapter 11), that occurs inside a molecule. While Lewis structures can represent the connectivity between atoms in a molecule, a good Lewis structure does not necessarily convey any information about the three-dimensional structure of a molecule. In this chapter, we will consider the basics of stereochemistry (three-dimensional chemical structure) and the representation of molecular structure.

COMPARING LEWIS STRUCTURES AND MOLECULAR MODELS

A Lewis structure does not have to convey the three-dimensional shape of the molecule or polyatomic ion it represents. If actual molecules are like a globe, with three-dimensional structure, a typical Lewis structure is akin to a map, a two-dimensional representation of a three-dimensional structure. For ammonium, for example, it is wholly correct to draw a Lewis structure like that in Figure 12.1.

```
     H
     |⊕
  H—N—H
     |
     H
```

FIGURE 12.1 Lewis structure of ammonium.

From X-ray crystallography, however, it is known that ammonium is not flat, as the Lewis structure in Figure 12.1 would suggest.[1] Figure 12.2 shows a molecular model, which highlights ammonium's distinctly three-dimensional shape.

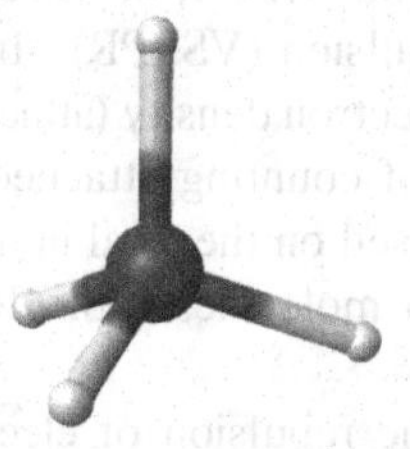

FIGURE 12.2 Three-dimensional molecular model of ammonium. The larger dark gray sphere represents the nitrogen atom, and the white spheres represent hydrogen atoms.

To properly account for this three-dimensional shape, then, we need to employ new ways of drawing bonds to supplement our existing symbology. Looking at the ammonium molecular model (Figure 12.2), we can see that there are two hydrogen atoms that lie in the plane of the page with the nitrogen atom. By convention, if something is in the plane of the page, we use a regular line (Figure 12.3).

The other two hydrogen atoms in the ammonium molecular model (Figure 12.2) are not in the plane of the page. One hydrogen atom is coming out of the plane towards the viewer. To show this, we draw a bond which starts at the central atom and broadens and boldens as it comes towards the atom (◀) that is meant to be above the plane of the page (Figure 12.4.). This is called a wedge bond.

DOI: 10.1201/9781003487210-12

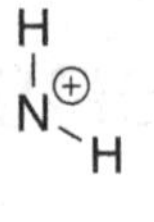

FIGURE 12.3 The nitrogen atom and two hydrogen atoms of ammonium that all lie in the same plane. These are drawn with regular, straight lines, indicating that the bonds are flat in the plane of the page.

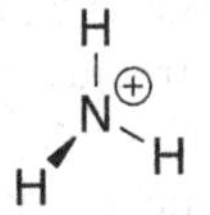

FIGURE 12.4 The wedge bond employed in the drawing of ammonium to represent the hydrogen atom that comes up, out of the plane, of the page.

The last hydrogen atom in the ammonium molecular model (Figure 12.2) is going back, into the plane of the page away from the viewer. To show this, we draw a bond, which starts at the central atom and broadens as a series of dashed lines (· ı ı I I |) going towards the atom that is meant to be behind the plane of the page. This is called a dashed bond. Using both wedge and dash notation, we can provide a Lewis structure that reflects the three-dimensional shape of the actual molecule. Compare, now, the model with the three-dimensional Lewis structure (Figure 12.5).

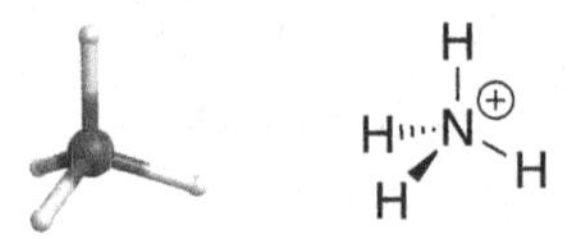

FIGURE 12.5 The molecular model (*left*) and three-dimensional Lewis structure (*right*) of ammonium.

VALENCE SHELL ELECTRON PAIR REPULSION (VSEPR)

For ammonium, we had a model to look at, but how could someone predict the shape of a molecule without having a molecular model at hand? One of the most used systems for predicting molecular shape is valence shell electron pair repulsion (VSEPR).[2] In VSEPR, the shape is predicted based upon counting the number of areas of electron density (attached atoms or lone pair) around a central atom. Table 12.1 shows the examples of counting attached atoms and lone pair around an atom. VSEPR theory then predicts a shape based on the total number of areas of electron density around the central atom (Table 12.2). For larger molecules, VSEPR can be applied to each atom in turn to determine the geometry at each point.

Note that VSEPR considers only the repulsion of electrons around an atom as the basis for electronic geometry around that atom. In actuality, the electronic geometry around an atom is also impacted by several factors, including delocalization (Chapter 11) and hydrogen bonding (Chapter 13). It is worth noting that VSEPR predictions for oxygen and other atoms with multiple lone pairs strongly diverge from those results from quantum mechanics. As such, the results from VSEPR should be considered as predictions and not necessarily as empirically factual.

The molecular models in Table 12.2 may seem at odds with the Lewis structure. For example, consider Figure 12.6, which shows a Lewis structure for ClF_3 and the molecular model for ClF_3. The molecular model seems to show what would be the equivalent of a hypervalent structure (with three lines connecting the central chlorine to the fluorides).

There are two points to note about Lewis structures compared to the molecular models. First, the connections between atoms in molecular models only imply a bond and not the nature of that bond. That is, a molecular model connection does not identify whether the connection is covalent, polar covalent, or ionic, and for covalent bonds, it does not convey the degree of covalency between the atoms. Second, the Lewis structure shown in Figure 12.6 for chlorine trifluoride is only one of

TABLE 12.1
Counting Areas of Electron Density

Lewis structure	Number of connected atoms	Number of lone pair	Total number of areas of electron density
⊖:O–N≡N:⊕	:O–N≡N: = 2	:O–N≡N: = 0	2
:N≡N:	:N≡N: = 1	:N≡N: = 1	2
⊖:O–N⊕(=O:)–O:⊖	= 3	= 0	3
⊖:O–N=O:	:O–N=O: = 2	:O–N=O: = 1	3
H–N⊕–H (H above and below)	H–N–H = 4	H–N–H = 0	4
:F–P–F: (:F: above)	= 3	= 1	4
:F–P⊕ :F:⊖ (F::F above, :F: below)	= 5	= 0	5
:F–Cl⊕ :F:⊖ (:F: below)	= 3	= 2	5
S2+ with six F (two F⊖)	= 6	= 0	6

the three contributing structures we could draw (Figure 12.7) for chlorine trifluoride. If we consider the three contributing structures, we can see that the bonding between each fluoride and the central chlorine atom is a mixture of covalent and ionic bonding. And so, when considering molecular models, it is important to realize that connections in the model imply a bond, but the nature of the bond is unclear without the accompanying Lewis structure (and any necessary contributing structures).

Bent's Rule

We will consider chlorine trifluoride a bit further to answer a common question about trigonal bipyramidal structures: Why are the lone pairs at the equatorial positions (those around the middle, equator, of a trigonal bipyramidal atom) and not at the axial positions (those at the top and bottom

TABLE 12.2
VSEPR Reference for 2–6 Areas of Electron Density

Areas of electron density	Electronic geometry	Idealized molecular model (Angles shown are ideal, lone pair cause a compression of idealized angles)	Example structures	Molecular models
2	Linear	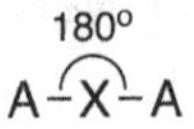 180° A–X–A	 :N≡N:	
			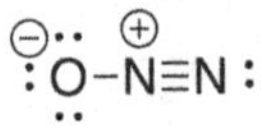 :O–N≡N:	
3	Trigonal planar	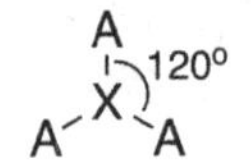 A, X, A, A 120°	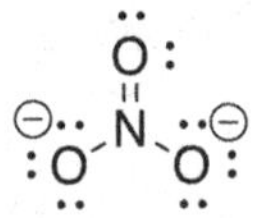 O=N(O)O	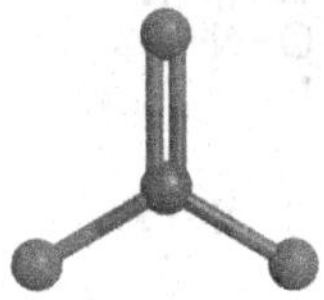
			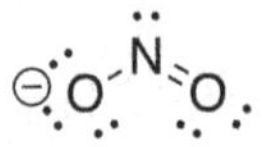 O–N=O	
4	Tetrahedral	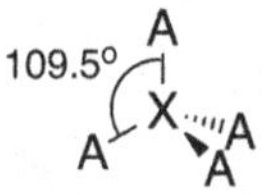 109.5° X, A, A, A, A	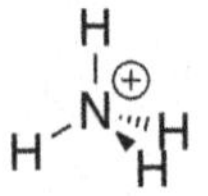 NH_4^+	
			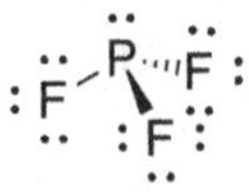 PF_3	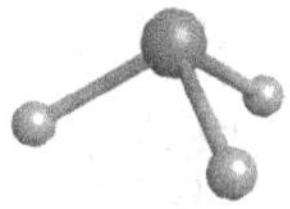
5	Trigonal bipyramidal	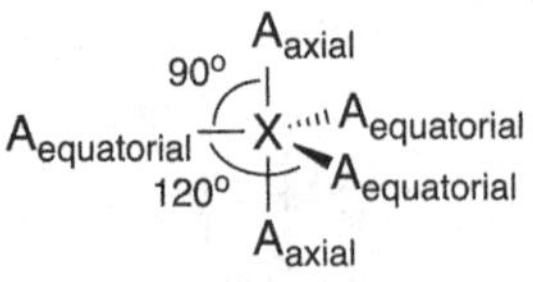 90° 120° X, A_{axial}, A_{axial}, $A_{equatorial}$, $A_{equatorial}$, $A_{equatorial}$	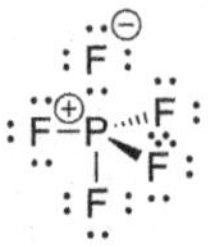 PF_5	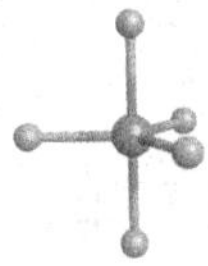
			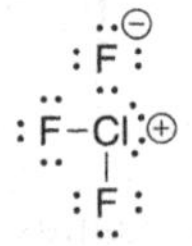 ClF_3	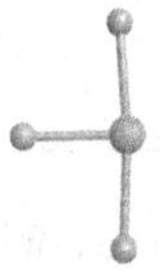
6	Octahedral	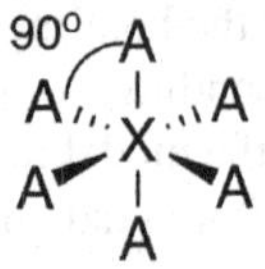 90° X, A, A, A, A, A, A	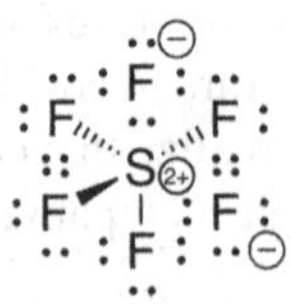 SF_6	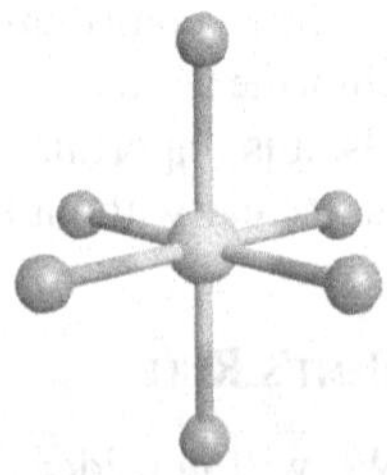

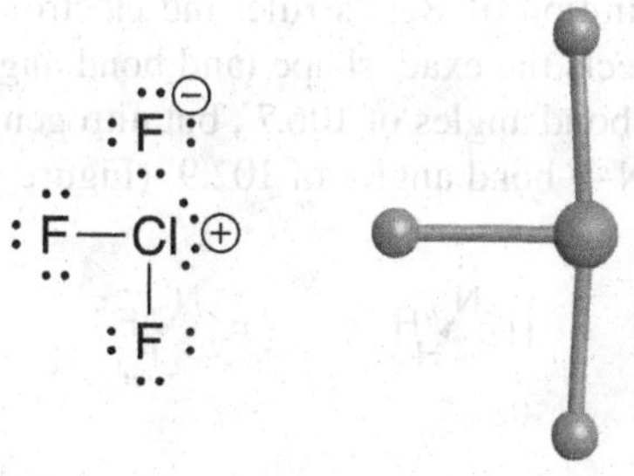

FIGURE 12.6 Lewis structure (*left*) and molecular model (*right*) of chlorine trifluoride.

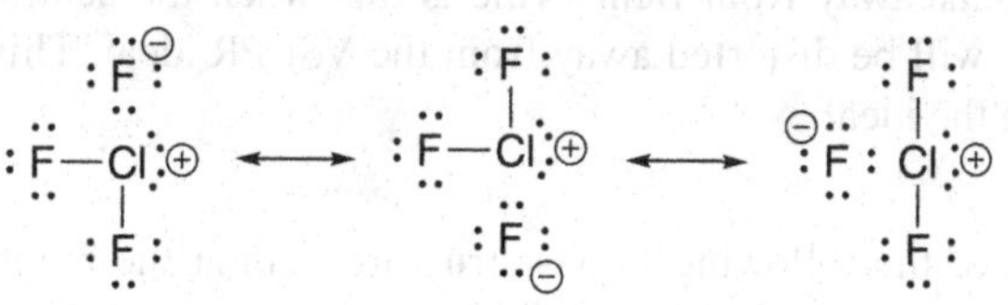

FIGURE 12.7 Contributing structures of chlorine trifluoride, showing a mixture of covalent and ionic bonding between each fluoride and the central chlorine atom.

of a trigonal bipyramidal atom)? For example, we could imagine and might have at first drawn the wrong chlorine trifluoride structure shown in Figure 12.8. In Figure 12.8, the lone pairs are axial, rather than equatorial.

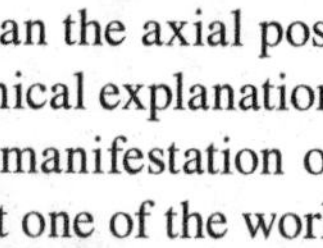

FIGURE 12.8 Incorrect geometry of chlorine trifluoride.

For a trigonal bipyramidal molecule, the lone pairs always occupy equatorial positions. The equatorial positions are lower in energy than the axial positions. An explanation for this difference would require a thorough quantum mechanical explanation, which is beyond the scope of this book. But it is worth highlighting that this is a manifestation of what is called Bent's rule.[3] A working definition of Bent's rule has two parts. Part one of the working definition of Bent's rule: lone pair(s) will always occupy the lowest energy position(s) around a central atom. For all molecules, molecules will modify their shape and compress bond angles, which will allow for electrons to occupy the lowest energy position. As mentioned, for trigonal bipyramidal chlorine trifluoride, this means that the electrons occupy equatorial positions. In the case of ammonium (NH_4^+) and ammonia (NH_3), they both have four areas of electron density. Ammonium adopts a more or less perfect tetrahedral shape (average bond angles 109.2°),[4] while ammonia distorts (Figure 12.9) its shape to compress the H–N–H bond angles to 106.7°,[5] which makes the lone pair lower in energy than it would be in a perfectly tetrahedral electron geometry.

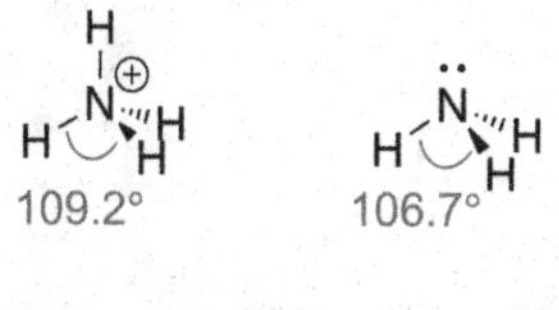

109.2° 106.7°

FIGURE 12.9 H–N–H bond angles in ammonium (*left*) and ammonia (*right*).

Part two of the working definition of Bent's rule: the electronegativity of surrounding atoms (relative to the central atom) affects the exact shape (and bond angles) that a molecule adopts. For example, ammonia has H–N–H bond angles of 106.7°, but nitrogen trifluoride (with highly electronegative fluorine atoms) has F–N–F bond angles of 102.9° (Figure 12.10).[6]

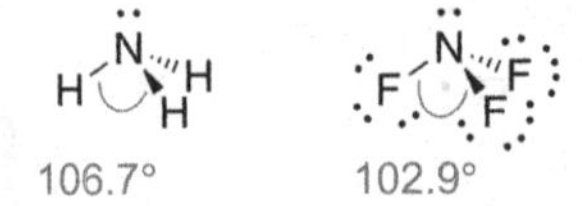

106.7° 102.9°

FIGURE 12.10 H–N–H bond angles in ammonia (*left*) and F–N–F bond angles in nitrogen trifluoride (*right*).

The most important takeaway from Bent's rule is that when the central atom has a lone pair, the molecular geometry will be distorted away from the VSEPR ideal. This is measurable in bond angles that are less than the ideal.

Problem 12.1. For each of the following Lewis structures, count the number of areas of electron density and redraw the Lewis structure (using Table 12.2) to appropriately show the correct three-dimensional shape. Provide the electron geometry shape name and identify the bond angle(s).

a.

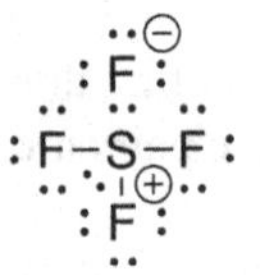

b. 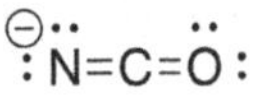 N=C=O

c.

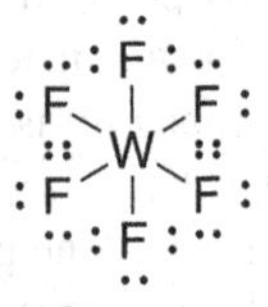

d.

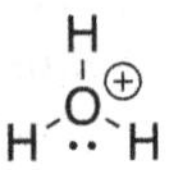

e.

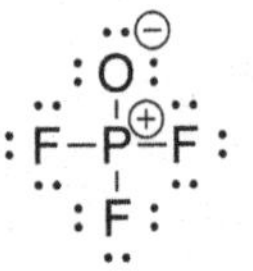

f. 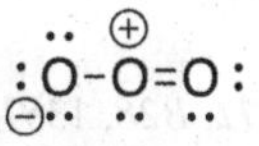 $:\overset{..}{\underset{\ominus ..}{O}}-\overset{\oplus}{\underset{..}{O}}=\underset{..}{O}:$

g. 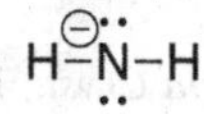 $H-\overset{\ominus ..}{\underset{..}{N}}-H$

Problem 12.2. For each of the following chemical formulae, draw a Lewis structure that shows the correct three-dimensional shape of the molecule.

a. nitrogen dioxide

b. iodine pentafluoride

c. boron tribromide

d. silicon tetraiodide

e. sulfite

f. beryllium chloride

g. bromine trifluoride

THREE-DIMENSIONAL SHAPE: MINIMIZING REPULSION

Molecules adopt a variety of three-dimensional electronic geometries. The exact geometry – linear, trigonal planar, tetrahedral, trigonal bipyramidal, or octahedral – depends on the number of areas of electron density, and the shape is the result of electron density areas spreading out as far as possible, which minimizes electron–electron repulsion. We investigated this through the lens of valence shell electron pair repulsion (VSEPR) theory. We also considered Bent's rule as a framework for understanding how lone pairs and electronegativity affect the molecular shape. The three-dimensional shape of molecules can help us to understand the ways that molecules interact (Chapter 13).

NOTES

1. Choi, C.S.; Mapes, J.E. The Structure of Ammonium Nitrate (IV). *Acta Cryst.*, **1972**, *B28*, 1357–1361. DOI: 10.1107/s0567740872004303.
2. Gillespie, R.J.; Nyholm, R.S. Inorganic Stereochemistry. *Q. Rev. Chem. Soc.*, **1957**, *11*, 339–380. DOI: 11.1039/QR9571100339.
3. Bent, H.A. An Appraisal of Valence-Bond Structures and Hybridization in Compounds of the First-Row Elements. *Chem. Rev.*, **1961**, *61* (3), 275–311. DOI: 10.1021/cr60211a005.
4. Robertson, J.H. Ammonium Oxalate Monohydrate: Structure Refinement at 30 °K. *Acta Cryst.*, **1965**, *18* (3), 410–417. DOI: 10.1107/s0365110x65000919.
5. Mulliken, R.S. Bond Angles in Water-Type and Ammonia-Type Molecules and Their Derivatives[1]. *J. Am. Chem. Soc.*, **1955**, *77* (4), 887–891. DOI: 10.1021/ja01609a021.
6. Mulliken, R.S. Bond Angles in Water-Type and Ammonia-Type Molecules and Their Derivatives[1]. *J. Am. Chem. Soc.*, **1955**, *77* (4), 887–891. DOI: 10.1021/ja01609a021.

13 Intermolecular Interactions

We have now looked at the structure of atoms and compounds, and we have established the basic tenets of atomic and molecular structure. But nothing so far explains how or why atoms and molecules aggregate into solid or liquid states or why chemicals interact at all. In this chapter, we will look at the physical basis of the interactions between molecules, that is, intermolecular interactions.

INTERMOLECULAR FORCES

When molecules come together, they release energy (an exothermic process); for molecules to move away from each other, energy must be added (an endothermic process). Together, then, there is an attractive energy between particles, which holds them together. This attractive energy is what chemists call intermolecular forces (IMFs). Put another way, intermolecular forces, van der Waals interactions and hydrogen bonds, are stabilizing interactions between molecules. We will consider noncovalent interactions (van der Waals interactions) and hydrogen bonds in turn and look at the impact of each on the physical properties of different substances.

INTERACTIONS BETWEEN MOLECULES: VAN DER WAALS (NONCOVALENT) INTERACTIONS

We will utilize the relationship between molecular size and the strength of intermolecular interactions (Figure 13.1) to develop some insights into intermolecular forces.[1] Looking at Figure 13.1, the particle size and the strength of interaction show a significant correlation ($R^2 = 0.9705$): the larger a molecule is, the stronger its interactions with other particles. This general trend is the most universal of intermolecular forces: dispersion interactions (London interactions).[2,3]

DISPERSION INTERACTIONS

Dispersion interactions are attractive interactions between molecules that result from the spreading out or concentration of electrons on the molecular surface, which leads to polarization as two molecules approach one another. This means that some areas are electron-rich (indicated with ∂^-, which signifies a partial negative charge) and some areas are electron-poor (indicated with ∂^+,

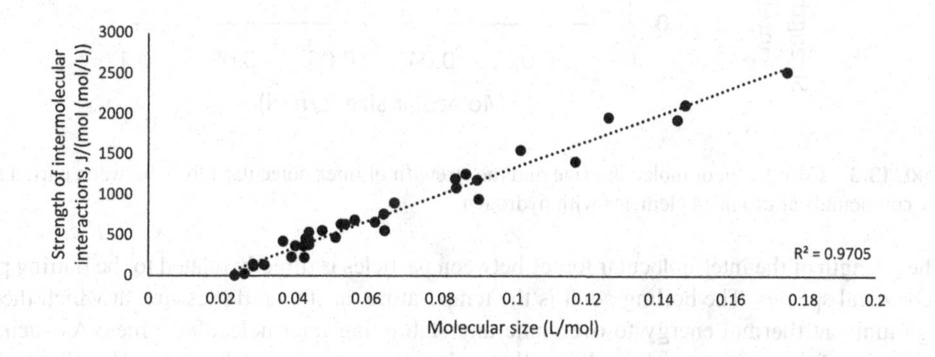

FIGURE 13.1 Correlation of particle size and the strength of intermolecular forces between particles for an average sample of chemical species.

DOI: 10.1201/9781003487210-13

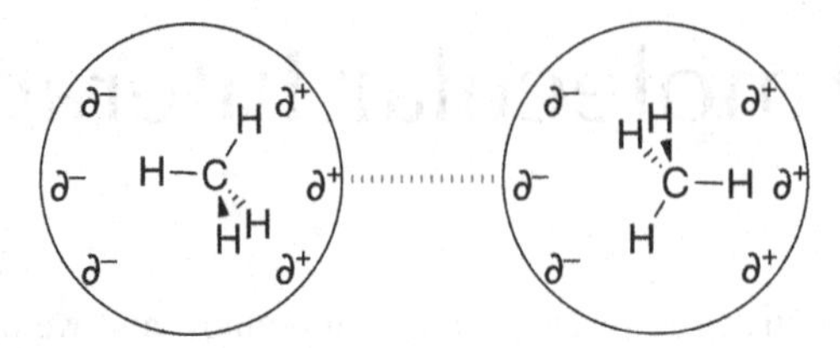

FIGURE 13.2 Model of dispersion interaction in methane, showing the formation of an induced dipole – due to the dispersion (deformation) of the electron cloud – in neighboring methane molecules, showing the resulting attraction (*hash marks*).

which signifies a partial positive charge).[4] These partially negative and partially positive areas then attract each other as shown in Figure 13.2. Note that the interaction energy is inversely proportional to distance. The attraction is strongest as gas particles approach during a collision or when particles are in a solid or liquid phase.

Because dispersion interactions rely on the deformation of the electron cloud, the strength of the interactions increases (Figure 13.1) with molecular size, the size of the electron cloud. A larger molecule has a greater ability to deform its larger electron cloud upon interaction with another molecule. For example, methane, CH_4 (van der Waals volume of 0.0258 nm^3), has a dispersion energy of −9.8 kJ/mol.[5] That is, it takes 9.8 kJ/mol to move methane molecules apart. Tetrachloromethane, CCl_4 (van der Waals volume of 0.0854 nm^3), with its greater size, has a greater dispersion interaction energy of −32.6 kJ/mol.[4] It is important to keep in mind that dispersion interactions are the most common form of intermolecular attraction interaction. All other intermolecular forces that will be considered are add-ons that some molecules have and others do not.

If we look at the binary compounds of group-14 elements with hydrogen (CH_4, SiH_4, GeH_4, and PbH_4) we can see exceptionally good correlation between molecular size and the strength of their intermolecular interactions (Figure 13.3). This can be attributed to the increasing size of the molecules and their greater ability to engage in stabilizing dispersion interactions.

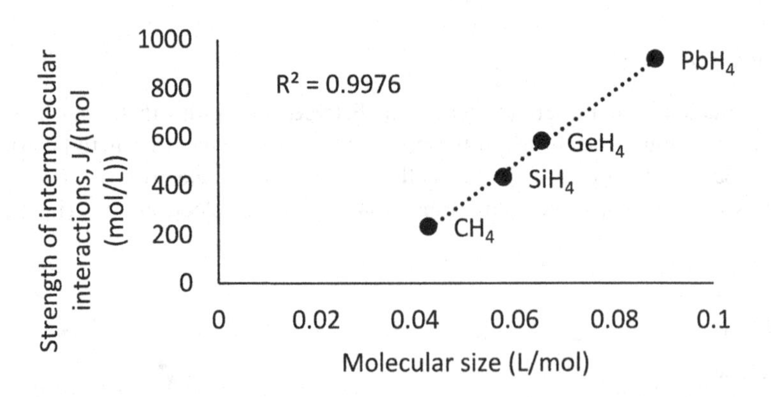

FIGURE 13.3 Correlation of molecular size and the strength of intermolecular forces between particles for binary compounds of group-14 elements with hydrogen.

The strength of the intermolecular forces between particles is directly related to the boiling point of a chemical species. The boiling point is the temperature, at standard pressure, at which there is enough ambient thermal energy to overcome any stabilizing intermolecular forces. As such, the boiling point of these compounds is also well correlated with the molecular size value (Figure 13.4) because bigger molecules have stronger dispersion interactions, and stronger dispersion interactions

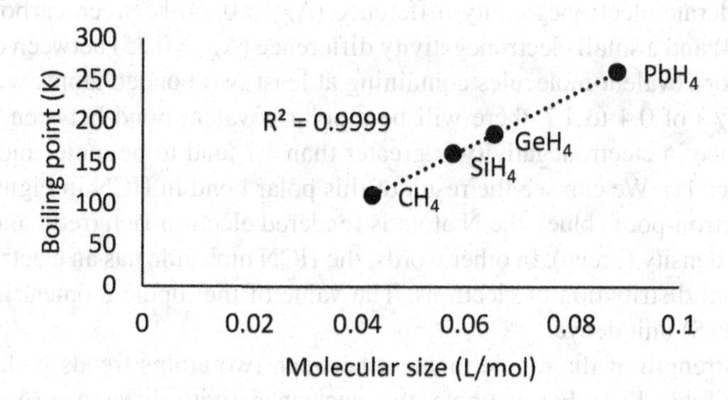

FIGURE 13.4 Correlation of molecular size and boiling point for binary compounds of group-14 elements with hydrogen.

require more thermal energy for the molecules to separate into the gas phase. As a liquid, molecules are in close contact, which allows for strong intermolecular interactions between particles; as a gas, however, molecules are separated by great distances and the intermolecular interactions are negligible except when they collide (Figure 2.6).

Problem 13.1. Considering the intermolecular forces involved, can you provide an explanation for the difference in the boiling point between these two compounds?

C_3H_8	C_4H_{10}
Molecular size = 0.09 L/mol	Molecular size = 0.15 L/mol
Boiling point = -42 ˚C	Boiling point = -1 ˚C

Dipole–Dipole Interactions[6]

Along with dispersion interactions, dipole–dipole interactions are the other common type of noncovalent (van der Waals) interaction. Dipole–dipole interactions are a stronger, more fixed version of dispersion interactions. While dispersion relies on the induced deformation of a molecule's electron cloud, dipole–dipole interactions occur between molecules that have a fixed, unequal distribution of charge (polar molecules). Consider, for example, HCN in Figure 13.5.

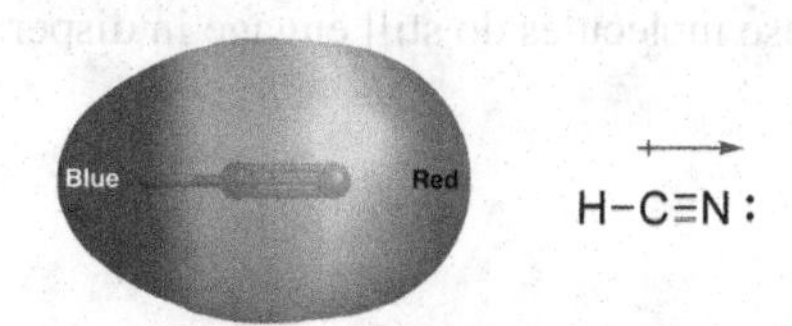

FIGURE 13.5 Electrostatic potential map (*left*) and Lewis structure with dipole moment arrow (*right*) of HCN. In the electrostatic potential map, red indicates areas of high electron density (∂^-), blue indicates areas of low electron density (∂^+), and green indicates a middling amount of electron density. On a Lewis structure, the dipole moment arrow (+——▸) conveys the same information as it points from the electron-poor end of a molecule towards the electron-rich end of the molecule.

There is a moderate electronegativity difference ($\Delta\chi_p$ = 0.49) between carbon (χ_p = 2.55) and nitrogen (χ_p = 3.04) and a small electronegativity difference ($\Delta\chi_p$ = 0.35) between carbon and hydrogen (χ_p = 2.20). For covalent molecules containing at least two bonded atoms with electronegativity differences ($\Delta\chi_p$) of 0.4 to 1.7, there will be a polar covalent bond between the atoms. Bonds where the difference in electronegativity is greater than 1.7 tend to be ionic, and the electrons are not shared (Chapter 11). We can see the result of this polar bond in HCN in Figure 13.5, where the H atom is left electron-poor (blue), the N atom is rendered electron-rich (red), and the C atom is of middling electron density (green). In other words, the HCN molecule has an electric dipole moment, a fixed and unequal distribution of electrons. The value of the dipole moment in HCN is 2.90 D, where D is the non-SI unit debye.

Note that the strength of the dipole moment between two atoms trends with the difference in electronegativity (Table 13.1). For example, the electronegativity difference ($\Delta\chi_p$ = 1.78) between hydrogen (χ_p = 2.20) and fluorine (χ_p = 3.98) gives hydrogen fluoride a dipole moment of 1.82 D, while the smaller electronegativity difference ($\Delta\chi_p$ = 0.46) between hydrogen (χ_p = 2.20) and iodine (χ_p = 2.66) gives hydrogen iodide a dipole moment of 0.44 D.

TABLE 13.1
Trend in the Dipole Moment of Hydrogen Halides

Hydrogen halide	$\Delta\chi_P$	Dipole moment (D)
HF	1.78	1.82
HCl	0.96	1.08
HBr	0.76	0.82
HI	0.46	0.44

The HCN molecules will then interact through the attraction of the opposite ends of the dipole (electron-poor to electron-rich) in what is called a dipole–dipole interaction (Figure 13.6).

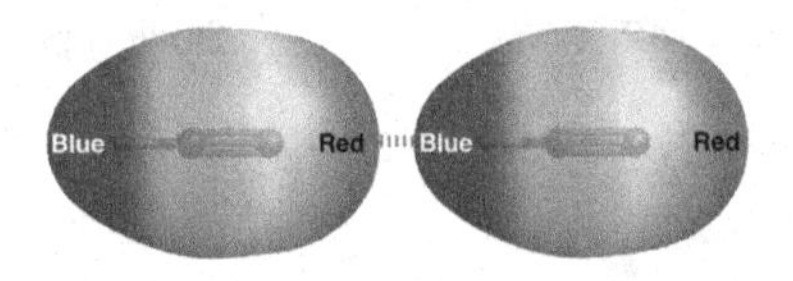

FIGURE 13.6 Dipole–dipole interaction (*hash marks*) between HCN molecules.

For most molecules with polar bonds, they will have a fixed, unequal distribution of charge (an electric dipole) and will engage in dipole–dipole interactions. There are some molecules, like sulfur trioxide (Figure 13.7) that have polar bonds but that are not polar. These molecules are highly symmetric, and thus there is no electron-poor end and no electron-rich end and so they cannot engage in dipole–dipole interactions. These molecules do still engage in dispersion interactions.

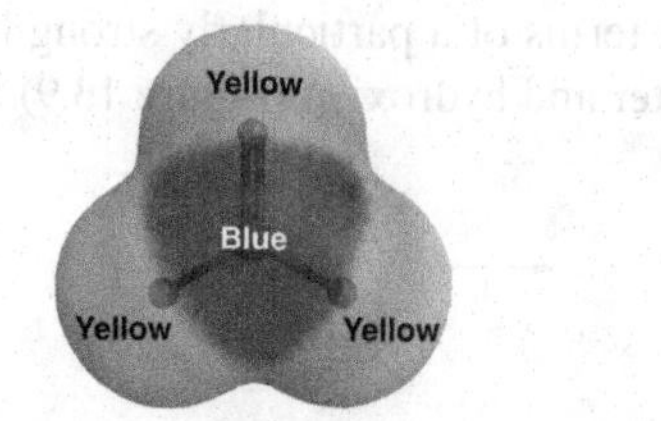

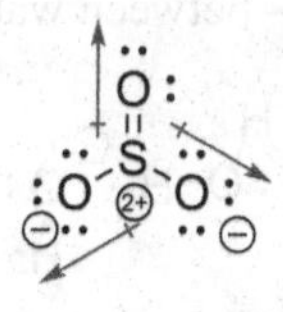

FIGURE 13.7 Electrostatic potential map (*left*) and Lewis structure with dipole moment arrows (*right*) of sulfur trioxide, a molecule with polar bonds, but which is nonpolar due to its high symmetry.

Problem 13.2. Considering the intermolecular forces involved, can you provide an explanation for the difference in the boiling point of the two compounds below?

C_4H_{10}
Molecular size = 0.15 L/mol
Boiling point = -1 °C

C_4H_8O
Molecular size = 0.13 L/mol
Boiling point = 79 °C

HYDROGEN BONDS (H-BONDS)

While van der Waals interactions (dispersion and dipole–dipole) are a common mode of intermolecular interaction, they are not the only ones. We will now turn our attention to an intermolecular interaction that is much more exclusive: hydrogen bonds (H-bonds).[7] First, let's see the significance of this interaction on the graph of molecular size and the strength of intermolecular forces for binary compounds of group-16 elements and hydrogen (Figure 13.8).

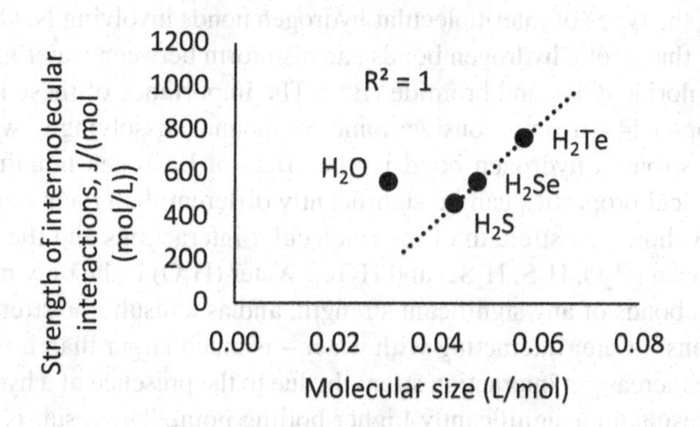

FIGURE 13.8 Correlation of molecular size and the strength of intermolecular interactions for binary compounds of group-16 elements and hydrogen.

Looking at Figure 13.8, we can see that water, which can hydrogen bond, has significantly stronger intermolecular interactions than we would expect from size alone. Hydrogen bonding is a concept that has attracted significant debate since it was first proposed. This is in part because a hydrogen bond is more complicated than a purely electrostatic, noncovalent (van der Waals) interaction and involves a degree of covalency (shared, electron-pair bonding).[8] A hydrogen bond is defined as a fractional chemical bond due to partial intermolecular delocalization (A–H···:B ↔ A:···

H–B⁺). Let's consider this a bit further in terms of a particularly strong hydrogen bond – −109 kJ/mol of stabilization energy – between water and hydroxide (Figure 13.9).[9]

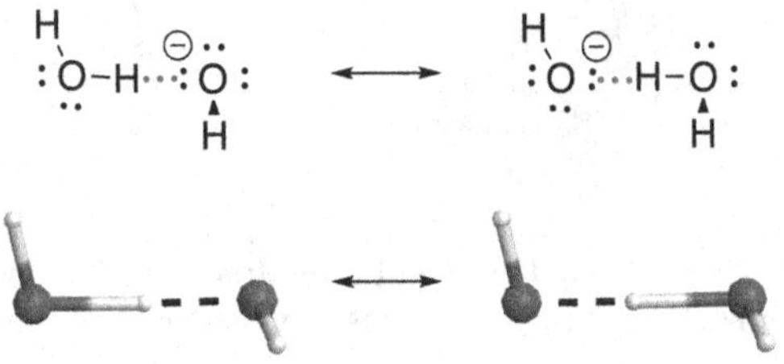

FIGURE 13.9 Contributing structures for the hydrogen bond between water and hydroxide with Lewis structures (*top*) and molecular models (*bottom*).

If we consider the contributing structures (Figure 13.9) we can see that the hydrogen bond in the water–hydroxide pair ($HO{-}H\cdots{:}OH^- \leftrightarrow HO{:}^-\cdots H{-}OH$) consists of a partial bond from each oxygen to the central hydrogen atom. We can see this further if we quantify the amount of bonding present, using what is called the bond order. When not interacting, in the gas phase, the bond order between each hydrogen and oxygen in water and in hydroxide is 1.0. That is, as is shown in the Lewis structure in Figure 13.9, there is a single bond between each hydrogen and oxygen. After the hydrogen (from water) and oxygen (from hydroxide) come together and interact, a hydrogen bond forms between the hydroxide ion and water molecule. The bond order (of the hydrogen bond) is 0.35. That is, the hydrogen bond is a partial (<1.0) covalent bond. Forming this bond weakens the O–H bond in water, and the bond order decreases from 1.0 to 0.65.

Hydrogen bonds are possible between any molecule that contains hydrogen bonded to a nonmetal interacting with a nonmetal lone pair. The most significant (stabilization energy values of greater than 10 kJ/mol) hydrogen bonds between molecules occur when a hydrogen atom attached to N, O, or F interacts with an N, O, or F lone pair. As such, we tend to focus on O–H, N–H, and F–H groups (moieties) when thinking about and identifying hydrogen bonding abilities between molecules. Examples of the types of intermolecular hydrogen bonds involving N, O, and F are shown in Figure 13.10.[10] Note that strong hydrogen bonds can also form between water molecules and anions like fluoride (F^-), chloride (Cl^-), and bromide (Br^-). The importance of these hydrogen bonds will be discussed in Chapter 14 when we consider ionic compounds dissolving in water.

Because of how strong a hydrogen bond is, the effect of hydrogen bonding can be dramatic, and molecules' physical properties can be significantly different than their size (dispersion) would suggest. Figure 13.8 shows the strength of intermolecular interactions and the size of the group-16 elements with hydrogen (H_2O, H_2S, H_2Se, and H_2Te). Water (H_2O) is the only member of this series that forms hydrogen bonds of any significant strength, and as a result, the strength of water's intermolecular interactions – water interacting with water – is much larger than it would be without the hydrogen bond. The increase in interaction strength, due to the presence of a hydrogen bond in addition to dispersion, results in a significantly higher boiling point. The result is a dramatic increase in the boiling point for water both in absolute terms (water boils at 373 K but would boil at 156 K if it did not hydrogen bond) and relative to the other, larger group-16 elements with hydrogen: H_2S (boiling point 214 K), H_2Se (boiling point 232 K), and H_2Te (boiling point 217 K) (Figure 13.11).

We see a similar effect for the binary compounds of group-15 elements with hydrogen (Figure 13.12) and group-17 with hydrogen (Figure 13.13). That is, ammonia (NH_3) and hydrogen fluoride (HF) both have boiling points substantially higher than we would expect based on their size because NH_3 hydrogen bonds to other NH_3 molecules and HF hydrogen bonds to other HF molecules. These intermolecular hydrogen bonds substantially increase the boiling points of NH_3 (Figure 13.12) and HF (Figure 13.13).

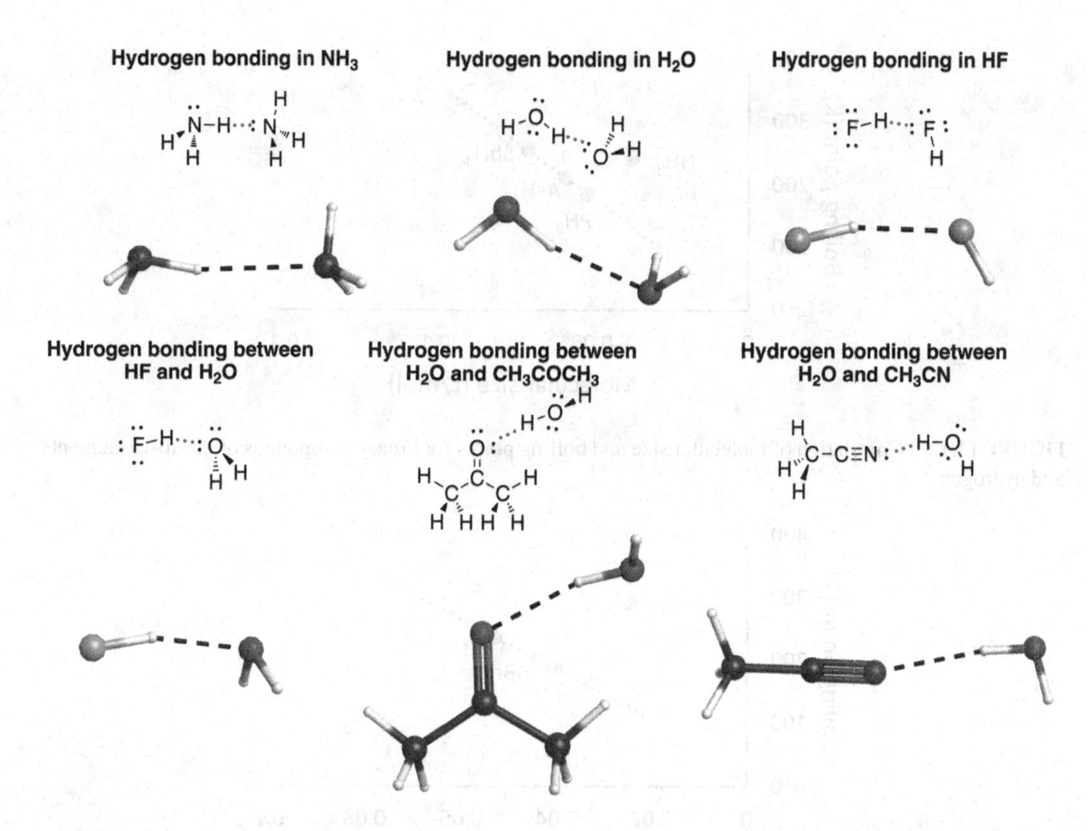

FIGURE 13.10 Examples of hydrogen bonds between two of the same molecular entities (*top*) and two different molecular entities (*bottom*).

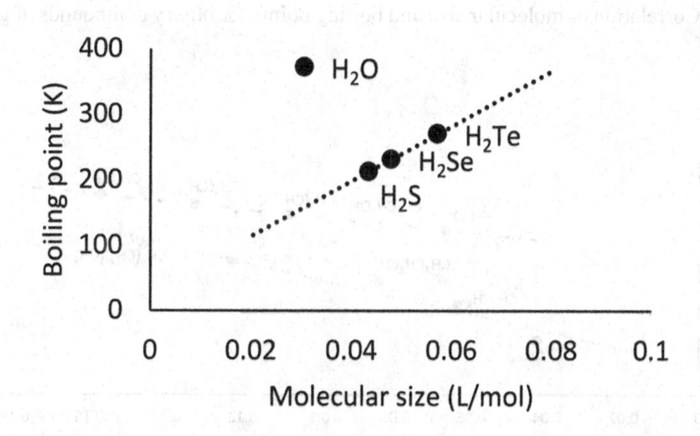

FIGURE 13.11 Correlation of molecular size and boiling points for binary compounds of group-16 elements and hydrogen.

The impact of hydrogen bonding on the strength of intermolecular interactions or molecular properties is greatest for small molecules. This is because in large molecules a single hydrogen-bonding N–H or O–H only affects a small portion of the structure and not the whole molecule, in contrast to dispersion that affects the whole structure. Figure 13.14 shows the diminishing effect that a single hydroxyl group (OH) has on the boiling point of saturated hydrocarbons (compounds consisting of carbon and hydrogen).

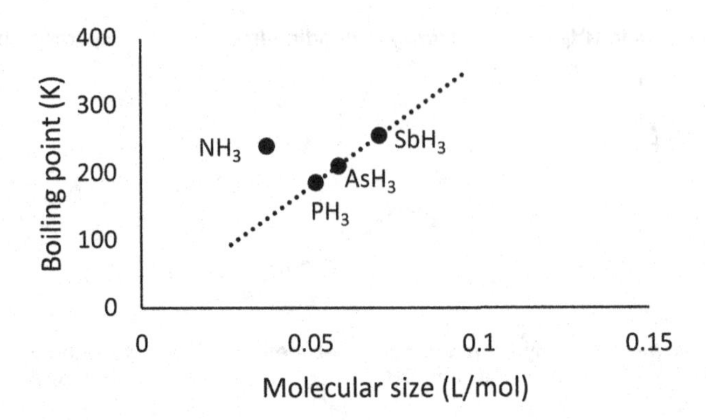

FIGURE 13.12 Correlation of molecular size and boiling points for binary compounds of group-15 elements and hydrogen.

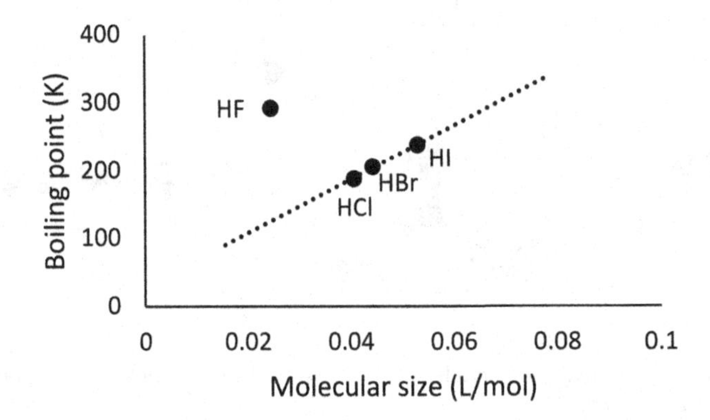

FIGURE 13.13 Correlation of molecular size and boiling points for binary compounds of group-17 elements and hydrogen.

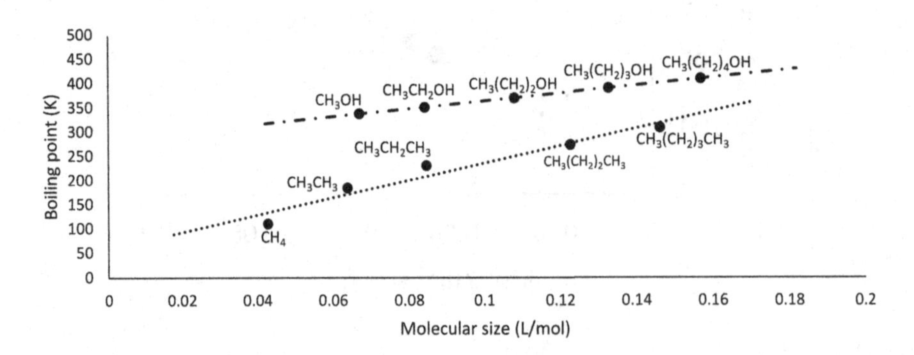

FIGURE 13.14 Correlation of molecular size and boiling points for saturated hydrocarbons (alkanes) and hydrocarbons containing an OH group (alcohols), which shows the diminished impact of hydrogen bonding on larger molecules.

Problem 13.3. Considering the intermolecular interactions involved, can you provide an explanation for the difference in the boiling point of the two isomers below? Isomers are compounds that have the same formula but a different arrangement of atoms.

Methyl acetate	Propionic acid
$C_3H_6O_2$	$C_3H_6O_2$
Molecular size = 0.12 L/mol	Molecular size = 0.13 L/mol
Boiling point = 57 ˚C	Boiling point = 140 ˚C

HOMOGENEOUS MIXTURES: INTERMOLECULAR FORCES AND SOLUBILITY AND MISCIBILITY

When a solid and a liquid combine to form a homogeneous mixture (a solution), we say that the solid is soluble in the liquid. Similarly, when two liquids combine to form a solution, we say that the liquids are miscible. Whether two compounds will combine to produce a solution depends on the intermolecular forces involved. For example, you may have seen that oil and water do not mix (they are immiscible). Oil is a mixture of hydrocarbons, fatty acids, and triglycerides, which interact predominantly through van der Waals interactions (dispersion and dipole–dipole interactions). Water is a highly polar molecule that can hydrogen bond. Hydrogen bonding is a stronger, more stabilizing intermolecular force than van der Waals interactions. For water molecules to mix with oil molecules, they would have to exchange strong hydrogen-bonding interactions with other water molecules for weaker (less stabilizing) van der Waals interactions with oil molecules. At normal temperatures and pressures, this will not happen: Water will prefer to stay unmixed, forming stronger hydrogen bonds with other water molecules. In contrast, methanol (CH_3OH) can hydrogen bond and is therefore fully miscible with water. We tend to use the phrase "like dissolves like" to encapsulate the general patterns of solubility and miscibility, but at its core, the important fact to remember is that hydrogen-bonding compounds mix preferentially with other hydrogen-bonding compounds, and compounds that interact through van der Waals interactions mix preferentially with other compounds that interact through van der Waals interactions.

Problem 13.4. For the following questions, we will consider two different vitamins – vitamin C and vitamin E – and their physical properties.

a. Consider the Lewis structure for vitamin C.

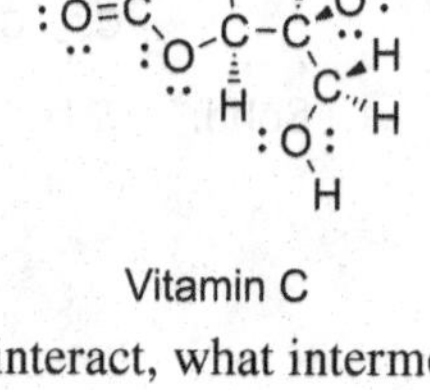

Vitamin C

When two vitamin C molecules interact, what intermolecular forces hold the two vitamin C molecules together?

b. Consider the Lewis structure for Vitamin E.

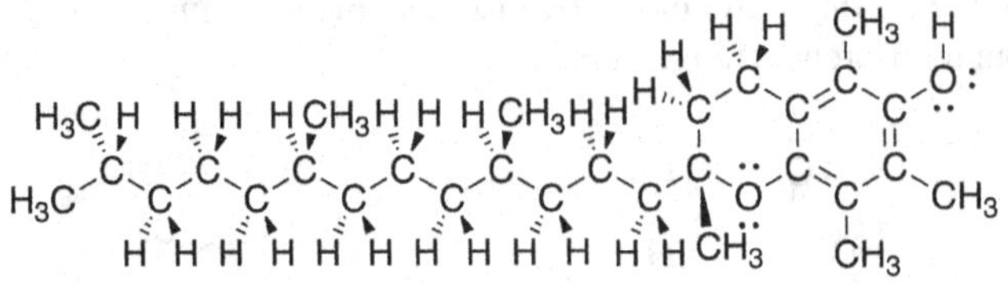

Vitamin E

When two vitamin E molecules interact, what intermolecular forces hold the two vitamin E molecules together?

c. Vitamin C is a water-soluble vitamin and vitamin E is a fat-soluble vitamin, which means that vitamin C will dissolve in water while vitamin E will not. Provide an explanation for the difference in water solubility of vitamin C and vitamin E.

d. At normal temperature and pressure, vitamin C is a solid and vitamin E is a liquid. Provide an explanation for this difference in their states of matter.

Problem 13.5. Whether or not two compounds will mix together depends on whether they have similar intermolecular forces. This is often summarized as "like dissolves like." For each of the following, the first molecule will be the liquid, solvent, and the second will be a solid, which may or may not dissolve in the liquid, a potential solute. State whether the solid will, or will not, dissolve in the liquid and explain your reasoning.

a. Liquid: H O H Solid:

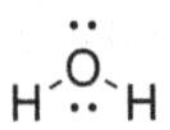

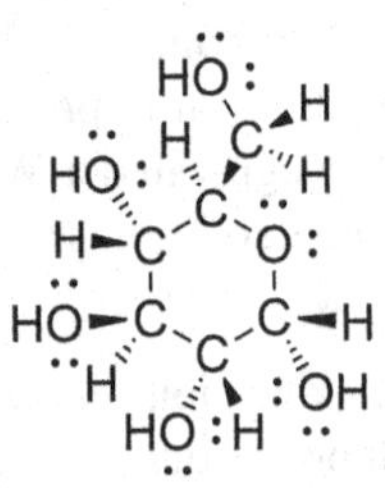

b. Liquid: H O H Solid:

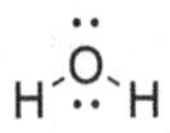

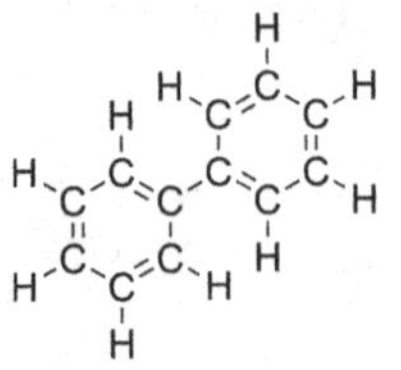

c. Liquid: 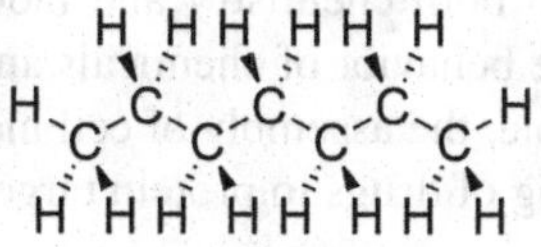 Solid:

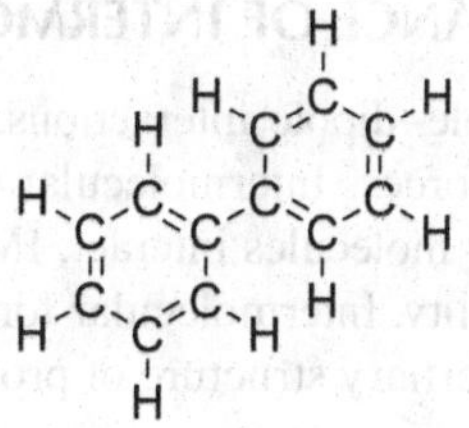

d. Liquid: 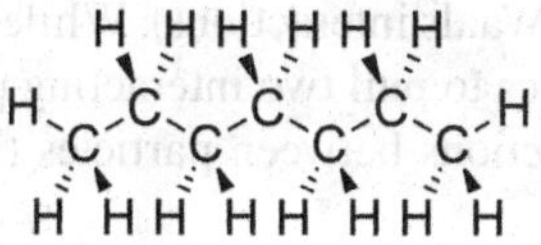 Solid:

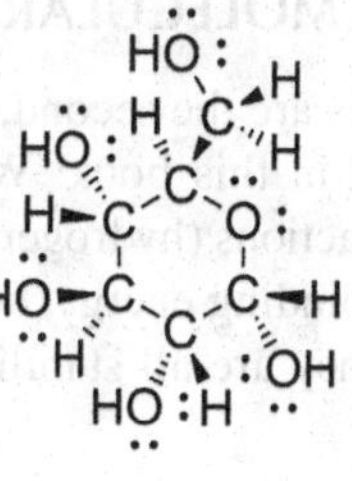

Problem 13.6. Ethanol and dimethyl ether are isomers.

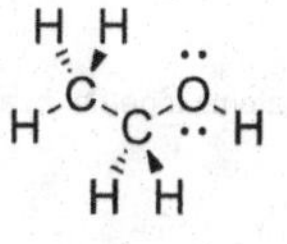

Ethanol
Formula C_2H_6O
Boiling point 78 ˚C

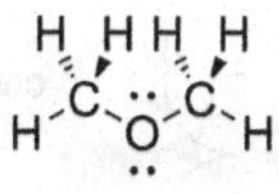

Dimethyl ether
Formula C_2H_6O
Boiling point -11 ˚C

Provide an explanation for the difference in the boiling points of ethanol and dimethyl ether.

Problem 13.7. As we will see in the future, molecules mix more readily if they share non-covalent interactions. Why is it that dimethyl ether is miscible (mixes with) with water, but dimethyl sulfide is not miscible with water?

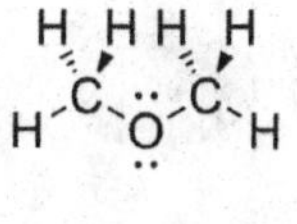

Dimethyl ether
C_2H_6O
Miscible with water

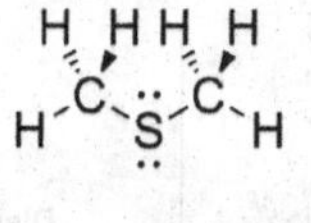

Dimethyl sulfide
C_2H_6S
Immiscible with water

THE IMPORTANCE OF INTERMOLECULAR FORCES

Dispersion, dipole–dipole interactions, and hydrogen bonds are the three most common types of intermolecular forces. Intermolecular forces are central to both chemistry and biology. Because they govern how molecules interact, IMFs dictate the phase behavior of chemicals and their miscibility and solubility. Intermolecular forces drive, for example, the assembly of cell membranes, the secondary and tertiary structure of proteins, and the binding of drugs to protein targets.

BONDING AND MOLECULAR INTERACTIONS

Intermolecular forces are the second, and last, type of stabilizing interaction between particles that will be investigated in this book. We have discussed intramolecular interactions (bonding) and intermolecular interactions (hydrogen bonding and van der Waals interactions). While these interactions differ in their binding energy, how much energy it takes to pull two interacting particles apart from one another, they are all stabilizing, attractive interactions between particles (Figure 13.15).

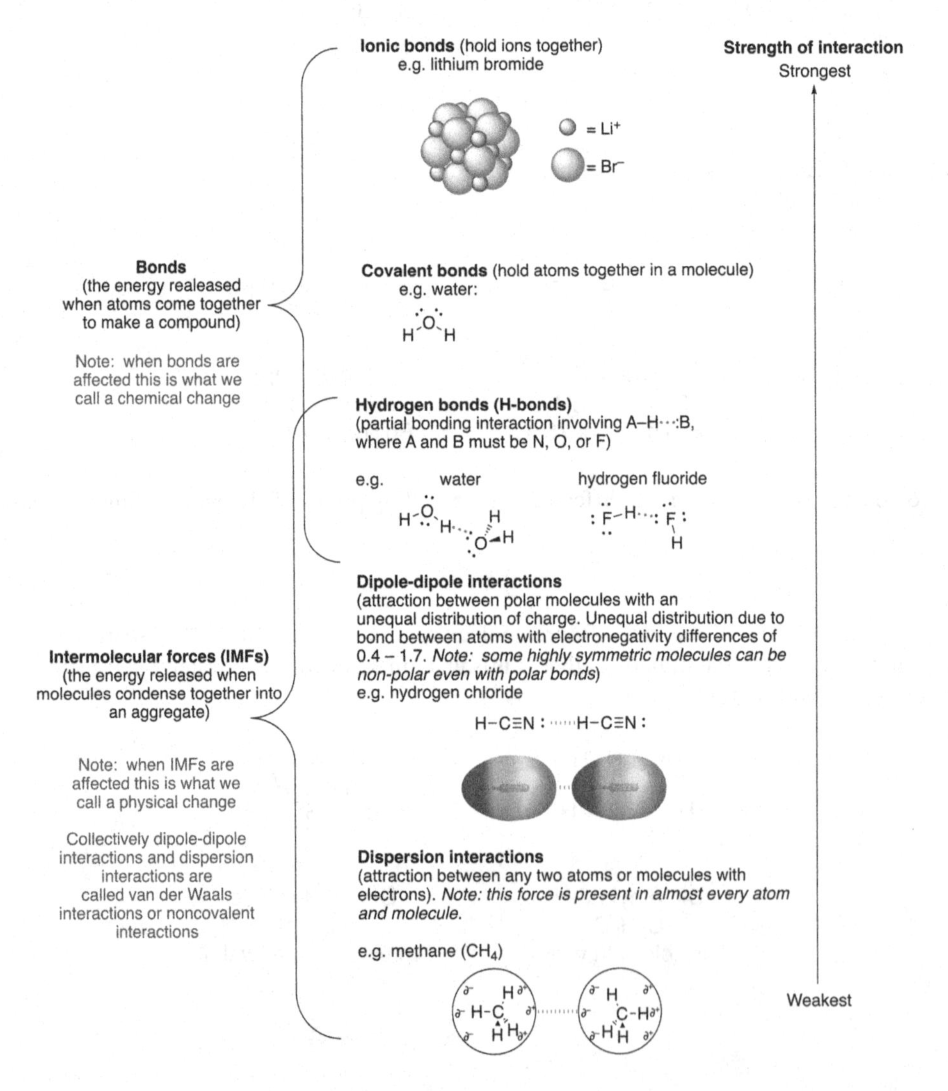

FIGURE 13.15 Synopsis and comparison of the binding energy of ionic bonding, covalent bonding, and intermolecular forces.

That is, the atoms and molecules are all lower in energy when engaged in these interactions, and it requires added energy to disrupt these forces. When the weaker intermolecular forces are changed, we call that a physical change (Chapter 14), and when the stronger bonds are changed, we call that a chemical reaction (Chapter 16).

NOTES

1. This relationship comes from an analysis of the van der Waals equation *a* and *b* values.
2. i) Eisenschitz, R.; London F. Über das Verhältnis der van der Waalschen Kräfte zu den homöopolaren Bindungskräften. *Z. Physik*, **1930**, *60* (7–8), 491–527. DOI: 10.1007/BF01341258.
   ii) London, F. The General Theory of Molecular Forces. *Trans. Faraday Soc.*, **1937**, *33*, 8b–26. DOI: 10.1039/TF937330008B.
3. Note that dispersion interactions are often also called van der Waals interactions, but dispersion is only one type of van der Waals interactions along with dipole--dipole interactions, which are considered below, and dipole–dispersion interactions, which are not considered here.
4. Use of the ∂ symbol first used to indicate partial charges: Ingold, C.K.; Ingold, E.H. CLXIX. The Nature of the Alternating Effect in Carbon Chains. Part V. A Discussion of Aromatic Substitution with Special Reference to the Respective Roles of Polar and Non-polar Dissociation; and a Further Study of the Relative Directive Efficiencies of Oxygen and Nitrogen. *J. Chem. Soc.*, **1926**, *129*, 1310–1328. DOI: 10.1039/JR9262901310.
5. Israelachvili, J.N. *5 – Interactions Involving the Polarization of Molecules. In Intermolecular and Surface Forces* (3rd ed.). Academic Press: Burlington, MA, **2011**, pp. 91–106. DOI: 10.1016/B978-0-12-375182-9.10005-3.
6. Keesom, W.M. The Second Virial Coefficient for Rigid Spherical Molecules Whose Mutual Attraction Is Equivalent to That of a Quadruplet Placed at Its Center. *Proc. Sect. Sci. K. Ned. Akad. Wet. Amst.*, **1915**, *18*, 636–646.
7. i) Lewis, G.N. *Valence and the Structure of Atoms and Molecules.* Chemical Catalog Co.: New York, **1923**.
   ii) Pauling, L. The Shared-electron Chemical Bond. *Proc. Natl. Acad. Sci. USA.*, **1928**, *14* (4), 359–362. DOI: 10.1073/pnas.14.4.359.
8. For the full definition and its explanation, see: Weinhold, F.; Klein, R.A. What Is a Hydrogen Bond? Resonance Covalency in the Supramolecular Domain. *Chem Educ. Res. Pract.* **2014**, *51* (3), 276–285. DOI: 10.1039/C4RP00030G.
9. Cao, W.; Xantheas, S.S.; Wang, X.-B. Cryogenic Vibrationally Resolved Photoelectron Spectroscopy of $OH^-(H_2O)$: Confirmation of Multidimensional Franck-Condon Simulation Results for the Transition State of the $OH + H_2O$ Reaction. *J. Phys. Chem. A*, **2021**, *125* (10), 2154–2162. DOI: 10.1021/acs.jpca.1c00848.
10. References for the hydrogen bonding structures. Note that the structure of the hydrogen bonding shown for the ammonia dimer is not the only minimum as it is highly fluxional with a very flat potential energy surface.
    i) HF dimer and H_2O dimer: Park, C.-Y.; Kim, Y.; Kim, Y. The Multi-coefficient Correlated Quantum Mechanical Calculations for Structures, Energies, and Harmonic Frequencies of HF and H_2O Dimers. *J. Chem. Phys.*, **2001**, *115* (7), 2926–2935. DOI: 10.1063/1.1386416.
    ii) NH_3 dimer: Jing, A.; Szalewicz, K.; van der Avoird, A. Ammonia Dimer: Extremely Fluxional But Still Hydrogen Bonded. *Nat. Commun.*, **2022**, *13*, DOI: 10.1038/s41467-022-28862-z.
    iii) H_2O and acetone dimer: Zhang, X.K.; Lewards, E.G.; March, R.E.; Parnis, J.M. Vibrational Spectrum of the Acetone-water Complex: A Matrix Isolation FTIR and Theoretical Study. *J. Phys. Chem.*, **1993**, *97* (17), 4320–4325. DOI: 10.1021/j100119a012.
    iv) H_2O and HF dimer: Sexton, T.M.; Howard, J.C.; Tschumper, G.S. Dissociation Energy of the $H_2O\cdots HF$ Dimer. *J. Phys. Chem. A*, **2018**, *122* (21), 4902–4908. DOI: 10.1021/acs.jpca.8b03397.
    v) H_2O and acetonitrile: Reimers, J.R.; Hall, L.E. The Solvation of Acetonitrile. *J. Am. Chem. Soc.*, **1999**, *121* (15), 3730–3744. DOI: 10.1021/ja983878n.

14 States and Physical Changes

This chapter builds off from the discussion of intermolecular forces from Chapter 13. One of the most important impacts of intermolecular forces is that IMFs are a key factor in dictating the state of matter of a chemical. This chapter represents a bridge as we will transition from describing and understanding matter – in terms of the states of matter (solids, liquids, and gases) – to describing and understanding change – in terms of changes in state (physical changes).

STATE VERSUS PHASE

Before diving into this topic, we first need to establish a distinction in two commonly used terms: state and phase. The three common states of matter are solids, liquids, and gases. The states of matter differ depending on the density of particle packing, the arrangement and degree of order, and the type(s) of particle motion.[1] In contrast to a state, a phase is a system that is uniform in physical state and in chemical composition.[2] Let us consider a biphasic heterogeneous mixture (Figure 14.1) as an example. A mixture of heptane, azulene, water, and iron(3+) is in the liquid state. Within the liquid, there are two phases, which differ in chemical composition: an organic phase (a homogeneous mixture of heptane and azulene) and an aqueous phase (a homogeneous mixture of water and iron(3+) chloride).

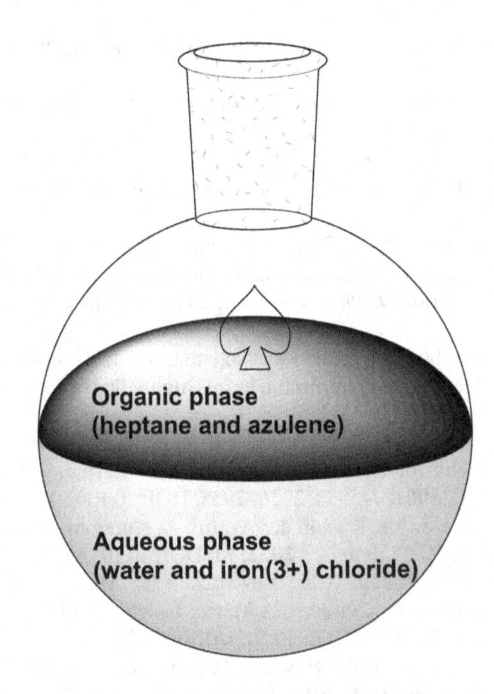

FIGURE 14.1 Diagram of a biphasic, liquid-state mixture with an organic phase (*top*) and aqueous phase (*bottom*) in a round-bottom flask. The organic phase is a homogeneous mixture of heptane (C_7H_{16}) and azulene ($C_{10}H_8$) and the aqueous phase is a homogeneous mixture of water (H_2O) and iron(3+) chloride.

DOI: 10.1201/9781003487210-14

Problem 14.1. For each of the following, identify the state(s) of matter (solid, liquid, or gas) and whether the material is a singular phase or more than one phase.

a. Seltzer water

b. A typical school desk

SOLID, LIQUID, AND GAS PROPERTIES

Now that we have firmly established some terminology, let's reconsider solids, liquids, and gases in terms of the density of particle packing, the arrangement and degree of order, and the type(s) of particle motion. Both solids and liquids are condensed phases of matter, which means they cannot be easily compressed and there is very little space between particles (Figure 14.2). In contrast, gases are dispersed phases with significant distance between the particles (Figure 14.2), which makes gases compressible. Unlike liquids or gases, solids are distinguished by having long-range order in the arrangement of particles. The persistence of long-range order is because solid particles vibrate (move around a fixed point) but do not easily rotate (spin around their center of mass) nor do they translate (move through three-dimensional space). In contrast, liquids and gases are fluid phases that take on the shapes of their containers as the particles not only vibrate but also rotate and translate. The greater extent of motion for liquids and gases is observable in terms of the lack of long-range order (lack of regular, repeating arrangement) among the particles (Figure 14.2). The phase that a given chemical has at normal temperature and pressure (25.0 °C, 101 kPa) is dictated by the strength of the intermolecular forces present in the compound.

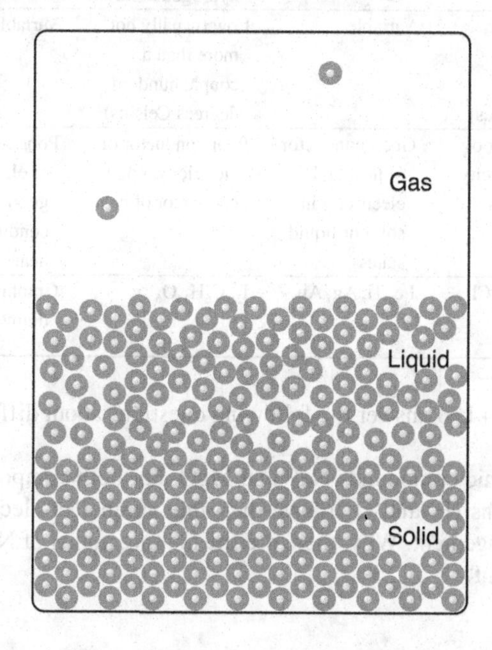

FIGURE 14.2 Density and arrangement of argon atoms in a sample with solid argon (*bottom*), liquid argon (*middle*), and gaseous argon (*top*). The atoms occupy 64%, 57%, and 0.064% of the space respectively. van Witzenburg, W. and Stryland, J.C. Density measurements of compressed solid and liquid argon. *Can. J. Phys.*, **1968**, *46* (7), 811–816. DOI: 10.1139/p68-102.

Problem 14.2. Consider Figure 14.2. Rank a solid (ice) versus a liquid (water) versus a gas (steam) from highest to lowest energy. Explain your answer.

SOLIDS

Solids are distinguished by having long-range order in the arrangement of particles. The persistence of long-range order is because solid particles vibrate (move around a fixed point) but do not easily rotate (spin around their center of mass) nor do they translate (move through three-dimensional space). Solids differ in terms of the types of particles and the forces that hold the particles together. The five most common types of solids – ionic, metallic, molecular, network, and amorphous – and their properties are given in Table 14.1.

TABLE 14.1
Descriptions of the Common Types of Solids

Type	Ionic	Metallic	Molecular	Network	Amorphous
Force holding units together	Ionic bonds	Metal bonding, ordered metal cations surrounded by dispersed electrons	Intermolecular forces	Covalent bonds	Covalent bonds
Soluble in water?	Yes, often	No	Yes, if molecules are polar or can hydrogen bond	No	No
Melting point	Very high (hundreds to thousands of degrees Celsius)	Variable	Low (usually not more than a couple hundred degrees Celsius)	Variable	Variable
Conductor?	Poor conductor of heat or electricity, only conducts electricity in liquid state	Good conductor of heat and electricity in solid or liquid states	Poor conductor of electricity, good conductor of heat	Poor conductor of electricity, good conductor of heat	Poor conductor of electricity, good conductor of heat
Examples	NaF, MgO, NaCl	Fe, Ti, Ag, Au	I_2, $C_6H_{12}O_6$	Graphite, diamond	Glass, plastic, nylon

Problem 14.3. Use Table 14.1 to answer the following questions about different types of solids.

a. Consider the two ionic compounds NaF and MgO. For both compounds, the cations (Na^+ and Mg^{2+}) and anions (F^- and O^{2-}) have the same number of electrons and both have a similar atomic arrangement. Why, then, are the melting points of NaF (992 °C) and MgO (2825 °C) so different?

b. Titanium(IV) bromide is a brown, crystalline solid with a melting point of 39 °C and a boiling point of 230 °C. It does not conduct electricity as a solid or liquid. Chromium(III) bromide is a lustrous black, crystalline solid with a melting point of 1130 °C and it is soluble in water and does not conduct electricity, but molten chromium(III) bromide does conduct electricity.
   i. Is titanium(IV) bromide an ionic, molecular, metallic, network, or amorphous solid? Explain your selection.

   ii. Is chromium(III) chloride an ionic, molecular, metallic, network, or amorphous solid? Explain your selection.

Liquids

A liquid is, like a solid, still a condensed phase, which means that they are not easily compressed as the particles are in close proximity. Unlike a solid, the liquid particles are moving rapidly through space and rotating, which means that the liquid particles are more disordered (Figure 14.2). An important physical property of liquids (and solids) is density in terms of mass (g) per volume (mL). The density of a solid or liquid is relatively constant over a wide temperature range. For example, liquid water's density does not vary by more than 5% from 0.0 °C to 100.0 °C.

TABLE 14.2
The Density of Water as a Function of Temperature.[3]

Phase	Temperature (°C)	Density (g/mL)
Solid	0.0	0.9167
	0.0	0.9999
	4.0	1.0000
	15.6	0.9991
	26.7	0.9967
	37.8	0.9932
Liquid	48.9	0.9887
	60.0	0.9834
	71.1	0.9773
	82.2	0.9706
	93.3	0.9633
	100.0	0.9587
Gas	100.0	0.0006

Problem 14.4. Water is unique because it is the only substance where its density as a solid is less than its density as a liquid. Think about life on Earth and provide an example of how this unique property is important for life.

Problem 14.5. The density of solid ice and liquid water differ by less than 10%, but there is a dramatic change in density for steam. Provide an explanation as to why the density of steam is so much less than the density of solid ice and liquid water.

GASES

Compared to solids and liquids, gases are unique because they are a dispersed phase. That is the gas particles are widely separated from each other. For example, an argon atom, in air, is separated from any other gas particles by a distance that is about 250 times greater than its own diameter.[4] Gas particles are very high energy and moving rapidly in every direction. In addition, the density of a gas is highly dependent on both the temperature and pressure of the gas. The pressure, volume, temperature, and amount of a gas are related together by the conversion factor (Chapter 4) known as the gas constant (R). In the most used units in chemistry (liter for volume and atmosphere for pressure), the value of the gas constant is $\frac{0.082\ 06\,\mathrm{L\,atm}}{\mathrm{mol\,K}}$. The gas constant is used in the Ideal Gas Law (Equation 14.1), which relates pressure (p), volume (V), amount (n), and temperature (T).

$$pV = nRT \tag{14.1}$$

p is the pressure (atm)
V is the volume (L)
n is the amount (mol)
R is the gas constant $\left(\frac{0.082\ 06\,\mathrm{L\,atm}}{\mathrm{mol\,K}}\right)$
T is the temperature (K)

For example, let's use the ideal gas constant to calculate the volume of a 1.00 mol of gas at 273.15 K and 0.987 atm. Given this information, we know the pressure (p) is 0.987 atm, the amount (n) is 1.00 mol, and the temperature is 273.15 K. We can plug these in and solve for the volume to find that it is 22.7 L.

$$(0.987\ \mathrm{atm})\mathrm{V} = (1.00\ \mathrm{mol})\left(\frac{0.082\ 06\,\mathrm{L\,atm}}{\mathrm{mol\,K}}\right)(273.15\ K)$$

$$V = \left((1.00\ \mathrm{mol})\left(\frac{0.082\ 06\,\mathrm{L}\,\cancel{\mathrm{atm}}}{\cancel{\mathrm{mol}}\,\mathrm{K}}\right)(273.15\ \mathrm{K})\right)/(0.987\,\cancel{\mathrm{atm}}$$

$$V = 22.7\ \mathrm{L}$$

Problem 14.6. A researcher has three containers of gas at 27.0 °C are connected by a closed valve:

	Container 1	Container 2	Container 3
Gas	O_2	N_2	Ar
Volume	0.00300 m^3	200 mL	5.00 L
Pressure	146 kPa	0.908 atm	630 mmHg
Gas constant	$\frac{8.314\,\mathrm{Pa\,m^3}}{\mathrm{mol\,K}}$	$\frac{0.082\ 06\,\mathrm{L\,atm}}{\mathrm{mol\,K}}$	$\frac{62.36\,\mathrm{L\,mmHg}}{\mathrm{mol\,K}}$

Which container has the most amount (mol) of gas?

Problem 14.7. If you breathe out 240 mL of CO_2 per minute, what is the mass (in kg) of CO_2 you exhale in a 24.0-hour day? Assume normal conditions (1.00 atm, 25.0 °C).

Solids, liquids, and gases are the most common phases that one encounters. There are several other phases that matter can exist in, including: Bose–Einstein condensate, superfluid, supercritical fluid, and plasma. We will not consider all these phases here, but it is worth highlighting two from this list. The first is plasma, which is a gas that is so high energy that electrons are no longer bound to their nuclei. Plasma is a dispersed phase with free ions and free electrons. The flame of a very hot fire is a plasma. The other phase to note is a supercritical fluid. A supercritical fluid has the density of a fluid but the energy and cavity-filling properties of a gas. Supercritical fluids have gained increasing attention for their applications as solvents in chemical reactions, decaffeinating coffee, and dry cleaning.

PRESSURE–TEMPERATURE DIAGRAMS AND PHASE CHANGES

The state of matter for a substance can be changed (Figure 14.3) by altering the pressure and/or by altering the temperature. The amount of thermal energy, or the change in pressure, needed to affect the phase of a substance depends on the strength of the interactions (bonds or intermolecular forces) between the particles (Chapter 13). The pressure–temperature impact on the phase behavior of a substance is represented by a pressure–temperature diagram, also called a phase diagram.[5]

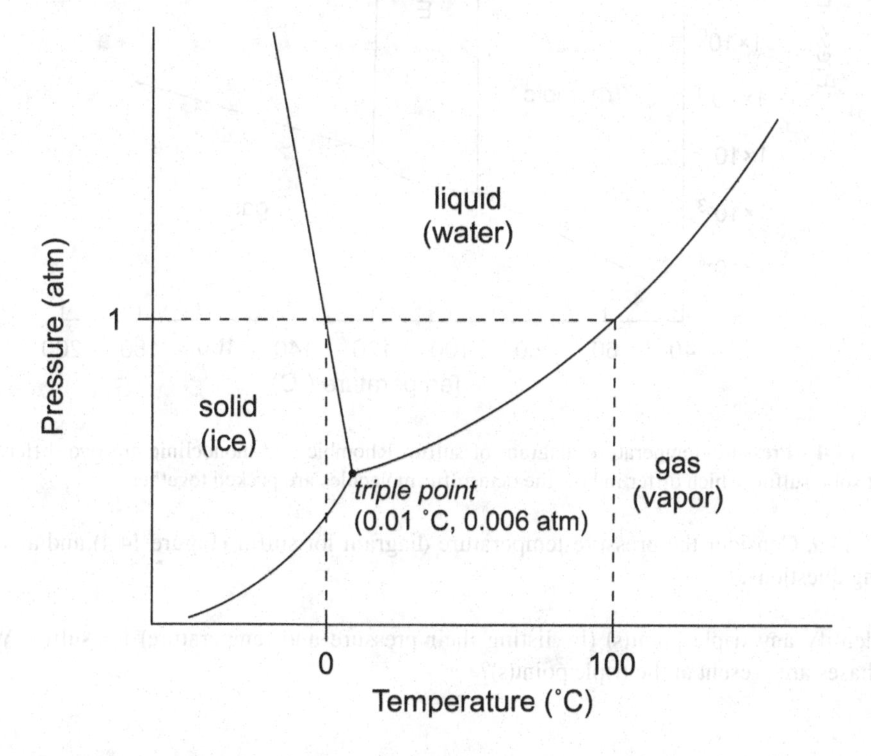

FIGURE 14.3 Pressure–temperature diagram of water showing the normal melting point (0 °C and 1 atm) and boiling point (100 °C and 1 atm).

Figure 14.3 shows a pressure–temperature diagram for water. Each line represents a phase boundary where two phases exist simultaneously. At 1 atm, normal pressure, the phase boundaries for liquid water are 0 °C (the melting point) and 100 °C (the boiling point). The triple point is the point on any pressure–temperature diagram where three phases exist simultaneously. For water, the triple point (0.01 °C and 0.006 atm) is the point where ice, water, and steam exist all at the same time.

Problem 14.8. If you have ice at −1 °C and 1 atm and increase the pressure, what happens to the solid water? What does this tell you about the density of liquid water compared to solid water?

The pressure–temperature diagram shown for water (Figure 14.3) is a relatively simple phase diagram. If we consider Figure 14.4, we can see a more complicated phase diagram for sulfur. Here there are two solid phases (allotropes) of sulfur that can exist: rhombic and monoclinic. These solid phases differ in how the particles of octasulfur are packed together in the solid. Rhombic sulfur is the dominant phase at STP.

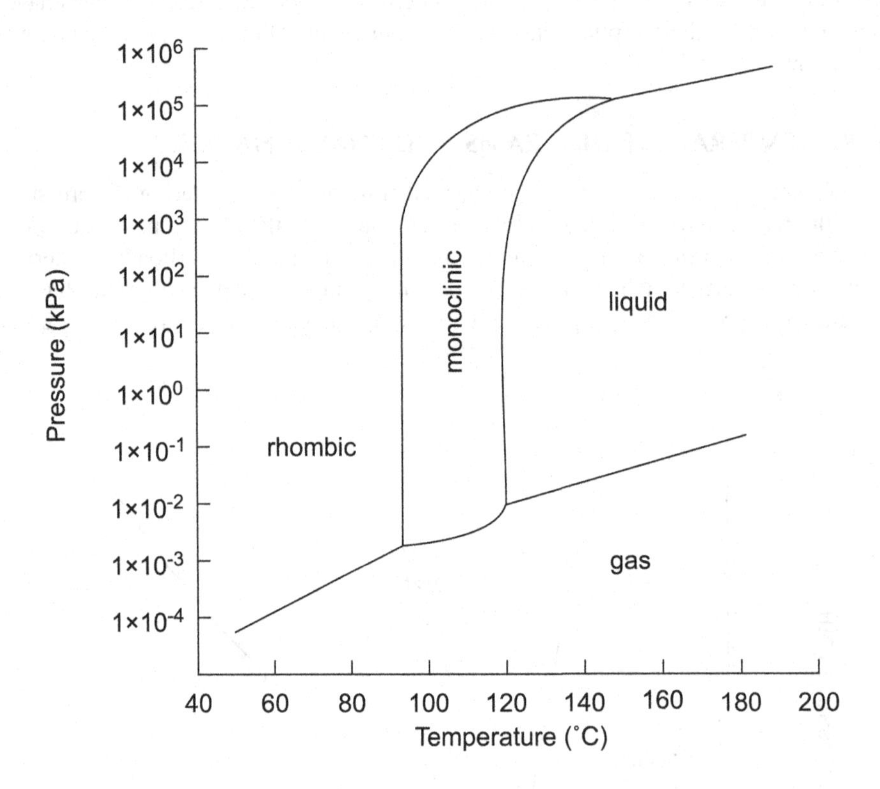

FIGURE 14.4 Pressure–temperature diagram of sulfur. Rhombic and monoclinic are two different allotropes of solid sulfur, which differ in how the octasulfur molecules are packed together.

Problem 14.9. Consider the pressure-temperature diagram for sulfur (Figure 14.4) and answer the following questions.

a. Identify any triple point(s) (by listing their pressure and temperature) for sulfur. What phases are present at the triple point(s)?

b. If a sample of sulfur gas at 130 °C and 10^{-4} atm is slowly pressurized up to 10^{4} atm, list the sequence of phases that would be visible as the pressure increased.

c. Which, if either, is denser monoclinic or rhombic sulfur? Explain your answer.

Figure 14.5 shows a pressure–temperature diagram for hydrogen sulfide. Here the diagram has been annotated to show the terminology associated with each type of phase change. Moving from solid to liquid (melting) and then liquid to gas (vaporization), at constant pressure, requires increasing the kinetic energy of the particles by increasing the temperature.

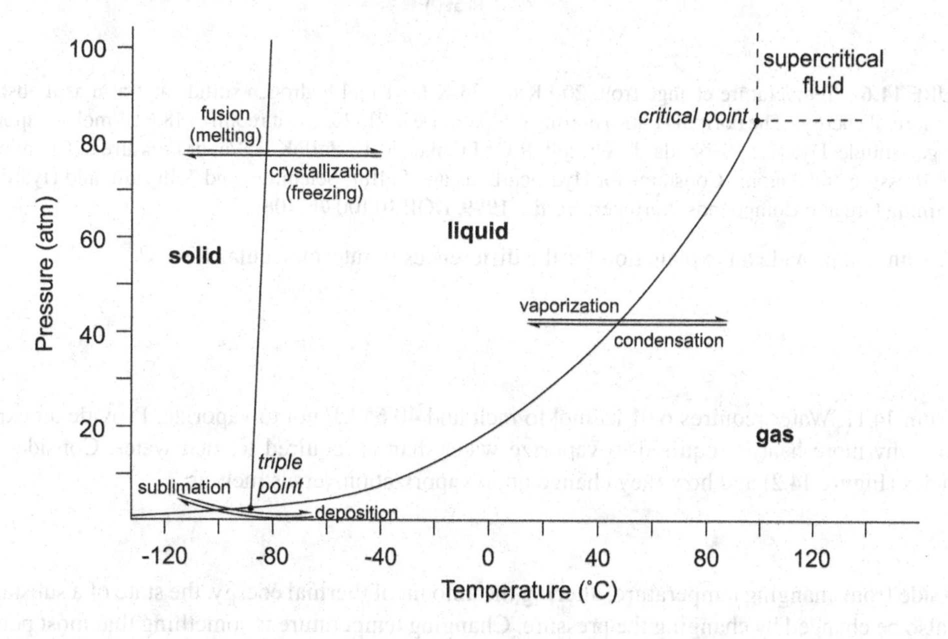

FIGURE 14.5 Pressure–temperature diagram for hydrogen sulfide showing the phase change terminology. The triple point is at −85.49 ˚C and 0.23 atm. The critical point is at 100.3 ˚C and 88.52 atm.

At a phase boundary, energy must be added or removed to cause the phase transition, but the temperature of the substance does not change (Figure 14.6). The discontinuity in Figure 14.6 corresponds to the phase transition.[6] For a plot of temperature versus heat (energy), the magnitude of the discontinuity is proportional to the strength of the intermolecular forces between the particles. For the example of hydrogen sulfide, the jump in heat corresponds to the energy required (18.6 kJ/mol) to disrupt the van der Waals interactions between particles.

Problem 14.10. If we compare hydrogen sulfide (boils at −60.28 ˚C at 1 atm) and hydrogen selenide (boils at −41.25 ˚C at 1 atm), we can see that hydrogen selenide must be given more thermal energy to vaporize.

a. What can we say about the strength of the intermolecular forces between hydrogen sulfide molecules versus the strength of the intermolecular forces between hydrogen selenide?

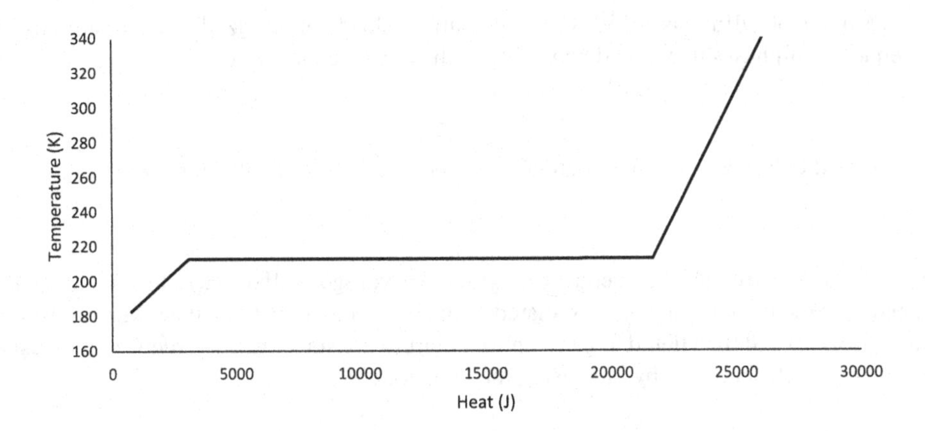

FIGURE 14.6 Temperature change from 200 K to 335 K for 1 mol hydrogen sulfide at 1 atm as it absorbs heat (thermal energy). The normal (1 atm) boiling temperature is 213 K, and it requires 18.6 kJ/mol to vaporize hydrogen sulfide. Dykyj, J.; Svoboda, J.; Wilhoit, R.C.; Frenkel, M.L; Hall, K.R. Vapor Pressure of Chemicals. Vapor Pressure and Antoine Constants for Hydrocarbons and Sulfur, Selenium, and Tellurium and Hydrogen Containing Organic Compounds. Springer: Berlin. **1999**. DOI: 10.1007/b71086.

b. Can you provide an explanation for the differences in intermolecular forces?

Problem 14.11. Water requires 6.01 kJ/mol to melt and 40.65 kJ/mol to vaporize. Provide an explanation why more heat is required to vaporize water than is required to melt water. Consider the particles (Figure 14.2) and how they change upon vaporization versus melting.

Aside from changing temperature, altering the amount of thermal energy, the state of a substance can also be changed by changing the pressure. Changing temperature is something that most people have lived experience with, but most people typically do not have lived experience with altering the pressure of a substance. Let's consider what's at play in changing pressure. To reduce the pressure, a vacuum is used. Vacuums mechanically reduce the amount of gas present in a container, which will in turn reduce the pressure (force per area) on a sample. A sufficient decrease in pressure can cause a substance to melt and to vaporize as the particles are not held together by force of gas particles hitting the surface.

While decreasing the pressure can be done with a vacuum, there are two different ways to increase pressure, depending on the sample and the pressures desired. One way to increase pressure is by adding an inert gas, like nitrogen or argon, to a container. As the amount of gas in the container increases, the pressure inside the container will also increase. It should be noted that pressurizing a container with gas can lead to explosive results, if not careful. The other way to increase pressure is through the application of a force using some sort of mechanical press. For example, a diamond anvil cell can pressurize a material up to 7.8×10^{11} Pa (7.7×10^{6} atm).

In terms of the intermolecular forces between particles, a material with strong intermolecular forces will require very little added pressure to condense from a gas, and it will, conversely, require very low pressures to vaporize. A substance with weak intermolecular forces will require high pressures to condense and only a small reduction in pressure to vaporize.

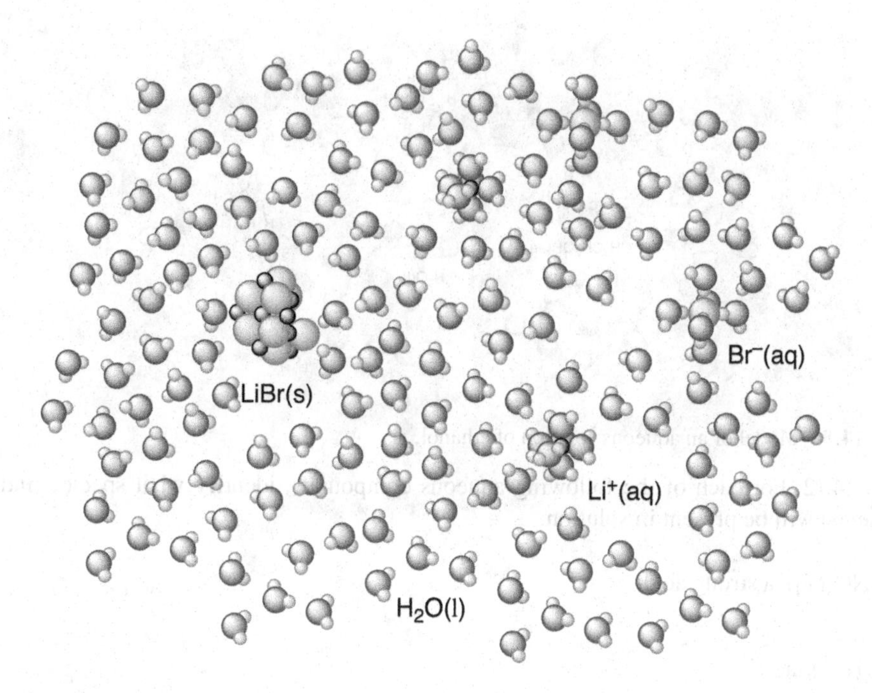

FIGURE 14.7 Dissociation of an ionic compound (lithium bromide) into separately solvated lithium cations and bromide anions in aqueous solution.

AQUEOUS PHASE

To conclude this chapter, we will briefly consider aqueous solutions, that is homogenous mixtures with water. When ionic compounds dissolve in water (Figure 14.7), the ionic solid, which we call a crystal lattice, see the LiBr(s) in Figure 14.7, dissociates into separately solvated cations (Li^+(aq)) and anions (Br^-(aq)). That means when ionic compounds dissolve, the ionic bonds are being broken. In the absence of water this would require large amounts of energy, but the polar water molecules are able to stabilize the ions by solvating them. In water, cations are stabilized by cation–dipole interactions, where the electron-rich oxygen of water stabilizes the positively charged ion. Water also stabilizes the anions in two ways. First, the nonmetal anions are stabilized by electrostatic attraction to the electron-poor hydrogen atoms. Second, the water stabilizes the anions by forming partial covalent bonds (hydrogen bonds, Chapter 13).

When ionic compounds dissolve, they typically dissociate into separate ions. In contrast, for molecular compounds the molecules will be separated by the water, disrupting their intermolecular forces, but the compounds' covalent bonds do not break (Figure 14.8).

The exception to this general difference in aqueous behavior between ionic (ions dissociate) and molecular compounds (atoms stay together) are acids. Acids are molecular compounds, but like ionic compounds they dissociate in water to give H^+(aq), a hydron (also called a hydrogen cation), and an anion. For example, HCl(aq) dissociates to give H^+(aq) and Cl^-(aq). The strength of an acid is dictated by the extent to which it dissociates in water. Hydrogen chloride fully dissociates in water (making it a strong acid) and so only H^+(aq) and Cl^-(aq) are in solution, while acetic acid does not fully dissociate in water (making it a weak acid) and so in solution there will be CH_3CO_2H(aq), H^+(aq), and $CH_3CO_2^-$(aq).

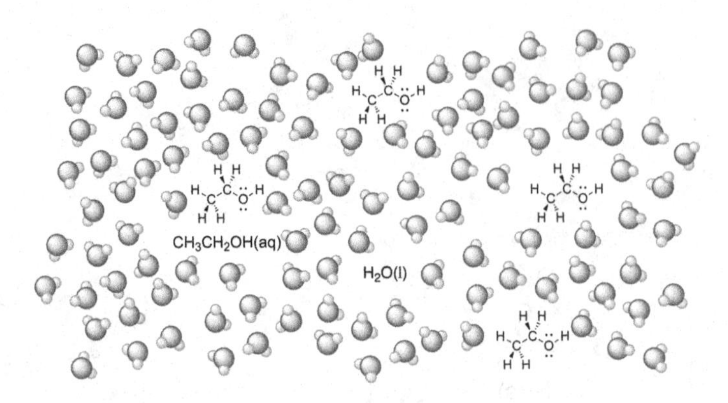

FIGURE 14.8 Model of an aqueous solution of ethanol.

Problem 14.12. For each of the following aqueous compounds, identify what species (individual components) will be present in solution.

a. HNO_3(aq), a strong acid:

b. $C_6H_{12}O_6$(aq):

c. CH_3OCH_3(aq):

d. Na_2SO_4(aq):

e. H_2SO_4(aq), a strong acid and a weak acid:

f. HBr, a strong acid:

g. $CaCl_2$(aq):

Problem 14.13. Sodium bromide is soluble in water, methanol (CH_3OH), and ethanol (CH_3CH_2OH). The solubility of sodium bromide in these three solvents varies: water (94 g/mL at 25 °C), methanol (17 g/mL at 25 °C), and ethanol (2 g/mL at 25 °C). Provide an explanation for the differences in solubility of sodium bromide in these three solvents.

PHASE AND PHYSICAL CHANGE

The phase of a substance is the result of the balance of intermolecular forces, temperature, and pressure. A physical change is the change in phase of substance, not the chemical formula, by altering its temperature and/or pressure. Physical changes can be understood through quantifying the macroscopic variables of temperature and pressure. In the following chapters, we will move onto studying chemical reactions, which can involve a change in temperature and/or pressure but must involve a change in the chemical formula of the substance.

NOTES

1. Barker, J.A.; Henderson, D. What Is "liquid"? Understanding the States of Matter. *Rev. Mod. Phys.*, **1976**, *48* (4), 587–671. DOI: 10.1103/RevModPhys.48.587.
2. "Phase." IUPAC. *Compendium of Chemical Terminology,* 2nd ed. (the "Gold Book"). Compiled by A.D. McNaught; A. Wilkinson. Blackwell Scientific Publications: Oxford, **1997**. Online version (2019-) created by S.J. Chalk. DOI: 10.1351/goldbook.P04528.
3. Data from: i) Rumble, J.R.; Lide, D.R.; Bruno, T.J. *CRC Handbook of Chemistry and Physics : A Ready-Reference Book of Chemical and Physical Data. 2019th-2020th,* 100th ed. CRC Press: Boca Raton, FL, **2019**.
   ii) United States Geological Survey, "Water Density." https://www.usgs.gov/special-topics/water-science-school/science/water-density (Date Accessed: 08/13/23).
4. Jennings, S.G. The Mean Free Path in Air. *J. Aerosol Sci.*, **1988**, *19* (2), 159–166. DOI: 10.1016/0021-8502(88)90219-4.
5. Pressure–temperature diagrams were first introduced by Gibbs: Gibbs, J.W. A Method of Geometrical Representation of the Thermodynamic Properties of Substances by Means of Surfaces. *Trans. Conn. Acad. Arts Sci.*, **1873**, *2*, 382–404.
6. Sauer, T. A Look Back at the Ehrenfest Classification. *Eur. Phys. J. Special Topics,* **2017**, *226*, 539–549. DOI: 10.1140/epjst/e2016-60344-y.

15 Describing and Quantifying Chemical Reactions

In this chapter, we begin our last topic for this book: chemical change. Chemical change, recall, is any process where the chemical composition of a substance is altered. The evidence of chemical change, which can include a change in color, bubbling (without added heat), the appearance of a solid (without cooling), the appearance of light, changes in smell, changes in taste, changes in texture or luster without external forces, and temperature changes.

Chemical change is represented with chemical equations. A generic chemical equation is shown in Figure 15.1.

$$\text{Reactants} \xrightarrow[\text{Conditions (time, temperature, solvent)}]{\text{Reagents}} \text{Products}$$

FIGURE 15.1 A generic chemical equation identifying where various components and details appear.

On the left side of the arrow are listed all reactants: the chemical formula (formulae) for the substance(s) that are undergoing a change. There is then an arrow, which means "yields" or "produces," and products are shown on the right: the chemical formula (formulae) for the substance(s) that are the outcome of a change. We also indicate the phase of a substance utilizing the abbreviations: (s) for solid, (l) for liquid, (g) for gas, and (aq) for aqueous or dissolved in water. For example, $TiCl_4(l)$ is liquid titanium tetrachloride.

The arrow in a chemical equation can also be annotated with information about the reaction. Anything above the arrow is a reagent, which is any substance that is necessary to cause the chemical reaction to occur, but the atom(s) of the reagent are not in any of the product formulae. Common reagents are catalysts, often metals, which are used to facilitate the reaction but are not consumed by the reaction. Below the arrow are any conditions for the reaction, including time (how long it takes for the reaction), temperature (if no temperature is included it is assumed to be at room temperature), and any solvents (the liquid media in which reactions often take place). The temperature may not be a specific value, like 50.0 °C, but it may simply say heat (or use the symbol Δ) if the reaction is heated, or it may say ice bath if the reaction is cooled. An increasingly popular way to heat up reactions is with chemical reaction microwaves, which can be listed as microwave or as the abbreviation μW. Finally, there are many solvents that are common, especially in organic chemistry, and their manifold abbreviations will not be included here. Most reactions that we will consider are run in water if any solvent is used.

EMPIRICALLY DERIVING CHEMICAL EQUATIONS

Baking soda is used as a leavening agent in baking. Let us consider the thermal decomposition of baking soda, sodium hydrogencarbonate, at 170 °C. Sodium hydrogencarbonate has the chemical formula $NaHCO_3$ and it is a solid powder, so the start of our chemical equation for the decomposition would be

$$NaHCO_3(s) \xrightarrow[170\,°C]{}$$

DOI: 10.1201/9781003487210-15

The reaction produces a solid powder, a gas, and steam. The solid powder is shown, through elemental analysis (Chapter 10), to have the empirical formula Na_2CO_3, which is sodium carbonate. The gas is shown to lower the pH when mixed with water, which is a property of several gases including carbon dioxide, and its elemental analysis gives a chemical formula of CO_2, which is carbon dioxide. Finally, the steam is inferred to be water, H_2O, which is confirmed by elemental analysis. Altogether, then, the chemical components of this reaction are as follows:

$$NaHCO_3(s) \xrightarrow[170\,^{\circ}C]{} Na_2CO_3(s) + CO_2(g) + H_2O(g)$$

This decomposition reaction, as written, is invalid because it does not obey the law of conservation of mass. We call such equations unbalanced.

Problem 15.1. Provide an explanation of why the equation, as written, violates the law of conservation of mass.

To fix – that is to balance – this equation, we will investigate the amount of each substance before and after the reaction. The initial sample of sodium bicarbonate had a mass of 0.25 g (about two teaspoons). Using the molar mass (84.006 g/mol), we can calculate that the reaction started with 3.0 mmol $NaHCO_3$:

0.25 ~~g~~	1 ~~mol~~	1 mmol	= 3.0 mmol $NaHCO_3$
	84.006 ~~g~~	1×10^{-3} ~~mol~~	

In this decomposition, 0.15 g of sodium carbonate is produced. Using the molar mass of sodium carbonate (105.99 g/mol), we can determine that the reaction produces 1.4 mmol Na_2CO_3:

0.15 ~~g~~	1 ~~mol~~	1 mmol	= 1.4 mmol Na_2CO_3
	105.99 ~~g~~	1×10^{-3} ~~mol~~	

The steam and gas are collected in a syringe (that measures the volume of gas) that has a molecular sieve filter (which absorbs the water). The syringe collects 37.0 mL of gas at 25 °C (298 K) and 0.987 atm, and the molecular sieves start at 5.250 g and end at 5.280 g. These data can be used to determine the amount of each product. Using the gas constant $\left(\frac{0.08206\,\text{L atm}}{\text{mol K}}\right)$, we find that there are 1.5 mmol CO_2 produced:

37.0 ~~mL~~	1×10^{-3} ~~L~~	mol K	= 0.451 mol K/atm
	1 mL	0.082 06 ~~L~~ atm	

0.987 ~~atm~~	0.451 mol K	= 0.445 mol K
	~~atm~~	

0.445 mol K	1	1 mmol	$= 1.5$ mmol CO_2
	298 K	1×10^{-3} mol	

And finally, the difference in mass of the molecular sieves (0.030 g) corresponds to the mass of water absorbed. Using the molar mass of water (18.015 g/mol), this gives 1.7 mmol of water:

0.030 ~~g~~	1 ~~mol~~	1 mmol	$= 1.7$ mmol H_2O
	18.015 ~~g~~	1×10^{-3} ~~mol~~	

Considering the amount of each chemical, we can improve our unbalanced chemical equation by showing the relationship among the various chemicals:

$$3.0 \text{ mmol } NaHCO_3(s) \xrightarrow[170°C]{} 1.4 \text{ mmol } Na_2CO_3(s) + 15 \text{ mmol } CO_2(g) + 1.7 \text{ mmol } H_2O(g)$$

These specific amounts (mol) are, however, only for the given reaction that we investigated. If we divide each amount by the smallest:

$$\frac{3.0\,\text{mmol}}{1.4\,\text{mmol}} NaHCO_3(s) \xrightarrow[170°C]{} \frac{1.4\,\text{mmol}}{1.4\,\text{mmol}} Na_2CO_3(s) + \frac{1.5\,\text{mmol}}{1.4\,\text{mmol}} CO_2(g) + \frac{1.7\,\text{mmol}}{1.4\,\text{mmol}} H_2O(g)$$

This will determine the whole number, amount (mole) ratios of the reactions involved. Note that this approach should feel familiar from our work determining empirical formulae (Chapter 10). Note that if the coefficient is 1, then the number is typically omitted from in front of the formula:

$$2\ NaHCO_3(s) \xrightarrow[170°C]{} Na_2CO_3(s) + CO_2(g) + H_2O(g)$$

Altogether, this is now a balanced chemical formula for the thermal decomposition of sodium bicarbonate. The number of atoms on the reactant and product side are equal, which means that the mass should be conserved between the reactants and products.

Problem 15.2. The reaction of thermite – iron(3+) oxide powder with aluminium powder – to produce solid aluminium oxide and solid iron metal is incredibly exothermic and dramatic to watch.

a. Write a chemical equation that shows all the reactants, products, and the appropriate phase notation.

b. Initially 10.0 g of iron(3+) oxide is mixed with 3.45 g of aluminium to produce 6.40 g of aluminium oxide and 6.95 g of iron. Determine the amount (mol) of each reactant and product.

c. Use the amounts determined in 15.2.b to find the appropriate coefficients to balance the reaction.

The coefficients, called stoichiometric coefficients, in a balanced chemical reaction show the amount (mol) ratios of chemicals in a reaction. Aside from making sure that the chemical equations conform to the law of conservation of mass, these amount ratios are also useful conversion factors. The use of amount ratios from chemical reactions for unit conversions is called stoichiometry.

WRITING BALANCED CHEMICAL REACTIONS

Balanced chemical equations can be derived empirically following the sodium hydrogencarbonate decomposition example. But it is often useful to know what the balanced chemical equation for a given reaction is without having to do the empirical work. Writing and balancing a chemical equation follows a regular and prescriptive pattern.

The first step is to take the text provided and convert each IUPAC name into a chemical formula. For this example, we will look at the decomposition of potassium chlorate. Potassium chlorate is placed in a beaker and is heated to yield potassium chloride and oxygen gas. Using the rules of IUPAC nomenclature (Chapter 10), we can produce the following chemical formula:

$$KClO_3(s) \xrightarrow[\Delta]{} KCl(s) + O_2(g)$$

Next, we take an inventory of the reaction to identify the number of each type of atom or ion present.

$$KClO_3(s) \xrightarrow[\Delta]{} KCl(s) + O_2(g)$$

Reactant	Product
1 K	1 K
1 Cl	1 Cl
3 O	2 O

Looking at the inventory of atoms and ions, we can see that the reaction is not balanced because the number of oxygen atoms differs between the reactant and product sides. We will now introduce whole-number coefficients to balance the equation. To balance the number of oxygens, we add a 2 to $KClO_3$ and a 3 to O_2, which balances the oxygen atoms with six (the lowest common multiple) on each side.

$$2\ KClO_3(s) \xrightarrow[\Delta]{} KCl(s) + 3\ O_2(g)$$

Reactant	Product
2 ~~1~~ K	1 K
2 ~~1~~ Cl	1 Cl
6 ~~3~~ O	6 ~~2~~ O

In balancing the oxygen atoms, we unbalanced the potassium and chlorine. Unbalancing one element to balance another is not an uncommon outcome in this work. This means that we will need to add additional coefficients, now, to balance the potassium and chlorine by adding a 2 to the right side.

$$2\ KClO_3(s) \xrightarrow[\Delta]{} 2\ KCl(s) + 3\ O_2(g)$$

Reactant	Product
2 ~~1~~ K	2 ~~1~~ K
2 ~~1~~ Cl	2 ~~1~~ Cl
6 ~~3~~ O	6 ~~2~~ O

Altogether, we now have a balanced chemical reaction for the decomposition of potassium chlorate. We will see an additional balancing example before turning to practice.

This reaction is an example of combustion: solid sucrose ($C_{12}H_{22}O_{11}$) and dioxygen gas react to yield carbon dioxide gas and water vapor. Using the rules of IUPAC nomenclature (Chapter 10), we can produce the following chemical formula:

$$C_{12}H_{22}O_{11}(s) + O_2(g) \rightarrow CO_2(g) + H_2O(g)$$

Next, we take an inventory of the reaction to identify the number of each type of atom or ion present.

$$C_{12}H_{22}O_{11}(s) + O_2(g) \rightarrow CO_2(g) + H_2O(g)$$

Reactant	Product
12 C	1 C
22 H	2 H
13 O	3 O

Looking at the inventory of atoms and ions, we can see that the reaction is not balanced because the number of carbon atoms, hydrogen atoms, and oxygen atoms all differ between the reactant and product sides. In general, it is best to balance elements that only appear in compounds (here carbon and hydrogen) before balancing elements that appear as the pure element (oxygen on the reactant side). It does not matter whether we start with carbon or hydrogen, so let us start with carbon by adding a 12 to CO_2

$$C_{12}H_{22}O_{11}(s) + O_2(g) \rightarrow 12\ CO_2(g) + H_2O(g)$$

Reactant	Product
12 C	12 ~~1~~ C
22 H	2 H
13 O	25 ~~3~~ O

Now if we add an 11 to H_2O, we will balance the number of hydrogen atoms:

$$C_{12}H_{22}O_{11}(s) + O_2(g) \rightarrow 12\ CO_2(g) + 11\ H_2O(g)$$

Reactant	Product
12 C	12 ~~1~~ C
22 H	11 ~~2~~ H
13 O	35 ~~25~~ ~~3~~ O

Finally, we turn to balancing the oxygen atoms. There are 35 oxygen atoms in the products. On the reactant side, the oxygen atoms are split between 11 in sucrose and 2 in dioxygen. The 11 in sucrose are fixed (unless we add a coefficient to sucrose, which will unbalance all of the balancing we have done so far), which means there are 24 oxygen atoms on the right that are not balanced by O_2 (the other source of oxygen). If we add a 12 to dioxygen, then we will balance the number of oxygen atoms on both sides and have an overall balanced equation.

$$C_{12}H_{22}O_{11}(s) + 12\ O_2(g) \rightarrow 12\ CO_2(g) + 11\ H_2O(g)$$

Reactant	Product
12 C	12 ~~1~~ C
22 H	11 ~~2~~ H
35 ~~13~~ O	35 ~~25~~ ~~3~~ O

Problem 15.3. For each problem, write a balanced chemical equation (making sure to include phase notation).

a. Liquid octane (C_8H_{18}) combines with oxygen gas to burn and yield carbon dioxide gas and water vapor.

b. Dihydrogen reacts with octasulfur to produce hydrogen sulfide gas.

c. Solid aluminium chloride decomposes to produce aluminium metal and chlorine gas.

d. Solid magnesium burns in the presence of oxygen gas to produce magnesium oxide.

e. Dichlorine gas reacts with an aqueous solution of sodium bromide to produce liquid bromine and an aqueous solution of sodium chloride.

f. When lithium is added to solid sodium chloride, solid lithium chloride and sodium are produced.

g. Aqueous rubidium hydroxide solution is combined with beryllium fluoride solid to produce aqueous rubidium fluoride and solid beryllium hydroxide.

h. A solution of aqueous silver nitrate combines with solid copper to produce elemental silver and copper(2+) nitrate solution.

i. Lithium metal reacts with solid magnesium bromide to produce solid lithium bromide and pure magnesium metal.

j. Chlorine gas reacts with aqueous sodium bromide solution to produce liquid bromine and aqueous sodium chloride.

k. Lead dihydroxide solid reacts with aqueous hydrogen chloride to produce water and solid lead dichloride.

l. Aqueous sodium phosphate and aqueous calcium chloride react to produce aqueous sodium chloride and solid calcium phosphate.

STOICHIOMETRY[1]

The stoichiometric coefficients in a balanced chemical equation represent amount ratios. More importantly, these coefficients represent conversion factors that can be used in calculations. For example, in the combustion of sucrose, the balanced chemical equation is as follows:

$$C_{12}H_{22}O_{11}(s) + 12\ O_2(g) \rightarrow 12\ CO_2(g) + 11\ H_2O(g)$$

In this reaction, then, we can derive the conversion factors 1 mol $C_{12}H_{22}O_{11}(s)$ = 12 mol $O_2(g)$ or 1 mol $C_{12}H_{22}O_{11}(s)$ = 12 mol $CO_2(g)$ or an amount ratio for any two chemicals. How is this useful? With these relationships, we can calculate, for example, how many liters of carbon dioxide, at STP, will be produced when 10.0 g of sucrose is combusted:

10.0 $\cancel{g}$ $\cancel{C_{12}H_{22}O_{11}}$	1 $\cancel{mol}$	12 $\cancel{mol}$ CO_2	22.7 L	= 7.96 L CO_2
	342.297 $\cancel{g}$	1 $\cancel{mol}$ $\cancel{C_{12}H_{22}O_{11}}$	1 $\cancel{mol}$	

In this, and all stoichiometry problems, a useful mnemonic is the stoichiometry motto:[2]

Take what's given
Convert to moles
Use the mole-to-mole ratio
Convert to final appropriate units

In our example with sucrose, the given is 10.0 g (the initial measured quantity). We converted that to amount (mol) using the molar mass (342.297 g/mol) and converted from amount (mol) of sucrose to amount (mol) of carbon dioxide (12 mol CO_2 = 1 mol $C_{12}H_{22}O_{11}$). Finally, the problem asked for liters of gas at STP and so 22.7 L/mol was used to convert to the final appropriate units. Note that in this calculation we assume that there is more than enough oxygen gas to react with 10.0 g of sucrose. Problems highlight that other reactants are there in abundance (and therefore should not be considered in the problem) by stating that a reactant is present "in excess."

Problem 15.4. Consider the reaction of ClO_2 with water.

$$6\ ClO_2(g) + 3\ H_2O(l) \rightarrow 5\ HClO_3(aq) + HCl(aq)$$

a. How many milliliters of water (density = 1.00 g/mL) are needed to react with 2.00 g of ClO_2?

b. How many grams of $HClO_3$ are formed from 2.00 g of ClO_2?

Problem 15.5. Answer the following questions about the production of iron.

$$\_\_\_\_\_\ Fe_2O_3(s) + \_\_\_\_\_\ C(s) \rightarrow \_\_\_\_\_\ Fe(s) + \_\_\_\_\_\ CO_2(g)$$

a. Balance the above equation.

b. How many grams of iron(III) oxide do you need to produce 500.0 g Fe(s)?

c. How many liters of CO_2 are produced (at STP) when 3.45 kg iron(III) oxide is reacted with excess carbon?

Problem 15.6. Answer the following questions about the production of ammonia (NH_3).

a. Balance the following combination reaction:

$$\_\_\_\_\_\ N_2(g) + \_\_\_\_\_\ H_2(g) \rightarrow \_\_\_\_\_\ NH_3(g)$$

b. How many grams of nitrogen gas are needed to produce 10.0 g NH_3?

c. How many liters of hydrogen gas (at STP) are needed to produce 135 g NH_3?

Problem 15.7. Answer the following questions about the reaction of aluminium and copper(II) chloride.

$$2\ Al(s) + 3\ CuCl_2(aq) \rightarrow 3\ Cu(s) + 2\ AlCl_3(aq)$$

a. If 4.5 mol Al reacts with excess copper(II) chloride, what amount of copper (mol) will be produced?

b. If 69.5 g copper(II) chloride is reacted with excess aluminium, how many grams of aluminium chloride would be produced?

Problem 15.8. Answer the following question about aluminium chloride.

$$\_\_\_\_\_\_ AlCl_3(s) \rightarrow \_\_\_\_\_\_ Al(s) + \_\_\_\_\_\_ Cl_2(g)$$

a. Balance the above equation.

b. If 43.0 g of aluminium chloride is decomposed according to the equation above, how many chlorine atoms will be produced?

Problem 15.9. Using the following balanced chemical equation, what volume (in L) of 0.790 mol/L lead(II) nitrate is required to react with 1.25 L of 0.550 mol/L potassium chloride solution?

$$2\ KCl(aq) + Pb(NO_3)_2(aq) \rightarrow PbCl_2(s) + 2\ KNO_3(aq)$$

LIMITING REAGENT

So far, we have only considered stoichiometry problems where one of the reactants is said to be in excess. How do we determine the amount of product produced when we are not told a reactant is in excess? In these scenarios, we must determine which reactant is the limiting reagent, that is, the chemical that determines the maximum amount of product that could be produced from the particular measures given. Let's consider the example of iron(III) chloride reacting with magnesium metal.

$$2\ FeCl_3(aq) + 3\ Mg(s) \rightarrow 2\ Fe(s) + 3\ MgCl_2(aq)$$

If we wished to know what mass of iron, in grams, will be produced when 10.0 mL of 0.100 mol/L iron(III) chloride solution is combined with 0.050 g of magnesium metal, we would first determine the mass of iron that each reactant could produce. With 10.0 mL of 0.100 mol/L iron(III) chloride, 0.0558 g Fe could be produced:

10.0 ~~mL $FeCl_3$~~	1×10^{-3} ~~L~~	0.100 ~~mol~~	2 ~~mol~~ Fe	55.845 g	= 0.0558 g Fe
	1 ~~mL~~	~~L~~	2 ~~mol $FeCl_3$~~	~~mol~~	

And with 0.050 g of magnesium metal, 0.077 g Fe could be produced.

0.050 g Mg	mol	2 mol Fe	55.845 g	= 0.077 g Fe
	24.305 g	3 mol Mg	mol	

Given these results, we can conclude that for this amount of iron(III) chloride and magnesium, the iron(III) chloride solution is the limiting reagent and that we will only produce 0.0558 g (55.8 mg) of iron using this combination of iron(III) chloride and magnesium metal. We would also say, in this example, that magnesium was present in excess.

Problem 15.10. 5.0 mL of 0.100 mol/L calcium chloride is combined with 7.5 mL of 0.060 mol/L lithium carbonate. Determine the mass of solid calcium carbonate produced (in grams) and identify the limiting reagent and the reagent present in excess.

$$CaCl_2(aq) + Li_2CO_3(aq) \rightarrow CaCO_3(s) + 2\ LiCl(aq)$$

ENTHALPY OF REACTION ($\Delta_R H^\circ$)

In Chapters 4 and 5, we discussed the concept of enthalpy (ΔH), which is a measure of the amount of heat produced or consumed, per amount of chemical. In the SI system of units, enthalpy is typically calculated as the kilojoules of heat energy per mole (kJ/mol). Here, we conclude our discussion of reactions by looking at the enthalpy of reaction ($\Delta_r H^\circ$), which is the heat produced or consumed per amount of reaction.

In terms of enthalpy, reactions can either be endothermic (consuming energy) or exothermic (producing energy). Endothermic reactions are reactions where the products are less stable energetically than the reactants, which is indicated, by convention, as a positive enthalpy of reaction is positive ($\Delta_r H^\circ > 0$). Exothermic reactions are reactions where the products are more stable energetically than the reactants, which is indicated, by convention, as a negative enthalpy of reaction is negative ($\Delta_r H^\circ < 0$).

We can use the enthalpy of reaction values to determine the amount of energy produced, for an exothermic reaction, or the amount of energy consumed, for an endothermic reaction. In stoichiometric calculations, the enthalpy values are energy (kJ) per amount (mol) of reaction, and so a balanced chemical equation is used to determine the amount of reaction. Consider, for example, the combustion of octane (C_8H_{18}).

$$2\ C_8H_{18}(l) + 25\ O_2(g) \rightarrow 16\ CO_2(g) + 18\ H_2O(g),\ \Delta_r H^\circ = -10\,860\ kJ/mol$$

If 2.81 kg octane, the mass in one gallon of gasoline, is combusted, the amount of octane is 24.6 mol.

2.81 kg C_8H_{18}	1000 g	1 mol	= 24.6 mol C_8H_{18}
	1 kg	114.23 g	

Now considering the balanced equation above, there are two moles of octane for every one mole of reaction (mol rxn). That is, the amount of reaction is half of the amount of octane.

24.6 mol C_8H_{18}	1 mol rxn	= 12.3 mol rxn
	2 mol C_8H_{18}	

With the moles of reaction, we can use the enthalpy of reaction to determine that combusting 2.81 kg of octane produces, the negative sign indicates energy given off, 134 000 kJ of energy.

12.3 mol rxn	−10 860 kJ	
	1 mol	= −134 000 kJ

Problem 15.11. For each of the following, use the information provided, the balanced reaction, and the enthalpy of reaction to determine the energy produced (or consumed).

a. $2\ S(s) + C(s, graphite) \rightarrow CS_2(l)$, $\Delta_r H^\circ = 89.0$ kJ/mol
   1.0 kg of graphite is converted into carbon disulfide.

b. $3\ Ag(s) + AuCl_3(aq) \rightarrow 3\ AgCl(s) + Au(s)$, $\Delta_r H^\circ = -288.6$ kJ/mol
   450 g of silver is converted into silver chloride.

CHEMICAL EQUATIONS

Chemical equations provide a means of representing the transformation of initial chemicals (reactants) into final chemicals (products). Balanced chemical equations – those obeying the law of conservation of mass – also represent the amount (mol) ratio of elements. These amount ratios allow chemists to quantify and relate the masses or volumes of chemicals in a reaction. When combined with enthalpy of reaction, these stoichiometric calculations can also determine the amount of energy produced or consumed.

NOTES

1. Richter, J.B. *Anfangsgründe der Stöchyometrie oder Meßkunst chymischer Elemente. Erster Theil.* Johann Friedrich Korn dem Aeltern: Breßlau und Hirschberg, **1792**.
2. This mnemonic device was imparted to the author by his high school chemistry teacher, the late Gary Osborne, when the author took high school chemistry in the 2004–05 academic year. The mnemonic has served well for nearly two decades, and Gary Osborne was the inspiration for the author's own interest in pursuing chemistry.

16 Chemical Reactions
Electron Transfer and Electron Sharing

In Chapter 15, we developed systems for representing chemical reactions, chemical equations, and relating the amounts of different substances, stoichiometry. We did not, however, address deeper questions of how or why reactions take place. In this chapter, we will turn our attention to these questions. A full explanation of how and why reactions occur involves a thorough study of thermodynamics and kinetics, which are beyond the scope of this book.

Chemical change is driven by the movement of electrons; this is how reactions happen. Given that reactions involve the flow of electrons, we need to track electrons and how they change. The movement of electrons is driven, in part, by the lowering of the energy of electrons (this is the reason why reactions happen). We can use electronegativity (Chapter 9) to help understand why a reaction is likely or unlikely.[1] Electronegativity, recall (Figure 16.1) is the power of an atom to attract electrons to itself.[2]

ELECTRON TRANSFER

A chemical reaction involves the movement of electrons. That movement could involve the transfer of electrons (the focus of this section) or a change in how electrons are shared. When electrons are transferred, this is called a redox reaction, where **red***ox* is a portmanteau of **red**uction[3] and

1	2	3	4	5	6	7	8	9	10	11	12	13	14	15	16	17	18
1 H 2.20																	2 He
3 Li 0.98	4 Be 1.57											5 B 2.04	6 C 2.55	7 N 3.04	8 O 3.44	9 F 3.98	10 Ne
11 Na 0.93	12 Mg 1.31											13 Al 1.61	14 Si 1.90	15 P 2.19	16 S 2.58	17 Cl 3.16	18 Ar
19 K 0.82	20 Ca 1.00	21 Sc 1.36	22 Ti 1.54	23 V 1.63	24 Cr 1.66	25 Mn 1.55	26 Fe 1.83	27 Co 1.88	28 Ni 1.91	29 Cu 1.90	30 Zn 1.65	31 Ga 1.81	32 Ge 2.01	33 As 2.18	34 Se 2.55	35 Br 2.96	36 Kr 3.0
37 Rb 0.82	38 Sr 0.95	39 Y 1.22	40 Zr 1.33	41 Nb 1.6	42 Mo 2.16	43 Tc 1.9	44 Ru 2.2	45 Rh 2.28	46 Pd 2.20	47 Ag 1.93	48 Cd 1.69	49 In 1.78	50 Sn 1.96	51 Sb 2.05	52 Te 2.30	53 I 2.66	54 Xe 2.6
55 Cs 0.79	56 Ba 0.89	*	72 Hf 1.3	73 Ta 1.5	74 W 2.36	75 Re 1.9	76 Os 2.0	77 Ir 2.20	78 Pt 2.28	79 Au 2.54	80 Hg 2.00	81 Tl 2.04	82 Pb 2.33	83 Bi 2.02	84 Po 2.0	85 At 2.2	86 Rn
87 Fr 0.7	88 Ra 0.9	**	104 Rf	105 Db	106 Sg	107 Bh	108 Hs	109 Mt	110 Ds	111 Rg	112 Cn	113 Nh	114 Fl	115 Mc	116 Lv	117 Ts	118 Og

*	57 La 1.1	58 Ce 1.12	59 Pr 1.13	60 Nd 1.14	61 Pm 1.2	62 Sm 1.17	63 Eu 1.1	64 Gd 1.20	65 Tb 1.2	66 Dy 1.22	67 Ho 1.23	68 Er 1.24	69 Tm 1.25	70 Yb 1.1	71 Lu 1.27
**	89 Ac 1.1	90 Th 1.0	91 Pa 1.5	92 U 1.38	93 Np 1.36	94 Pu 1.28	95 Am 1.30	96 Cm	97 Bk	98 Cf	99 Es	100 Fm	101 Md	102 No	103 Lr

FIGURE 16.1 Pauling electronegativity values.

DOI: 10.1201/9781003487210-16

oxidation.[4] Reduction is the complete transfer of one or more electrons to an atom, ion, or molecule.[5] Oxidation is the complete removal of one or more electrons from an atom, ion, or molecule.[6] A useful mnemonic for remembering these definitions is LEO the lion says GER. LEO stands for lose electrons oxidation, and GER stands for gain electrons reduction.

TRACKING ELECTRONS – CHARGE

To begin our study of electron transfer reactions, we will look at the tarnishing of silver. When silver tarnishes, silver metal reacts with dioxygen gas to produce silver oxide.

$$4\ Ag(s) + O_2(g) \rightarrow 2\ Ag_2O(s)$$

In the reactants, the elements are in their pure form, which means that silver has no charge, and the two oxygen atoms each have no charge. The product is an ionic compound composed of silver(1+) and oxide(2–) ions. This means that each silver atom has lost one electron, and each oxygen atom has gained two electrons. Following the definitions of reduction and oxidation, the silver atoms are oxidized, and the oxygen atoms are reduced. This transfer – from silver to oxygen – occurs freely because it follows the general trend that the less electronegative elements ($\chi_P(Ag) = 1.93$) tend to be oxidized and the more electronegative elements ($\chi_P(O) = 3.44$) tend to be reduced in chemical reactions.

Problem 16.1. Using the common charges of elements in ionic compounds (Appendix 5), identify the charge of each element before and after the chemical reaction. Then identify which element is reduced and which element is oxidized.

a. $MgI_2(aq) + Br_2(l) \rightarrow I_2(s) + MgBr_2(aq)$

b. $TiCl_4(l) \rightarrow Ti(s) + 2\ Cl_2(g)$

c. $Fe_2O_3(s) + 3\ Mg(s) \rightarrow 3\ MgO(s) + 2\ Fe(s)$

d. $2\ Cu(s) + S(s) \rightarrow Cu_2S(s)$

Problem 16.2. Answer the following questions about the synthesis of sodium chloride from solid sodium and dichlorine gas.

$2\ Na(s) + Cl_2(g) \rightarrow 2\ NaCl(s),\ \Delta_r H^\circ = -411$ kJ/mol

a. What element is oxidized, and what element is reduced?

b. This reaction is very favorable; favorable reactions tend to release energy, and this reaction releases 411 kJ/mol. Consider the electron configuration of sodium and the electron configuration of chlorine, and explain why the electron transfer you identified in 16.2.a is so favorable.

c. The reaction of sodium and chlorine is very exothermic. Would the reaction of caesium and dichlorine be more or less energetic than the reaction of sodium and dichlorine? Consider periodic trends at play for sodium versus caesium and how that might impact the reaction.

d. Would the reaction of sodium and diiodine be more or less energetic than the reaction of sodium and dichlorine? Consider periodic trends at play for iodine versus chlorine and how that might impact the reaction.

Tracking Electrons – Oxidation Numbers

Combustion reactions are common examples of redox reactions. The balanced reaction for the combustion of methane is as follows:

$$CH_4(g) + 2\ O_2(g) \rightarrow CO_2(g) + 2\ H_2O(g)$$

Here, we cannot track the electrons like we did for ionic compounds because there are no ionic compounds (no compounds consisting of metal and nonmetal elements) and therefore no ions are present. The transfer of electrons defines redox. The question then arises: How do we keep track of electrons for molecular compounds?[7] There are two methods used for quantifying the number of electrons around an atom: formal charges (Chapter 11) and oxidation numbers (discussed here). Ideally, we would be able to directly measure the charge of an atom in a molecular compound, but to date, there is no direct method to do so. Formal charges and oxidation numbers are two bookkeeping methods that chemists employ. Let's consider the atoms in ammonium (Figure 16.2).

H
H–N–H
H

H
H··N··H
H

FIGURE 16.2 A Lewis structure for ammonium (*left*) showing only atoms and bonds and a modified Lewis structure (*right*) where the bonds are shown as two dots to represent the two shared electrons.

In a molecular compound that involves different elements – nitrogen and hydrogen in ammonium – the electrons are not shared equally. Nitrogen (χ_P(N) = 3.04) pulls on the electrons more than hydrogen (χ_P(H) = 2.20), which means that the electron density around the nitrogen atom is greater than the electron density around the hydrogen atoms. This partially covalent (sharing) and partially ionic (unequal sharing) character of covalent bonds makes it challenging to determine the charge of an atom in a molecular compound without a computational chemistry program. We shall see an example of computationally derived charges, natural population analysis (NPA),[8] which is a result of a natural bond orbital (NBO) calculation.[9]

To simplify the complexity in a real bond, our bookkeeping tools – formal charge (Chapter 11) and oxidation number (briefly mentioned in Chapter 10)[10] – make simplifying assumptions. Formal charge assumes that all bonds are perfectly covalent, and all electrons are shared equally. The oxidation number formalism, in contrast, assumes that all bonds are perfectly ionic, and no electrons are shared. These two assumptions allow us to consider a structure (Figure 16.3) and assign the bonding electrons to each atom (formal charge) or to the more electronegative atom (oxidation number).

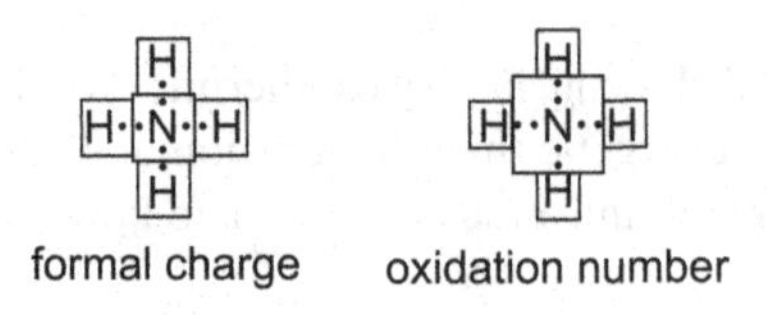

FIGURE 16.3 Assignment of electrons in bonds to each atom following the formal charge assumption (*left*) and only to the more electronegative atom following the oxidation number assumption (*right*).

Looking at Figure 16.3, we can see the differences between how formal charge and oxidation numbers consider bonding electrons. In the case of formal charge, the bonding electrons are split equally, which means that each hydrogen atom is assigned one electron, and the nitrogen atom is assigned four electrons. We then subtract the assigned number of electrons from the valence electron count for each atom – one for hydrogen and five for nitrogen – to give 0 formal charge for hydrogen and 1+ for nitrogen (Figure 16.4). In contrast, oxidation number determinations assign no electrons to the hydrogen atom and all eight electrons to the nitrogen atom. We then subtract the assigned number of electrons from the valence electron count for each atom – one for hydrogen and five for nitrogen – to give an oxidation number of +1 for hydrogen and –3 for nitrogen (Figure 16.4). In terms of the calculated natural population analysis (NPA) charge, we can see that oxidation number values give a better sense of the charge distribution, that is, it correctly predicts positive and negative but overestimates the charge magnitudes (nitrogen has an NPA value of –0.979 while the oxidation number is –3).

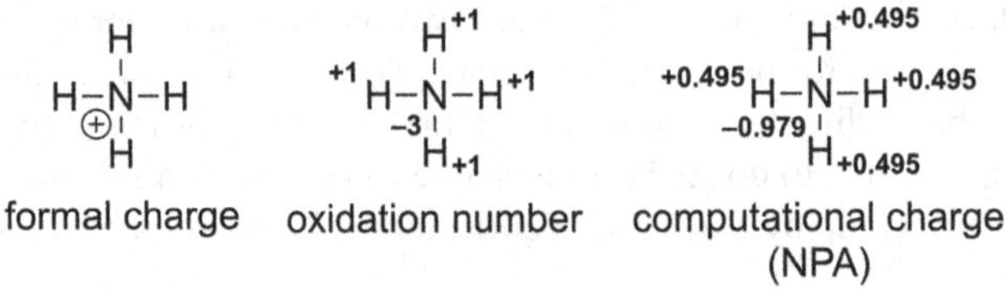

formal charge oxidation number computational charge (NPA)

FIGURE 16.4 Assignment of atomic charge using formal charge (*left*), oxidation numbers (*middle*), and calculated charge using natural population analysis (NPA).

While oxidation numbers can be determined with a Lewis structure, often we only have a chemical formula. In such instances, we can either draw a Lewis structure and assign oxidation numbers by assigning electrons based on electronegativity or we can use the following rules with the chemical formulae.

RULES FOR ASSIGNING OXIDATION NUMBERS TO ELEMENTS IN A CHEMICAL FORMULA

1. **For any atom in its elemental form, the oxidation number is always zero.**
   For example, H in H_2 has an oxidation number of 0. Each P atom in P_4 has an oxidation number of 0. Every Cu atom in copper metal has an oxidation number of 0.

2. **For any monatomic ion, the oxidation number always equals the charge.** For example, in the compound K_2S, the ions are K^+ and S^{2-}. The oxidation number of potassium is +1, and the oxidation number of sulfide is –2.
3. **For any polyatomic ion or molecular compound, assign oxidation numbers in order.**
   a. **Fluorine always has an oxidation number of –1.**
   b. **Oxygen usually has an oxidation number of –2.**
      The exceptions are when oxygen is bonded to fluorine or when oxygen is part of a peroxide (oxygen has an oxidation number of –1 in peroxides).
   c. **The oxidation number of hydrogen is +1 when combined with non-metals and –1 when combined with metals.**
      In NaH, sodium has an oxidation number of +1 and hydrogen has an oxidation number of –1.
      In HF, hydrogen has an oxidation number of +1 and fluorine has an oxidation number of –1.
   d. **Halogen atoms Cl, Br, and I usually have an oxidation number of –1, except when bonded to another element with higher electronegativity.**
      In HCl, hydrogen has an oxidation number of +1 and chlorine has an oxidation number of –1.
      In ClF, chlorine has an oxidation number of +1 and fluorine has an oxidation number of –1.
4. **The sum of all oxidation numbers must equal 0 for a neutral molecule or they must equal the charge of a polyatomic ion.**
   In ClF_3 the fluorine atoms each have an oxidation number of –1 and the chlorine atom has an oxidation number of +3: $1Cl + 3(-1) = 0$, $Cl = +3$.
   For CO_3^{2-} oxygen atoms each have an oxidation number of –2 and the carbon atom has an oxidation number of +4: $1C + 3(-2) = -2$, $C = +4$.

Problem 16.3. Assign the oxidation number for each atom or ion.

a. Mn(s)
b. Ca^{2+}(aq)
c. S_8(s)
d. N_2(g)
e. I^-(aq)

Problem 16.4. Assign the oxidation number for each element in the following binary compounds.

a. C_2H_6(g)
b. C_2H_4(g)
c. C_2H_2(g)
d. $MnCl_2$(s)
e. IF_7(g)

Problem 16.5. Assign the oxidation number for each element in the following polyatomic ions.

a. NO_2^-(aq)
b. NO_3^-(aq)
c. Hg_2^{2+}(aq)
d. OH^-(aq)
e. PO_4^{3-}(aq)

Problem 16.6. Assign the oxidation number for each element in the following compounds.

a. $Fe(NO_3)_3(s)$
b. $H_3PO_4(l)$
c. $KMnO_4(s)$
d. $KOH(s)$
e. $Al_2(SO_4)_3(s)$

Problem 16.7. What is the oxidation number of carbon in each of the following compounds?

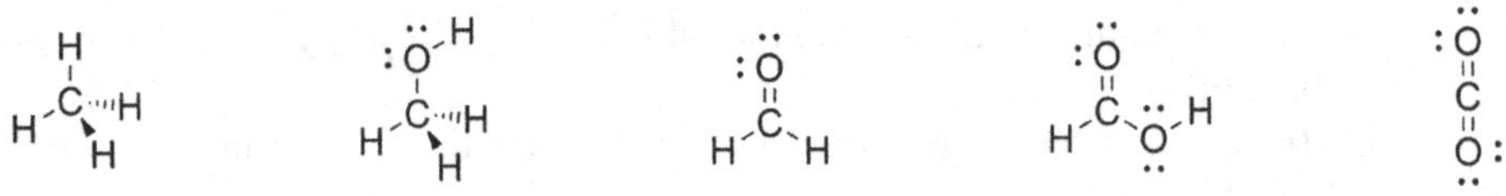

Problem 16.8. For each of the following, identify which element is oxidized and which element is reduced.

a. $Hg_2O(s) \rightarrow HgO(s) + Hg(l)$

b. $CH_3Cl(g) + 3\ Cl_2(g) \rightarrow CCl_4(g) + 3\ HCl(g)$

c. $2\ KMnO_4(aq) + 5\ HCO_2H(aq) + 6\ HCl(aq) \rightarrow 2\ MnCl_2(aq) + 8\ H_2O(l) + 5\ CO_2(g) + 2\ KCl(aq)$

Enthalpy of Reaction – A Tool for Identifying Whether a Reaction Is Favored

So far, we have considered tracking electrons and identifying whether a chemical species is oxidized or reduced. Before turning to electron sharing, we will consider the energy changes (Chapters 5 and 14) associated with redox reactions. Recall that enthalpy ($\Delta_r H^\circ$) for a process shows the amount of energy (J or kJ) per amount (mol). By convention, when a reaction is exothermic and releases heat, we indicate that with a negative enthalpy value ($\Delta_r H^\circ < 0$); when a reaction is endothermic and takes in heat, we indicate that with a positive enthalpy value ($\Delta_r H^\circ > 0$). Reactions that are exothermic are energetically favored—that is, they are more likely to proceed—while reactions that are endothermic are energetically disfavored—that is, they are less likely to proceed. For example, combustion (fire) is very exothermic and proceeds quite readily. The decomposition of iron(III) oxide into metallic iron and dioxygen gas, however, is very endothermic and will not proceed unless provided with a large amount of external energy.

The direction of electron transfer in redox reactions, that is, where electrons are most likely to go without added energy, can often be explained simply by electronegativity: Atoms with higher electronegativity values tend to get reduced, and atoms with lower electronegativity values tend to get oxidized. We can use enthalpy to help underscore this idea. Consider the synthesis reaction of dihydrogen and difluorine reacting to produce hydrogen fluoride.

$$H_2(g) + F_2(g) \rightarrow 2\ HF(l),\ \Delta_r H^\circ = -598.0\ \text{kJ/mol}$$

Here, the hydrogen atoms are oxidized (going from an oxidation state of 0 to an oxidation state of +1) and the fluorine atoms are reduced (going from an oxidation state of 0 to an oxidation state of −1). If we consider the Pauling electronegativity values (χ_P) of each, we can see that the electrons are transferred from the less electronegative hydrogen atoms ($\chi_P = 2.20$) to the more electronegative fluorine atoms (χ_P = 3.98). This is a favorable direction of transfer, which we see reflected in the highly exothermic reaction enthalpy, and such reactions proceed to make products.

Now compare and contrast the reaction of dihydrogen and difluorine with the reaction of dihydrogen with zinc oxide.

$$H_2(g) + ZnO(s) \rightarrow H_2O(l) + Zn(s),\ \Delta_r H^\circ = 64.7\ \text{kJ/mol}$$

Here, the hydrogen atoms are oxidized (going from an oxidation state of 0 to an oxidation state of +1) and the zinc is reduced (going from an oxidation state of +2 to an oxidation state of 0). If we consider the Pauling electronegativity values (χ_P) of each, we can see that the electrons are transferred from the more electronegative hydrogen atoms (χ_P = 2.20) to the less electronegative zinc atom (χ_P = 1.65). This is an unfavorable direction of transfer, which we see reflected in the endothermic reaction enthalpy. Reactions with endothermic reaction enthalpies tend not to produce products without the addition of external energy.

While other factors aside from electronegativity play a role in determining the outcome of a redox reaction, electronegativity is a great first approximation of whether a redox reaction is likely to proceed or not. Redox reactions are driven by electrons achieving a lower energy, which tends to happen when they move to more electronegative atoms.

Problem 16.9. For each of the following, predict whether you think the reaction enthalpy ($\Delta_r H^\circ$) would be positive (endothermic) or negative (exothermic).

a. $NH_2Cl(g) + Cl_2(g) \rightarrow NHCl_2(g) + HCl(g)$

b. $2\ NaCl(s) + Br_2(l) \rightarrow 2\ NaBr(s) + Cl_2(g)$

c. $3\ Ag(s) + AuCl_3(aq) \rightarrow 3\ AgCl(s) + Au(s)$

d. $2\ HCl(aq) + Pt(s) \rightarrow PtCl_2(s) + H_2(g)$

e. $CH_4(g) + 2\ O_2(g) \rightarrow CO_2(g) + 2\ H_2O(g)$

Redox Reactions in Context – Electrochemical Cells

Redox reactions play a central role in a number of areas: metabolism, metallurgy, energy generation, and energy storage. It is worth noting that electrochemical cells rely on redox reactions to function. In an electrochemical cell, the reaction happens indirectly, with electrons transferred through a wire. For a favorable reaction – where we see the enthalpy of reaction is exothermic – the electrochemical cell is termed a galvanic cell (or a battery in everyday language) and has a positive cell potential (measured in volts). For example, each lithium-ion battery in a laptop is typically 3.7 V. The value of 3.7 V indicates the likelihood of electrons moving, with the larger the voltage being the larger the potential for electrons to move. In contrast, a fully discharged electrochemical cell (a dead battery) would have a cell potential of 0.0 V as there is no longer a likelihood for electrons to move. Electrochemical cells in this state must be recharged with an external power connection, which increases the electrons energetically (the enthalpy of reaction is endothermic).

ELECTRON SHARING

To start our discussion of electron sharing, let us consider the reaction of hydrogen bromide with potassium hydroxide.

$$HBr(aq) + KOH(aq) \rightarrow H_2O(l) + KBr(aq)$$

If we assign oxidation numbers to each element, we find that potassium has an oxidation number of +1 before and after the reaction, bromine has an oxidation number of –1 before and after the reaction, potassium has an oxidation number of +1 before and after the reaction, and oxygen has an oxidation number of –2 before and after the reaction. No electrons have been transferred, and nothing has been oxidized or reduced. Yet, a reaction has taken place because a new molecular compound (H_2O) has formed from two aqueous solutions. In addition, the electrons in water and potassium bromide are shared differently than they are in hydrogen bromide and potassium hydroxide. We can see this more clearly if we consider the Lewis structures of the chemical species (Figure 16.5).

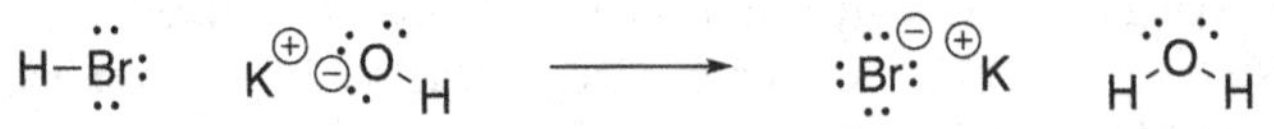

H–Br: K⊕⊖O–H ⟶ :Br:⊖ ⊕K H–O–H

FIGURE 16.5 Lewis structures showing the reaction of hydrogen bromide with potassium hydroxide to produce water and potassium bromide.

In the reactants, the oxygen atom is negatively charged, has one bond to hydrogen, and has three lone pairs. In the product, the oxygen is neutral, has two bonds to hydrogen, and has two lone pairs. In the reactants, bromine has three lone pairs and one bond to hydrogen. In the product, bromine is not sharing any electrons and now has four lone pair. Altogether, we can see the changes and we can talk about them, but there is not a single, quantifiable metric, like oxidation numbers, that we can use to measure the change. Chemists have, therefore, developed the terms "acid" and "base," which are based on definitions from acid–base theories. We will consider the Brønsted–Lowry and Lewis theories.

ACID–BASE DEFINITIONS

In Brønsted–Lowry acid–base theory,[11] an acid is a molecule that is a hydron, H^+, donor and a base is a molecule that is a hydron, H^+, acceptor. In Brønsted–Lowry theory, acids and bases are defined in relation to one another: Acids donate a hydron to a base, and this produces a conjugate base, while the base accepts the hydron and becomes a conjugate acid. In Figure 16.6, the acid hydrogen bromide (HBr) donates a hydron to the base potassium hydroxide (KOH), which results in the conjugate base bromide (Br^-) and the conjugate acid water (H_2O).

H–Br: K⊕⊖O–H ⟶ :Br:⊖ ⊕K H–O–H

acid base conjugate base conjugate acid

FIGURE 16.6 Analysis of the reaction according to Brønsted–Lowry theory: The hydron (circle) transfers from the acid (HBr) to the base (KOH) to yield the conjugate base (potassium bromide) and the conjugate acid (water).

Lewis acid–base theory is the most inclusive framework for defining acids and bases. In Lewis acid–base theory,[12] acids are defined as electron-pair acceptors and bases are defined as electron-pair donors. The flow of electrons is highlighted in Figure 16.7 with the arrows. We can see the

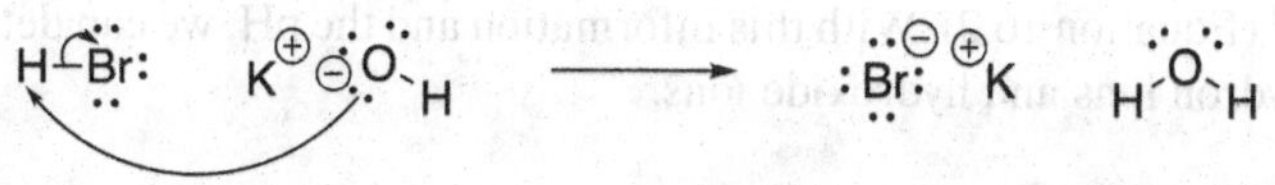

FIGURE 16.7 Analysis of the reaction according to Lewis theory: The base (KOH) donates an electron pair to the acid (HBr) as indicated by arrows.

oxygen lone pair is donated to hydrogen bromide, which accepts the lone pair and breaks the H–Br bond. Lewis acid–base assignments always agree with the Brønsted–Lowry assignments; the only difference between the two is the perspective. In Brønsted–Lowry theory, we focus on the movement of hydrons, and in Lewis theory, we focus on the movement of electrons.

Problem 16.10. For each of the following reactions, identify the acid and the base. For any reactions involving hydron transfer, also identify the conjugate acid and conjugate base.

a.

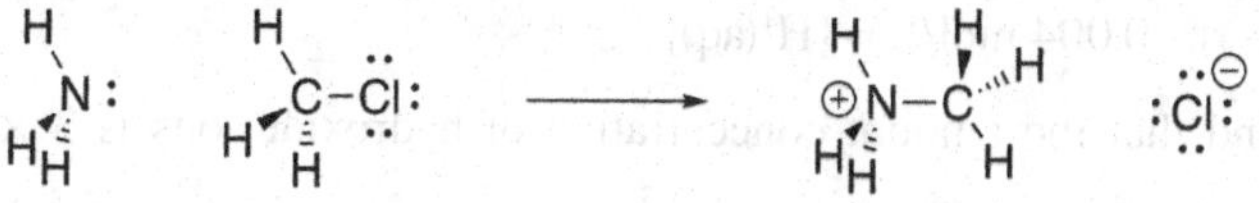

b.

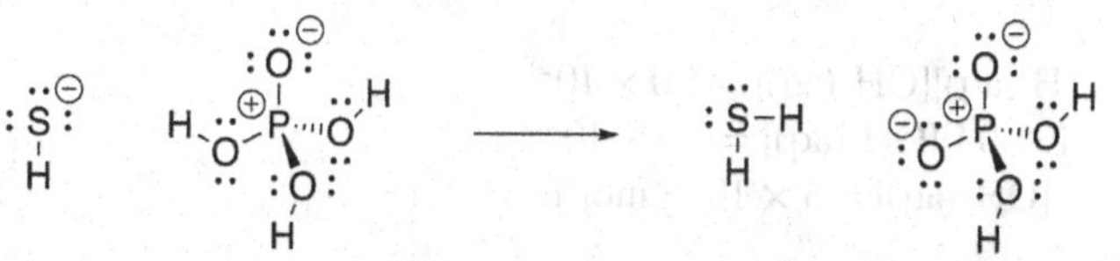

c.

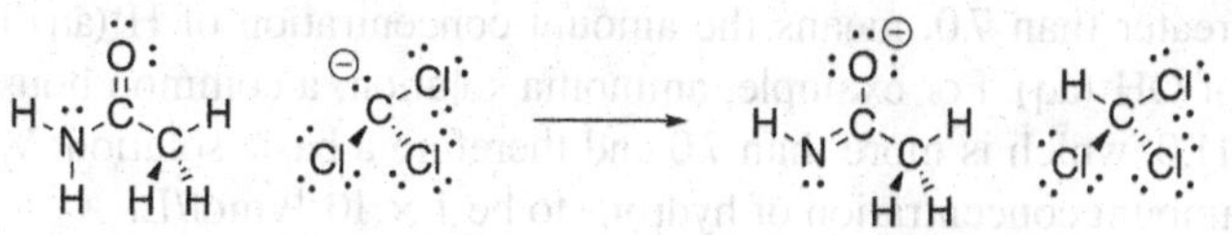

d.

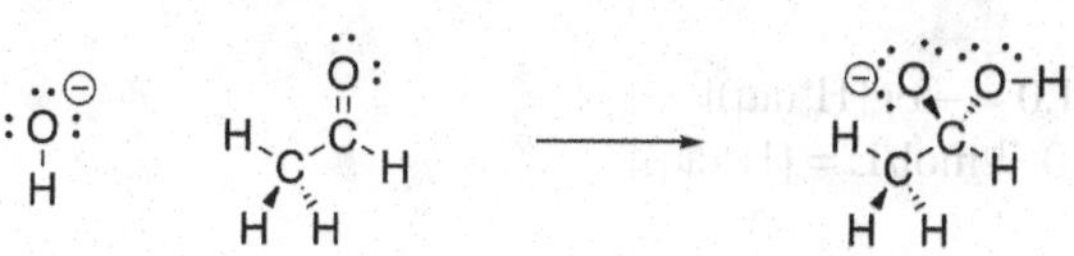

e.

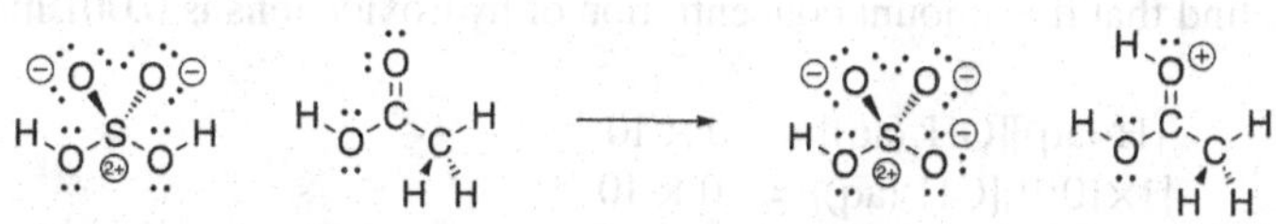

Measurement of Acidity – pH

When an acid is added to water, the aqueous concentration of hydrons, $H^+(aq)$, increases. Measuring the amount of $H^+(aq)$ in water allows us to quantify the acidity of a solution. We commonly measure acidity using pH, which is defined as the negative logarithm of $[H^+(aq)]$, the amount concentration (mol/L) of hydrons (Equation 16.1).[13]

$$pH = -\log[H^+(aq)] \quad (16.1)$$

$[H^+(aq)]$ is the amount concentration (mol/L) of $H^+(aq)$ in the solution.

For water at 25.0 °C, the neutral pH is 7.00, which is when the amount concentration of hydrons ($H^+(aq)$) is 1.0×10^{-7} mol/L and this equals the 1.0×10^{-7} mol/L concentration of hydroxide ions ($OH^-(aq)$). At 25.0 °C, the product of the hydron ion and hydroxide ion amount concentration values is always 1.0×10^{-14} (Equation 16.2). With this information and the pH, we can determine the amount concentration of hydron ions and hydroxide ions.

$$[H^+(aq)][OH^-(aq)] = 1.0 \times 10^{-14} \tag{16.2}$$

$[H^+(aq)]$ is the amount concentration (mol/L) of $H^+(aq)$ in the solution.
$[OH^-aq)]$ is the amount concentration (mol/L) of $OH^-(aq)$ in solution.

On the pH scale, an acidic pH, less than 7.0, means the amount concentration of $H^+(aq)$ is greater than the amount concentration of $OH^-(aq)$. For example, vinegar is typically pH 2.4, which is less than 7.0 and therefore an acidic solution. We can use this value to determine the amount concentration of hydrons to be 0.004 mol/L.

$$2.4 = -\log[H^+(aq)]$$
$$10^{-2.4} = [H^+(aq)]$$
$$0.004 \text{ mol/L} = [H^+(aq)]$$

And using Equation 16.2, we find that the amount concentration of hydroxide ions is 3×10^{-12} mol/L.

$$[H^+(aq)][OH^-(aq)] = 1.0 \times 10^{-14}$$
$$[0.004][OH^-(aq)] = 1.0 \times 10^{-14}$$
$$[OH^-(aq)] = 3 \times 10^{-12} \text{ mol/L}$$

On the pH scale, a basic pH, greater than 7.0, means the amount concentration of $H^+(aq)$ is less than the amount concentration of $OH^-(aq)$. For example, ammonia solution, a common household cleaner, is typically around pH 11.0, which is more than 7.0 and therefore a basic solution. We can use this value to determine the amount concentration of hydrons to be 1×10^{-11} mol/L.

$$11.0 = -\log[H^+(aq)]$$
$$1 \times 10^{-11} \text{ mol/L} = [H^+(aq)]$$

And using Equation 16.2, we find that the amount concentration of hydroxide ions is 0.001 mol/L.

$$[H^+(aq)][OH^-(aq)] = 1.0 \times 10^{-14}$$
$$[1\times10^{-11}][OH^-(aq)] = 1.0 \times 10^{-14}$$
$$[OH^-(aq)] = 0.001 \text{ mol/L}$$

Problem 16.11. Given the following pH values, determine the amount concentration of hydrons, $H^+(aq)$, and hydroxide, $OH^-(aq)$. Identify each as acidic or basic.

a. pH = 1.00

b. pH = 2.00

c. pH = 8.18

Problem 16.12. Given the following hydroxide amount concentrations, determine the amount concentration of hydrons and the pH of the solution. Identify each as acidic or basic.

a. $[OH^-(aq)] = 1.8 \times 10^{-5}$

b. $[OH^-(aq)] = 4.1 \times 10^{-2}$

c. $[OH^-(aq)] = 9.8 \times 10^{-8}$

Problem 16.13. Given the following hydron amount concentrations, determine the amount concentration of hydroxide and the pH of the solution. Identify each as acidic or basic.

a. $[H^+(aq)] = 1.0$

b. $[H^+(aq)] = 5.0 \times 10^{-7}$

c. $[H^+(aq)] = 5.0 \times 10^{-6}$

Precipitation Reactions

Reactions that involve the formation of a solid product from aqueous starting materials are called precipitation reactions, which are an example of an acid–base reaction. Consider the reaction of aqueous silver nitrate with aqueous sodium chloride to produce aqueous sodium nitrate and solid silver chloride. Note that the downward-facing arrow next to silver chloride can be included or not, but it highlights the formation of a solid product that sinks to the bottom.

$$AgNO_3(aq) + NaCl(aq) \rightarrow NaNO_3(aq) + AgCl(s)\downarrow$$

We can get a clearer sense of what is occurring if we look at the ionic equation, where all the aqueous ionic compounds are shown as dissociated ions (Chapter 14). Note that the dissolving of ionic compounds in water is a chemical change as the ionic bonds are broken and replaced by hydrogen bonds and ion–dipole interactions between the ions and the water molecules. In an ionic equation, we can get a better visualization of what chemical species are actually present.

$$Ag^+(aq) + NO_3^-(aq) + Na^+(aq) + Cl^-(aq) \rightarrow Na^+(aq) + NO_3^-(aq) + AgCl(s)\downarrow$$

We can see that the sodium (1+) and nitrate ions are unchanged over the course of the reaction, which we term spectator ions. In the net ionic equation, we simplify the reaction further by removing the spectator ions and only show the chemical species undergoing a change.

$$Ag^+(aq) + Cl^-(aq) \rightarrow AgCl(s)\downarrow$$

Here, the silver(1+) cation is a Lewis acid and the chloride(1–) is a Lewis base, and they interact to form solid silver chloride. Silver chloride is an ionic compound, but it exhibits partial covalent bonding (30.6%),[14] which comes from the donation of electrons from the chloride to the silver(1+).

Problem 16.14. For each of the following precipitation reactions, write a net ionic equation and identify the Lewis acid and the Lewis base.

a. $K_2CO_3(aq) + SrCl_2(aq) \rightarrow 2\ KCl(aq) + SrCO_3(s)\downarrow$

b. $CrCl_3(aq) + 3\ LiOH(aq) \rightarrow 3\ LiCl(aq) + Cr(OH)_3(s)\downarrow$

c. $Co(CH_3CO_2)_2(aq) + 2\ HF(aq) \rightarrow 2\ CH_3CO_2H(aq) + CoF_2(s)\downarrow$

Enthalpy and Periodic Trends: Understanding Acidity

There are many factors that influence acidity (how likely a species is to donate a hydron). Here, we will only consider how two periodic trends – electronegativity and atomic size – affect acidity. We will consider each trend and its impact on acidity, and we will use enthalpy to see the impact on acid–base reactions.

Electronegativity can help us predict whether a reaction is likely, or not, to occur when the base and conjugate base are both in the same period. Consider the reaction of aqueous hydrogen fluoride and aqueous hydroxide to yield aqueous fluoride and water.

$$HF(aq) + OH^-(aq) \rightarrow F^-(aq) + H_2O(l),\ \Delta_r H^\circ = -61.6\ \text{kJ/mol}$$

Given the exothermic enthalpy of reaction ($\Delta_r H^\circ < 0$), this is a favorable reaction, which will proceed to make products. Here, the hydron is transferred from fluoride ($\chi_P = 3.98$) to the oxygen ($\chi_P = 3.44$) of hydroxide. Fluorine has a higher electronegativity value, which means that fluorine tends to share its electrons less than oxygen. Therefore, the fluoride ion is a weaker base than hydroxide because of the greater electronegativity of fluorine. In other words, hydrogen fluoride is a stronger acid than water. Regardless of whether we consider the acid or the base, reactions tend to go from strong (acid or base) to weak (acid or base).

Contrast the reaction between hydrogen fluoride and hydroxide with the reaction between methane (CH_4) and azanide (NH_2^-).

$$CH_4(g) + NH_2^-(g) \rightarrow CH_3^-(g) + NH_3(g),\ \Delta_r H^\circ = 15.3\ \text{kJ/mol}$$

Here, the reaction is endothermic ($\Delta_r H^\circ > 0$), which indicates that the reaction is unfavorable and will not proceed without the addition of external energy. If we consider the electronegativity values for carbon ($\chi_P = 2.55$) and nitrogen ($\chi_P = 3.04$), we can see why this reaction is disfavored. Here, a hydron is transferred from carbon – less electronegative and wants to share its electrons more – to nitrogen, which is more electronegative and wants to share its electrons less. In other words, NH_2^- is a weaker base than CH_3^- and NH_3 is a stronger acid than CH_4. This reaction is disfavored because it goes from the weaker acid (CH_4) and the weaker base (NH_2^-) to the stronger acid (NH_3) and the stronger base (CH_3^-).

Atomic size can help us predict whether a reaction is likely, or not, to occur when the base and conjugate base are both in the same group. Consider the reaction of gaseous hydrogen fluoride and gaseous chloride to yield gaseous fluoride and gaseous hydrogen chloride.

$$HF(g) + Cl^-(g) \rightarrow F^-(g) + HCl(g),\ \Delta_r H^\circ = 33.0\ kJ/mol$$

Given the endothermic enthalpy of reaction ($\Delta_r H^\circ > 0$), this is an unfavorable reaction, which will not proceed to make products without adding external energy. Comparing the covalent radii (of fluorine and chlorine), chlorine is bigger (r_{cov} = 102 pm) than fluorine (r_{cov} = 57 pm). We can infer from the enthalpy that the larger chloride tends to share its electrons less than the smaller fluoride. Therefore, the fluoride ion is a stronger base than chloride because of its smaller size. This reaction is disfavored because it goes from the weaker acid (HF) and the weaker base (Cl^-) to the stronger acid (HCl) and stronger base (F^-).

Contrast the reaction between hydrogen fluoride and chloride with the reaction between phosphine (PH_3) and azanide (NH_2^-).

$$PH_3(g) + NH_2^-(g) \rightarrow PH_2^-(g) + NH_3(g),\ \Delta_r H^\circ = -19.3\ kJ/mol$$

Here, the reaction is exothermic ($\Delta_r H^\circ < 0$), which indicates that the reaction is favorable and will proceed to make products. If we consider the covalent radii of phosphorus (r_{cov} = 107 pm) and of nitrogen (r_{cov} = 71 pm), we can see why this reaction is favored. Here a hydron is transferred from phosphorus – larger and wants to share its electrons less – to nitrogen, which is smaller and wants to share its electrons more. In other words, NH_2^- is a stronger base than PH_2^- and NH_3 is a weaker acid than PH_3. This reaction is favored because it goes from the stronger acid (PH_3) and the stronger base (NH_2^-) to the weaker acid (NH_3) and weaker base (PH_2^-).

Problem 16.15. For each of the following, predict whether you think the reaction enthalpy ($\Delta_r H^\circ$) would be positive (endothermic) or negative (exothermic).

a. $HS^-(aq) + H_2O(l) \rightarrow H_2S(aq) + OH^-(aq)$

b. $CH_3^-(g) + SiH_4(g) \rightarrow CH_4(g) + SiH_3^-(g)$

c. $Br^-(aq) + H_2Se(aq) \rightarrow HBr(aq) + HSe^-(aq)$

d. $HS^-(aq) + HCl(aq) \rightarrow H_2S(aq) + Cl^-(aq)$

ACTIVATION ENERGY

Chemical kinetics studies how fast reactions happen and what factors affect the rate of reaction. Chemical kinetics provides us with an understanding of how reactions happen at the atomic scale. Here we will only consider one aspect of chemical kinetics: activation energy. Activation energy is the minimum amount of energy required to initiate a reaction. Let's consider activation energy in a specific context, the combustion of benzene (C_6H_6), a component of gasoline (petrol).

$$2\ C_6H_6(l) + 15\ O_2(g) \rightarrow 12\ CO_2(g) + 6\ H_2O(g),\ \Delta_rH^\circ = -6535.4\ \text{kJ/mol}$$

This reaction is very exothermic ($\Delta_rH^\circ < 0$) and will proceed to make products. However, when an internal combustion engine vehicle is filled with gasoline, the gasoline can sit in the tank indefinitely without combusting. Why? Because the reaction of benzene (and gasoline generally) with air requires a spark. The spark provides the activation energy to initiate the reaction. Energetically favorable reactions with higher activation energies tend to be slower to start, if they initiate at all, at room temperature. In contrast, energetically favorable reactions with lower activation energies tend to start faster.

The key takeaway is that reactions, even if energetically favorable, will not start if they do not have the required activation energy. If you have ever done so, you can think of it like rolling down a hill. Your energy will be substantially lower when you reach the bottom of the hill, and once you have started rolling you will keep doing so. The activation energy to start rolling is quite high, however, and you will not start rolling unless you provide the initial push (energy) to make yourself start rolling.

ELECTRON MOVEMENT AND CHEMICAL REACTIONS

This chapter introduced the two frameworks for identifying how electrons move in chemical reactions: electron transfer (redox) and electron sharing (acid–base). We also briefly investigated why (thermodynamics) reactions do, or do not, occur. Brief mention was given to how (kinetics) of reactions occur in terms of activation energy, the energy necessary for a reaction to begin. A thorough treatment of thermodynamics and kinetics is beyond the scope of this book. For those interested readers, the author recommends continuing to study chemistry.

NOTES

1. Pauling, L. The Nature of the Chemical Bond. IV. The Energy of Single Bonds and the Relative Electronegativity of Atoms. *J. Am. Chem. Soc.*, **1932**, *54* (9), 3570–3582. DOI: 10.1021/ja01348a011.
2. "Electronegativity." IUPAC. *Compendium of Chemical Terminology,* 2nd ed. (the "Gold Book"). Compiled by A.D. McNaught; A. Wilkinson. Blackwell Scientific Publications: Oxford, **1997**. Online version (2019-) created by S.J. Chalk. DOI: 10.1351/goldbook.E01990.
3. Reduction was named in antithesis to oxidation and meant to "reduce the amount of oxygen."
4. Lavoisier, A.L. *Elements of Chemistry*. Creech: Edinburgh, **1790**, pp. 159–172.
5. "Reduction." IUPAC. *Compendium of Chemical Terminology,* 2nd ed. (the "Gold Book"). Compiled by A.D. McNaught; A. Wilkinson. Blackwell Scientific Publications: Oxford, **1997**. Online version (2019-) created by S.J. Chalk. DOI: 10.1351/goldbook.R05222.
6. "Oxidation." IUPAC. *Compendium of Chemical Terminology,* 2nd ed. (the "Gold Book"). Compiled by A.D. McNaught; A. Wilkinson. Blackwell Scientific Publications: Oxford, **1997**. Online version (2019-) created by S.J. Chalk. DOI: 10.1351/goldbook.O04362.
7. First proposed by Lavoisier: Lavoisier, A.L. *Elements of Chemistry*. Creech: Edinburgh, **1790**, pp. 159–172. Latimer, W.M. *The Oxidation States of the Elements and Their Potentials in Aqueous Solution*. Prentice-Hall: New York, NY, **1938**.
8. Reed, A.E.; Weinstock, R.B.; Weinhold, F. Natural Population Analysis. *J. Chem. Phys.* **1985**, *83* (2), 735–746. DOI: 10.1063/1.449486.
9. Glendening, E.D.; Badenhoop, J.K.; Reed, A.E.; Carpenter, J.E.; Bohmann, J.A.; Morales, C.M.; Landis, C.R.; Weinhold, F.A. *NBO 6.0*. Theoretical Chemistry Institute: University of Wisconsin: Madison, **2013**.
10. In English, oxidation number and oxidation state are functionally synonyms. Oxidation number etymologically derives from the now deprecated Stock Number: Stock, A. Einige Nomenklaturfragen der anorganischen Chemie. **1919**, *32* (98), 373–374. DOI: 10.1002/ange.19190329802.

11. i) Brönsted, J.N. Einige Bemerkungen über den Begriff der Säuren und Basen. *Recl. Trav. Chim. Pays-Bas.,* **1923**, *42* (8), 718–728. DOI: 10.1002/recl.19230420815.
   ii) Lowry, T.M. The Uniqueness of Hydrogen. *J. Soc. Chem. Ind. (London).,* **1923**, *42* (3), 43–47. DOI: 10.1002/jctb.5000420302.
12. Lewis, G.N. *Valence and the Structure of Atoms and Molecules.* The Chemical Catalog Co., Inc.: New York, NY, **1923**.
13. Sørensen, S.P.L. Über die Messung und die Bedeutung der Wasserstoffionenkonzentration bei enzymatischen Prozessen. *Biochem. Z.*, **1909**, *21*, 131–304.
14. Benmessabih, T.; Amrani, B.; El Haj Hassan, F.; Hamdache, F.; Zoaeter, M. Computational study of AgCl and AgBr semiconductors. *Phys. B: Condens.*, **2007**, *392* (1–2), 309–317. DOI: 10.1016/j.physb.2006.11.046.

Appendices

APPENDIX 1 – PHYSICAL CONSTANTS AND CONVERSION FACTORS[1]

Physical Constants

Quantity	Symbol	Value
Avogadro constant	N_A	$6.022\ 140\ 76 \times 10^{23}$ 1/mol
Electron mass	m_e	$9.109\ 383\ 701\ 5 \times 10^{-31}$ kg
Elementary charge	e	$1.602\ 176\ 634 \times 10^{-19}$ C
Gas constant	R	8.314 462 618 J/(mol K)
		0.082 057 366 (L atm)/(mol K)
Neutron mass	m_n	$1.674\ 927\ 498\ 04 \times 10^{-27}$ kg
Proton mass	m_p	$1.672\ 621\ 923\ 69 \times 10^{-27}$ kg
Speed of light in a vacuum	c	299 792 458 m/s

Conversion Factors

Dimension	Units	Conversion factor
Length	inch to meter	1 in = 0.0254 m
	foot to meter	1 ft = 0.3048 m
	yard to meter	1 yd = 0.9144 m
	mile to meter	1 mi = 1609.344 m
	astronomical unit to meter	1 au = 149 597 870 700 m
	angstrom to meter	1 Å = 1×10^{-10} m
Time	minute to second	1 min = 60 s
	hour to second	1 h = 3600 s
	day to second	1 d = 86 400 s
	year (annum) to second	1 a = 31 557 600 s
Mass	ounce to kilogram	1 oz = 0.028 349 52 kg
	pound to kilogram	1 lb = 0.453 592 37 kg
	unified atomic mass unit to kilogram	1 u = $1.660\ 539\ 066\ 60 \times 10^{-27}$ kg (inexact)
Volume	gallon to m^3	1 gal = 0.003 785 411 784 m^3
	liter to m^3	1 L = 0.001 m^3
Pressure	atmosphere to pascal	1 atm = 101 325 Pa
	bar to pascal	1 bar = 100 000 Pa
	millimeter mercury to pascal	1 mmHg = 133.322 Pa
	pound per square inch to pascal	1 psi = 6894.757 Pa (inexact)
Energy	calorie to joule	1 cal = 4.184 J
	kilowatt-hour to joule	1 kWh = 3 600 000 J
	megaton of TNT equivalent to joule	1 Mt = 4.184×10^{15} J
	electronvolt to joule	1 eV = $1.602\ 176\ 634 \times 10^{-19}$ J

APPENDIX 2 – TABLE OF RECOMMENDED RELATIVE ATOMIC MASS (STANDARD ATOMIC WEIGHT) VALUES, $A_R°(E)$, WHICH HAVE BEEN ABRIDGED TO FIVE SIGNIFICANT FIGURES[2]

Atomic number (Z)	Name	Symbol	$A_r°$ (E)	Atomic number (Z)	Name	Symbol	$A_r°$ (E)
1	hydrogen	H	1.0080	60	neodymium	Nd	144.24
2	helium	He	4.0026	61	promethium	Pm	–
3	lithium	Li	6.94	62	samarium	Sm	150.36
4	beryllium	Be	9.0122	63	europium	Eu	151.96
5	boron	B	10.81	64	gadolinium	Gd	157.25
6	carbon	C	12.011	65	terbium	Tb	158.93
7	nitrogen	N	14.007	66	dysprosium	Dy	162.50
8	oxygen	O	15.999	67	holmium	Ho	164.93
9	fluorine	F	18.998	68	erbium	Er	167.36
10	neon	Ne	20.180	69	thulium	Tm	168.93
11	sodium	Na	22.990	70	ytterbium	Yb	173.05
12	magnesium	Mg	24.305	71	lutetium	Lu	174.97
13	aluminium	Al	26.982	72	hafnium	Hf	178.49
14	silicon	Si	28.085	73	tantalum	Ta	180.95
15	phosphorus	P	30.974	74	tungsten	W	183.84
16	sulfur	S	32.06	75	rhenium	Re	186.21
17	chlorine	Cl	35.45	76	osmium	Os	190.23
18	argon	Ar	39.95	77	iridium	Ir	192.22
19	potassium	K	39.098	78	platinum	Pt	195.08
20	calcium	Ca	40.078	79	gold	Au	196.97
21	scandium	Sc	44.956	80	mercury	Hg	200.59
22	titanium	Ti	47.867	81	thallium	Tl	204.38
23	vanadium	V	50.942	82	lead	Pb	207.2
24	chromium	Cr	51.996	83	bismuth	Bi	208.98
25	manganese	Mn	54.938	84	polonium	Po	–
26	iron	Fe	55.845	85	astatine	At	–
27	cobalt	Co	58.933	86	radon	Rn	–
28	nickel	Ni	58.693	87	francium	Fr	–
29	copper	Cu	63.546	88	radium	Ra	–
30	zinc	Zn	65.38	89	actinium	Ac	–
31	gallium	Ga	69.723	90	thorium	Th	232.04
32	germanium	Ge	72.630	91	protactinium	Pa	231.04
33	arsenic	As	74.922	92	uranium	U	238.03
34	selenium	Se	78.971	93	neptunium	Np	–
35	bromine	Br	79.904	94	plutonium	Pu	–
36	krypton	Kr	83.798	95	americium	Am	–
37	rubidium	Rb	85.468	96	curium	Cm	–
38	strontium	Sr	87.62	97	berkelium	Bk	–
39	yttrium	Y	88.906	98	californium	Cf	–
40	zirconium	Zr	91.224	99	einsteinium	Es	–
41	niobium	Nb	92.906	100	fermium	Fm	–
42	molybdenum	Mo	95.95	101	mendelevium	Md	–
43	technetium	Tc	–	102	nobelium	No	–
44	ruthenium	Ru	101.07	103	lawrencium	Lr	–
45	rhodium	Rh	102.91	104	rutherfordium	Rf	–

Atomic number (Z)	Name	Symbol	A_r° (E)	Atomic number (Z)	Name	Symbol	A_r° (E)
46	palladium	Pd	106.42	105	dubnium	Db	–
47	silver	Ag	107.87	106	seaborgium	Sg	–
48	cadmium	Cd	112.41	107	bohrium	Bh	–
49	indium	In	114.82	108	hassium	Hs	–
50	tin	Sn	118.71	109	meitnerium	Mt	–
51	antimony	Sb	121.76	110	darmstadtium	Ds	–
52	tellurium	Te	127.60	111	roentgenium	Rg	–
53	iodine	I	126.90	112	copernicium	Cn	–
54	xenon	Xe	131.29	113	nihonium	Nh	–
55	caesium	Cs	132.91	114	flerovium	Fl	–
56	barium	Ba	137.33	115	moscovium	Mc	–
57	lanthanum	La	138.91	116	livermorium	Lv	–
58	cerium	Ce	140.12	117	tennessine	Ts	–
59	praseodymium	Pr	140.91	118	oganesson	Og	–

APPENDIX 3 – MODERN PERIODIC TABLE WITH THE RECOMMENDED RELATIVE ATOMIC MASS VALUE OF EACH ELEMENT (APPENDIX 2) ROUNDED TO THREE SIGNIFICANT FIGURES

1	2	3	4	5	6	7	8	9	10	11	12	13	14	15	16	17	18
1 **H** 1.01																	2 **He** 4.00
3 **Li** 6.94	4 **Be** 9.01											5 **B** 10.8	6 **C** 12.0	7 **N** 14.0	8 **O** 16.0	9 **F** 19.0	10 **Ne** 20.2
11 **Na** 23.0	12 **Mg** 24.3											13 **Al** 27.0	14 **Si** 28.1	15 **P** 31.0	16 **S** 32.1	17 **Cl** 35.5	18 **Ar** 40.0
19 **K** 39.1	20 **Ca** 40.1	21 **Sc** 45.0	22 **Ti** 47.9	23 **V** 50.9	24 **Cr** 52.0	25 **Mn** 54.9	26 **Fe** 55.8	27 **Co** 58.9	28 **Ni** 58.7	29 **Cu** 63.5	30 **Zn** 65.4	31 **Ga** 69.7	32 **Ge** 72.6	33 **As** 74.9	34 **Se** 79.0	35 **Br** 79.9	36 **Kr** 83.8
37 **Rb** 85.5	38 **Sr** 87.6	39 **Y** 88.9	40 **Zr** 91.2	41 **Nb** 92.9	42 **Mo** 96.0	43 **Tc** –	44 **Ru** 101	45 **Rh** 103	46 **Pd** 106	47 **Ag** 108	48 **Cd** 112	49 **In** 115	50 **Sn** 119	51 **Sb** 122	52 **Te** 128	53 **I** 127	54 **Xe** 131
55 **Cs** 133	56 **Ba** 137	*	72 **Hf** 178	73 **Ta** 181	74 **W** 184	75 **Re** 186	76 **Os** 190.	77 **Ir** 192	78 **Pt** 195	79 **Au** 197	80 **Hg** 201	81 **Tl** 204	82 **Pb** 207	83 **Bi** 209	84 **Po** –	85 **At** –	86 **Rn** –
87 **Fr** –	88 **Ra** –	**	104 **Rf** –	105 **Db** –	106 **Sg** –	107 **Bh** –	108 **Hs** –	109 **Mt** –	110 **Ds** –	111 **Rg** –	112 **Cn** –	113 **Nh** –	114 **Fl** –	115 **Mc** –	116 **Lv** –	117 **Ts** –	118 **Og** –

*	57 **La** 139	58 **Ce** 140	59 **Pr** 141	60 **Nd** 144	61 **Pm** –	62 **Sm** 150.	63 **Eu** 152	64 **Gd** 157	65 **Tb** 159	66 **Dy** 163	67 **Ho** 165	68 **Er** 167	69 **Tm** 169	70 **Yb** 173	71 **Lu** 175
**	89 **Ac** –	90 **Th** 232	91 **Pa** 231	92 **U** 238	93 **Np** –	94 **Pu** –	95 **Am** –	96 **Cm** –	97 **Bk** –	98 **Cf** –	99 **Es** –	100 **Fm** –	101 **Md** –	102 **No** –	103 **Lr** –

APPENDIX 4 – ELECTRON CONFIGURATION OF EACH ELEMENT[3]

Element	Electron configuration	Noble gas configuration
H	$1s^1$	$1s^1$
He	$1s^2$	$1s^2$
Li	$1s^22s^1$	$[He]2s^1$
Be	$1s^22s^2$	$[He]2s^2$
B	$1s^22s^22p^1$	$[He]2s^22p^1$
C	$1s^22s^22p^2$	$[He]2s^22p^2$
N	$1s^22s^22p^3$	$[He]2s^22p^3$
O	$1s^22s^22p^4$	$[He]2s^22p^4$
F	$1s^22s^22p^5$	$[He]2s^22p^5$
Ne	$1s^22s^22p^6$	$[He]2s^22p^6$
Na	$1s^22s^22p^63s^1$	$[Ne]3s^1$
Mg	$1s^22s^22p^63s^2$	$[Ne]3s^2$
Al	$1s^22s^22p^63s^23p^1$	$[Ne]3s^23p^1$
Si	$1s^22s^22p^63s^23p^2$	$[Ne]3s^23p^2$
P	$1s^22s^22p^63s^23p^3$	$[Ne]3s^23p^3$
S	$1s^22s^22p^63s^23p^4$	$[Ne]3s^23p^4$
Cl	$1s^22s^22p^63s^23p^5$	$[Ne]3s^23p^5$
Ar	$1s^22s^22p^63s^23p^6$	$[Ne]3s^23p^6$
K	$1s^22s^22p^63s^23p^64s^1$	$[Ar]4s^1$
Ca	$1s^22s^22p^63s^23p^64s^2$	$[Ar]4s^2$
Sc	$1s^22s^22p^63s^23p^63d^14s^2$	$[Ar]3d^14s^2$
Ti	$1s^22s^22p^63s^23p^63d^24s^2$	$[Ar]3d^24s^2$
V	$1s^22s^22p^63s^23p^63d^34s^2$	$[Ar]3d^34s^2$
Cr	$1s^22s^22p^63s^23p^63d^54s^1$	$[Ar]3d^54s^1$
Mn	$1s^22s^22p^63s^23p^63d^54s^2$	$[Ar]3d^54s^2$
Fe	$1s^22s^22p^63s^23p^63d^64s^2$	$[Ar]3d^64s^2$
Co	$1s^22s^22p^63s^23p^63d^74s^2$	$[Ar]3d^74s^2$
Ni	$1s^22s^22p^63s^23p^63d^84s^2$	$[Ar]3d^84s^2$
Cu	$1s^22s^22p^63s^23p^63d^{10}4s^1$	$[Ar]3d^{10}4s^1$
Zn	$1s^22s^22p^63s^23p^63d^{10}4s^2$	$[Ar]3d^{10}4s^2$
Ga	$1s^22s^22p^63s^23p^63d^{10}4s^24p^1$	$[Ar]3d^{10}4s^24p^1$
Ge	$1s^22s^22p^63s^23p^63d^{10}4s^24p^2$	$[Ar]3d^{10}4s^24p^2$
As	$1s^22s^22p^63s^23p^63d^{10}4s^24p^3$	$[Ar]3d^{10}4s^24p^3$
Se	$1s^22s^22p^63s^23p^63d^{10}4s^24p^4$	$[Ar]3d^{10}4s^24p^4$
Br	$1s^22s^22p^63s^23p^63d^{10}4s^24p^5$	$[Ar]3d^{10}4s^24p^5$
Kr	$1s^22s^22p^63s^23p^63d^{10}4s^24p^6$	$[Ar]3d^{10}4s^24p^6$
Rb	$1s^22s^22p^63s^23p^63d^{10}4s^24p^65s^1$	$[Kr]5s^1$
Sr	$1s^22s^22p^63s^23p^63d^{10}4s^24p^65s^2$	$[Kr]5s^2$
Y	$1s^22s^22p^63s^23p^63d^{10}4s^24p^64d^15s^2$	$[Kr]4d^15s^2$
Zr	$1s^22s^22p^63s^23p^63d^{10}4s^24p^64d^25s^2$	$[Kr]4d^25s^2$
Nb	$1s^22s^22p^63s^23p^63d^{10}4s^24p^64d^45s^1$	$[Kr]4d^45s^1$
Mo	$1s^22s^22p^63s^23p^63d^{10}4s^24p^64d^55s^1$	$[Kr]4d^55s^1$
Tc	$1s^22s^22p^63s^23p^63d^{10}4s^24p^64d^55s^2$	$[Kr]4d^55s^2$
Ru	$1s^22s^22p^63s^23p^63d^{10}4s^24p^64d^75s^1$	$[Kr]4d^75s^1$
Rh	$1s^22s^22p^63s^23p^63d^{10}4s^24p^64d^85s^1$	$[Kr]4d^85s^1$
Pd	$1s^22s^22p^63s^23p^63d^{10}4s^24p^64d^{10}$	$[Kr]4d^{10}$
Ag	$1s^22s^22p^63s^23p^63d^{10}4s^24p^64d^{10}5s^1$	$[Kr]4d^{10}5s^1$
Cd	$1s^22s^22p^63s^23p^63d^{10}4s^24p^64d^{10}5s^2$	$[Kr]4d^{10}5s^2$
In	$1s^22s^22p^63s^23p^63d^{10}4s^24p^64d^{10}5s^25p^1$	$[Kr]4d^{10}5s^25p^1$

Element	Electron configuration	Noble gas configuration
Sn	$1s^2 2s^2 2p^6 3s^2 3p^6 3d^{10} 4s^2 4p^6 4d^{10} 5s^2 5p^2$	$[Kr]4d^{10} 5s^2 5p^2$
Sb	$1s^2 2s^2 2p^6 3s^2 3p^6 3d^{10} 4s^2 4p^6 4d^{10} 5s^2 5p^3$	$[Kr]4d^{10} 5s^2 5p^3$
Te	$1s^2 2s^2 2p^6 3s^2 3p^6 3d^{10} 4s^2 4p^6 4d^{10} 5s^2 5p^4$	$[Kr]4d^{10} 5s^2 5p^4$
I	$1s^2 2s^2 2p^6 3s^2 3p^6 3d^{10} 4s^2 4p^6 4d^{10} 5s^2 5p^5$	$[Kr]4d^{10} 5s^2 5p^5$
Xe	$1s^2 2s^2 2p^6 3s^2 3p^6 3d^{10} 4s^2 4p^6 4d^{10} 5s^2 5p^6$	$[Kr]4d^{10} 5s^2 5p^6$
Cs	$1s^2 2s^2 2p^6 3s^2 3p^6 3d^{10} 4s^2 4p^6 4d^{10} 5s^2 5p^6 6s^1$	$[Xe]6s^1$
Ba	$1s^2 2s^2 2p^6 3s^2 3p^6 3d^{10} 4s^2 4p^6 4d^{10} 5s^2 5p^6 6s^2$	$[Xe]6s^2$
La	$1s^2 2s^2 2p^6 3s^2 3p^6 3d^{10} 4s^2 4p^6 4d^{10} 5s^2 5p^6 5d^1 6s^2$	$[Xe]5d^1 6s^2$
Ce	$1s^2 2s^2 2p^6 3s^2 3p^6 3d^{10} 4s^2 4p^6 4d^{10} 5s^2 5p^6 4f^1 5d^1 6s^2$	$[Xe]4f^1 5d^1 6s^2$
Pr	$1s^2 2s^2 2p^6 3s^2 3p^6 3d^{10} 4s^2 4p^6 4d^{10} 5s^2 5p^6 4f^3 6s^2$	$[Xe]4f^3 6s^2$
Nd	$1s^2 2s^2 2p^6 3s^2 3p^6 3d^{10} 4s^2 4p^6 4d^{10} 5s^2 5p^6 4f^4 6s^2$	$[Xe]4f^4 6s^2$
Pm	$1s^2 2s^2 2p^6 3s^2 3p^6 3d^{10} 4s^2 4p^6 4d^{10} 5s^2 5p^6 4f^5 6s^2$	$[Xe]4f^5 6s^2$
Sm	$1s^2 2s^2 2p^6 3s^2 3p^6 3d^{10} 4s^2 4p^6 4d^{10} 5s^2 5p^6 4f^6 6s^2$	$[Xe]4f^6 6s^2$
Eu	$1s^2 2s^2 2p^6 3s^2 3p^6 3d^{10} 4s^2 4p^6 4d^{10} 5s^2 5p^6 4f^7 6s^2$	$[Xe]4f^7 6s^2$
Gd	$1s^2 2s^2 2p^6 3s^2 3p^6 3d^{10} 4s^2 4p^6 4d^{10} 5s^2 5p^6 4f^7 5d^1 6s^2$	$[Xe]4f^7 5d^1 6s^2$
Tb	$1s^2 2s^2 2p^6 3s^2 3p^6 3d^{10} 4s^2 4p^6 4d^{10} 5s^2 5p^6 4f^9 6s^2$	$[Xe]4f^9 6s^2$
Dy	$1s^2 2s^2 2p^6 3s^2 3p^6 3d^{10} 4s^2 4p^6 4d^{10} 5s^2 5p^6 4f^{10} 6s^2$	$[Xe]4f^{10} 6s^2$
Ho	$1s^2 2s^2 2p^6 3s^2 3p^6 3d^{10} 4s^2 4p^6 4d^{10} 5s^2 5p^6 4f^{11} 6s^2$	$[Xe]4f^{11} 6s^2$
Er	$1s^2 2s^2 2p^6 3s^2 3p^6 3d^{10} 4s^2 4p^6 4d^{10} 5s^2 5p^6 4f^{12} 6s^2$	$[Xe]4f^{12} 6s^2$
Tm	$1s^2 2s^2 2p^6 3s^2 3p^6 3d^{10} 4s^2 4p^6 4d^{10} 5s^2 5p^6 4f^{13} 6s^2$	$[Xe]4f^{13} 6s^2$
Yb	$1s^2 2s^2 2p^6 3s^2 3p^6 3d^{10} 4s^2 4p^6 4d^{10} 5s^2 5p^6 4f^{14} 6s^2$	$[Xe]4f^{14} 6s^2$
Lu	$1s^2 2s^2 2p^6 3s^2 3p^6 3d^{10} 4s^2 4p^6 4d^{10} 5s^2 5p^6 4f^{14} 5d^1 6s^2$	$[Xe]4f^{14} 5d^1 6s^2$
Hf	$1s^2 2s^2 2p^6 3s^2 3p^6 3d^{10} 4s^2 4p^6 4d^{10} 5s^2 5p^6 4f^{14} 5d^2 6s^2$	$[Xe]4f^{14} 5d^2 6s^2$
Ta	$1s^2 2s^2 2p^6 3s^2 3p^6 3d^{10} 4s^2 4p^6 4d^{10} 5s^2 5p^6 4f^{14} 5d^3 6s^2$	$[Xe]4f^{14} 5d^3 6s^2$
W	$1s^2 2s^2 2p^6 3s^2 3p^6 3d^{10} 4s^2 4p^6 4d^{10} 5s^2 5p^6 4f^{14} 5d^4 6s^2$	$[Xe]4f^{14} 5d^4 6s^2$
Re	$1s^2 2s^2 2p^6 3s^2 3p^6 3d^{10} 4s^2 4p^6 4d^{10} 5s^2 5p^6 4f^{14} 5d^5 6s^2$	$[Xe]4f^{14} 5d^5 6s^2$
Os	$1s^2 2s^2 2p^6 3s^2 3p^6 3d^{10} 4s^2 4p^6 4d^{10} 5s^2 5p^6 4f^{14} 5d^6 6s^2$	$[Xe]4f^{14} 5d^6 6s^2$
Ir	$1s^2 2s^2 2p^6 3s^2 3p^6 3d^{10} 4s^2 4p^6 4d^{10} 5s^2 5p^6 4f^{14} 5d^7 6s^2$	$[Xe]4f^{14} 5d^7 6s^2$
Pt	$1s^2 2s^2 2p^6 3s^2 3p^6 3d^{10} 4s^2 4p^6 4d^{10} 5s^2 5p^6 4f^{14} 5d^9 6s^1$	$[Xe]4f^{14} 5d^9 6s^1$
Au	$1s^2 2s^2 2p^6 3s^2 3p^6 3d^{10} 4s^2 4p^6 4d^{10} 5s^2 5p^6 4f^{14} 5d^{10} 6s^1$	$[Xe]4f^{14} 5d^{10} 6s^1$
Hg	$1s^2 2s^2 2p^6 3s^2 3p^6 3d^{10} 4s^2 4p^6 4d^{10} 5s^2 5p^6 4f^{14} 5d^{10} 6s^2$	$[Xe]4f^{14} 5d^{10} 6s^2$
Tl	$1s^2 2s^2 2p^6 3s^2 3p^6 3d^{10} 4s^2 4p^6 4d^{10} 5s^2 5p^6 4f^{14} 5d^{10} 6s^2 6p^1$	$[Xe]4f^{14} 5d^{10} 6s^2 6p^1$
Pb	$1s^2 2s^2 2p^6 3s^2 3p^6 3d^{10} 4s^2 4p^6 4d^{10} 5s^2 5p^6 4f^{14} 5d^{10} 6s^2 6p^2$	$[Xe]4f^{14} 5d^{10} 6s^2 6p^2$
Bi	$1s^2 2s^2 2p^6 3s^2 3p^6 3d^{10} 4s^2 4p^6 4d^{10} 5s^2 5p^6 4f^{14} 5d^{10} 6s^2 6p^3$	$[Xe]4f^{14} 5d^{10} 6s^2 6p^3$
Po	$1s^2 2s^2 2p^6 3s^2 3p^6 3d^{10} 4s^2 4p^6 4d^{10} 5s^2 5p^6 4f^{14} 5d^{10} 6s^2 6p^4$	$[Xe]4f^{14} 5d^{10} 6s^2 6p^4$
At	$1s^2 2s^2 2p^6 3s^2 3p^6 3d^{10} 4s^2 4p^6 4d^{10} 5s^2 5p^6 4f^{14} 5d^{10} 6s^2 6p^5$	$[Xe]4f^{14} 5d^{10} 6s^2 6p^5$
Rn	$1s^2 2s^2 2p^6 3s^2 3p^6 3d^{10} 4s^2 4p^6 4d^{10} 5s^2 5p^6 4f^{14} 5d^{10} 6s^2 6p^6$	$[Xe]4f^{14} 5d^{10} 6s^2 6p^6$
Fr	$1s^2 2s^2 2p^6 3s^2 3p^6 3d^{10} 4s^2 4p^6 4d^{10} 5s^2 5p^6 4f^{14} 5d^{10} 6s^2 6p^6 7s^1$	$[Rn]7s^1$
Ra	$1s^2 2s^2 2p^6 3s^2 3p^6 3d^{10} 4s^2 4p^6 4d^{10} 5s^2 5p^6 4f^{14} 5d^{10} 6s^2 6p^6 7s^2$	$[Rn]7s^2$
Ac	$1s^2 2s^2 2p^6 3s^2 3p^6 3d^{10} 4s^2 4p^6 4d^{10} 5s^2 5p^6 4f^{14} 5d^{10} 6s^2 6p^6 6d^1 7s^2$	$[Rn]6d^1 7s^2$
Th	$1s^2 2s^2 2p^6 3s^2 3p^6 3d^{10} 4s^2 4p^6 4d^{10} 5s^2 5p^6 4f^{14} 5d^{10} 6s^2 6p^6 6d^2 7s^2$	$[Rn]6d^2 7s^2$
Pa	$1s^2 2s^2 2p^6 3s^2 3p^6 3d^{10} 4s^2 4p^6 4d^{10} 5s^2 5p^6 4f^{14} 5d^{10} 6s^2 6p^6 5f^2 6d^1 7s^2$	$[Rn]5f^2 6d^1 7s^2$
U	$1s^2 2s^2 2p^6 3s^2 3p^6 3d^{10} 4s^2 4p^6 4d^{10} 5s^2 5p^6 4f^{14} 5d^{10} 6s^2 6p^6 5f^3 6d^1 7s^2$	$[Rn]5f^3 6d^1 7s^2$
Np	$1s^2 2s^2 2p^6 3s^2 3p^6 3d^{10} 4s^2 4p^6 4d^{10} 5s^2 5p^6 4f^{14} 5d^{10} 6s^2 6p^6 5f^4 6d^1 7s^2$	$[Rn]5f^4 6d^1 7s^2$
Pu	$1s^2 2s^2 2p^6 3s^2 3p^6 3d^{10} 4s^2 4p^6 4d^{10} 5s^2 5p^6 4f^{14} 5d^{10} 6s^2 6p^6 5f^6 7s^2$	$[Rn]5f^6 7s^2$
Am	$1s^2 2s^2 2p^6 3s^2 3p^6 3d^{10} 4s^2 4p^6 4d^{10} 5s^2 5p^6 4f^{14} 5d^{10} 6s^2 6p^6 5f^7 7s^2$	$[Rn]5f^7 7s^2$
Cm	$1s^2 2s^2 2p^6 3s^2 3p^6 3d^{10} 4s^2 4p^6 4d^{10} 5s^2 5p^6 4f^{14} 5d^{10} 6s^2 6p^6 5f^7 6d^1 7s^2$	$[Rn]5f^7 6d^1 7s^2$
Bk	$1s^2 2s^2 2p^6 3s^2 3p^6 3d^{10} 4s^2 4p^6 4d^{10} 5s^2 5p^6 4f^{14} 5d^{10} 6s^2 6p^6 5f^9 7s^2$	$[Rn]5f^9 7s^2$
Cf	$1s^2 2s^2 2p^6 3s^2 3p^6 3d^{10} 4s^2 4p^6 4d^{10} 5s^2 5p^6 4f^{14} 5d^{10} 6s^2 6p^6 5f^{10} 7s^2$	$[Rn]5f^{10} 7s^2$
Es	$1s^2 2s^2 2p^6 3s^2 3p^6 3d^{10} 4s^2 4p^6 4d^{10} 5s^2 5p^6 4f^{14} 5d^{10} 6s^2 6p^6 5f^{11} 7s^2$	$[Rn]5f^{11} 7s^2$
Fm	$1s^2 2s^2 2p^6 3s^2 3p^6 3d^{10} 4s^2 4p^6 4d^{10} 5s^2 5p^6 4f^{14} 5d^{10} 6s^2 6p^6 5f^{12} 7s^2$	$[Rn]5f^{12} 7s^2$

Element	Electron configuration	Noble gas configuration
Md	$1s^22s^22p^63s^23p^63d^{10}4s^24p^64d^{10}5s^25p^64f^{14}5d^{10}6s^26p^65f^{13}7s^2$	$[Rn]5f^{13}7s^2$
No	$1s^22s^22p^63s^23p^63d^{10}4s^24p^64d^{10}5s^25p^64f^{14}5d^{10}6s^26p^65f^{14}7s^2$	$[Rn]5f^{14}7s^2$
Lr	$1s^22s^22p^63s^23p^63d^{10}4s^24p^64d^{10}5s^25p^64f^{14}5d^{10}6s^26p^65f^{14}6d^17s^2$	$[Rn]5f^{14}6d^17s^2$
Rf	$1s^22s^22p^63s^23p^63d^{10}4s^24p^64d^{10}5s^25p^64f^{14}5d^{10}6s^26p^65f^{14}6d^27s^2$	$[Rn]5f^{14}6d^27s^2$
Db	$1s^22s^22p^63s^23p^63d^{10}4s^24p^64d^{10}5s^25p^64f^{14}5d^{10}6s^26p^65f^{14}6d^37s^2$	$[Rn]5f^{14}6d^37s^2$
Sg	$1s^22s^22p^63s^23p^63d^{10}4s^24p^64d^{10}5s^25p^64f^{14}5d^{10}6s^26p^65f^{14}6d^47s^2$	$[Rn]5f^{14}6d^47s^2$
Bh	$1s^22s^22p^63s^23p^63d^{10}4s^24p^64d^{10}5s^25p^64f^{14}5d^{10}6s^26p^65f^{14}6d^57s^2$	$[Rn]5f^{14}6d^57s^2$
Hs	$1s^22s^22p^63s^23p^63d^{10}4s^24p^64d^{10}5s^25p^64f^{14}5d^{10}6s^26p^65f^{14}6d^67s^2$	$[Rn]5f^{14}6d^67s^2$
Mt	$1s^22s^22p^63s^23p^63d^{10}4s^24p^64d^{10}5s^25p^64f^{14}5d^{10}6s^26p^65f^{14}6d^77s^2$	$[Rn]5f^{14}6d^77s^2$
Ds	$1s^22s^22p^63s^23p^63d^{10}4s^24p^64d^{10}5s^25p^64f^{14}5d^{10}6s^26p^65f^{14}6d^87s^2$	$[Rn]5f^{14}6d^87s^2$
Rg	$1s^22s^22p^63s^23p^63d^{10}4s^24p^64d^{10}5s^25p^64f^{14}5d^{10}6s^26p^65f^{14}6d^97s^2$	$[Rn]5f^{14}6d^97s^2$
Cn	$1s^22s^22p^63s^23p^63d^{10}4s^24p^64d^{10}5s^25p^64f^{14}5d^{10}6s^26p^65f^{14}6d^{10}7s^2$	$[Rn]5f^{14}6d^{10}7s^2$
Nh	$1s^22s^22p^63s^23p^63d^{10}4s^24p^64d^{10}5s^25p^64f^{14}5d^{10}6s^26p^65f^{14}6d^{10}7s^27p^1$	$[Rn]5f^{14}6d^{10}7s^27p^1$
Fl	$1s^22s^22p^63s^23p^63d^{10}4s^24p^64d^{10}5s^25p^64f^{14}5d^{10}6s^26p^65f^{14}6d^{10}7s^27p^2$	$[Rn]5f^{14}6d^{10}7s^27p^2$
Mc	$1s^22s^22p^63s^23p^63d^{10}4s^24p^64d^{10}5s^25p^64f^{14}5d^{10}6s^26p^65f^{14}6d^{10}7s^27p^3$	$[Rn]5f^{14}6d^{10}7s^27p^3$
Lv	$1s^22s^22p^63s^23p^63d^{10}4s^24p^64d^{10}5s^25p^64f^{14}5d^{10}6s^26p^65f^{14}6d^{10}7s^27p^4$	$[Rn]5f^{14}6d^{10}7s^27p^4$
Ts	$1s^22s^22p^63s^23p^63d^{10}4s^24p^64d^{10}5s^25p^64f^{14}5d^{10}6s^26p^65f^{14}6d^{10}7s^27p^5$	$[Rn]5f^{14}6d^{10}7s^27p^5$
Og	$1s^22s^22p^63s^23p^63d^{10}4s^24p^64d^{10}5s^25p^64f^{14}5d^{10}6s^26p^65f^{14}6d^{10}7s^27p^6$	$[Rn]5f^{14}6d^{10}7s^27p^6$

APPENDIX 5 – COMMON CHARGE FOR EACH ELEMENT AS A MONATOMIC ION

1	2	3	4	5	6	7	8	9	10	11	12	13	14	15	16	17	18
1 H 1+/1−																	2 He 0
3 Li 1+	4 Be 2+											5 B 3+	6 C variable	7 N 3−	8 O 2−	9 F 1−	10 Ne 0
11 Na 1+	12 Mg 2+											13 Al 3+	14 Si variable	15 P 3−	16 S 2−	17 Cl 1−	18 Ar 0
19 K 1+	20 Ca 2+	21 Sc 3+	22 Ti variable	23 V variable	24 Cr variable	25 Mn variable	26 Fe variable	27 Co variable	28 Ni variable	29 Cu variable	30 Zn 2+	31 Ga 3+	32 Ge variable	33 As variable	34 Se 2−	35 Br 1−	36 Kr 0
37 Rb 1+	38 Sr 2+	39 Y 3+	40 Zr 4+	41 Nb variable	42 Mo variable	43 Tc variable	44 Ru variable	45 Rh variable	46 Pd variable	47 Ag 1+	48 Cd 2+	49 In variable	50 Sn variable	51 Sb variable	52 Te variable	53 I 1−	54 Xe 0
55 Cs 1+	56 Ba 2+	*	72 Hf 4+	73 Ta variable	74 W variable	75 Re variable	76 Os variable	77 Ir variable	78 Pt variable	79 Au variable	80 Hg variable	81 Tl variable	82 Pb variable	83 Bi variable	84 Po variable	85 At variable	86 Rn 0
87 Fr 1+	88 Ra 2+	**	104 Rf unknown	105 Db unknown	106 Sg unknown	107 Bh unknown	108 Hs unknown	109 Mt unknown	110 Ds unknown	111 Rg unknown	112 Cn unknown	113 Nh unknown	114 Fl unknown	115 Mc unknown	116 Lv unknown	117 Ts unknown	118 Og unknown

*	57 La 3+	58 Ce 3+	59 Pr 3+	60 Nd 3+	61 Pm 3+	62 Sm 3+	63 Eu 3+	64 Gd 3+	65 Tb 3+	66 Dy 3+	67 Ho 3+	68 Er 3+	69 Tm 3+	70 Yb 3+	71 Lu 3+
**	89 Ac 3+	90 Th variable	91 Pa variable	92 U variable	93 Np variable	94 Pu variable	95 Am 3+	96 Cm 3+	97 Bk 3+	98 Cf 3+	99 Es 3+	100 Fm 3+	101 Md 3+	102 No variable	103 Lr 3+

0 means that monatomic ions of this element do not naturally form.

APPENDIX 6 – PAULING ELECTRONEGATIVITY VALUES

1	2	3	4	5	6	7	8	9	10	11	12	13	14	15	16	17	18
1 H 2.20																	2 He
3 Li 0.98	4 Be 1.57											5 B 2.04	6 C 2.55	7 N 3.04	8 O 3.44	9 F 3.98	10 Ne
11 Na 0.93	12 Mg 1.31											13 Al 1.61	14 Si 1.90	15 P 2.19	16 S 2.58	17 Cl 3.16	18 Ar
19 K 0.82	20 Ca 1.00	21 Sc 1.36	22 Ti 1.54	23 V 1.63	24 Cr 1.66	25 Mn 1.55	26 Fe 1.83	27 Co 1.88	28 Ni 1.91	29 Cu 1.90	30 Zn 1.65	31 Ga 1.81	32 Ge 2.01	33 As 2.18	34 Se 2.55	35 Br 2.96	36 Kr 3.0
37 Rb 0.82	38 Sr 0.95	39 Y 1.22	40 Zr 1.33	41 Nb 1.6	42 Mo 2.16	43 Tc 1.9	44 Ru 2.2	45 Rh 2.28	46 Pd 2.20	47 Ag 1.93	48 Cd 1.69	49 In 1.78	50 Sn 1.96	51 Sb 2.05	52 Te 2.30	53 I 2.66	54 Xe 2.6
55 Cs 0.79	56 Ba 0.89	*	72 Hf 1.3	73 Ta 1.5	74 W 2.36	75 Re 1.9	76 Os 2.0	77 Ir 2.20	78 Pt 2.28	79 Au 2.54	80 Hg 2.00	81 Tl 2.04	82 Pb 2.33	83 Bi 2.02	84 Po 2.0	85 At 2.2	86 Rn
87 Fr 0.7	88 Ra 0.9	**	104 Rf	105 Db	106 Sg	107 Bh	108 Hs	109 Mt	110 Ds	111 Rg	112 Cn	113 Nh	114 Fl	115 Mc	116 Lv	117 Ts	118 Og
		*	57 La 1.1	58 Ce 1.12	59 Pr 1.13	60 Nd 1.14	61 Pm 1.2	62 Sm 1.17	63 Eu 1.1	64 Gd 1.20	65 Tb 1.2	66 Dy 1.22	67 Ho 1.23	68 Er 1.24	69 Tm 1.25	70 Yb 1.1	71 Lu 1.27
		**	89 Ac 1.1	90 Th 1.0	91 Pa 1.5	92 U 1.38	93 Np 1.36	94 Pu 1.28	95 Am 1.30	96 Cm	97 Bk	98 Cf	99 Es	100 Fm	101 Md	102 No	103 Lr

APPENDIX 7 – CHEMISTRY NOMENCLATURE REFERENCE

Homonuclear Species

Element	Chemical formula	Common name (IUPAC)	Element	Chemical formula	Common name (IUPAC)
bromine	Br_2	bromine (dibromine)	nitrogen	N_2	nitrogen (dinitrogen)
chlorine	Cl_2	chlorine (dichlorine)	oxygen	O_2	oxygen (dioxygen)
fluorine	F_2	Fluorine (difluorine)	oxygen	O_3	ozone (trioxygen)
hydrogen	H_2	hydrogen (dihydrogen)	phosphorus	P_4	phosphorus (tetraphosphorus)
iodine	I_2	iodine (diiodine)	sulfur	S_8	sulfur (octasulfur)

Polyatomic Ions

a. Cations

Common name (IUPAC)	Chemical formula	Common name (IUPAC)	Chemical formula
ammonium (azanium)	NH_4^+	mercury(I) (dimercury(2+))	Hg_2^{2+}

b. Anions

Common name (IUPAC)	**Chemical formula**	**Common name (IUPAC)**	**Chemical formula**
acetate	$CH_3CO_2^-$	hydroxide (oxidanide)	OH^-
amide (azanide)	NH_2^-	hypobromite	BrO^-
azide (trinitride(1–))	N_3^-	hypochlorite	ClO^-
bromate	BrO_3^-	hypoiodite	IO^-
bromite	BrO_2^-	iodate	IO_3^-
bicarbonate (hydrogencarbonate)	HCO_3^-	iodite	IO_2^-
bifluoride (difluoridohydrogenate(1–))	HF_2^-	nitrate	NO_3^-
bisulfate (hydrogensulfate)	HSO_4^-	nitrite	NO_2^-
bisulfide (sulfanide)	HS^-	perbromate	BrO_4^-
bisulfite (hydrogensulfite)	HSO_3^-	perchlorate	ClO_4^-
carbonate	CO_3^{2-}	periodate	IO_4^-
chromate	CrO_4^{2-}	permanganate	MnO_4^-
chlorate	ClO_3^-	peroxide (dioxide(2–))	O_2^{2-}
chlorite	ClO_2^-	phosphate	PO_4^{3-}
cyanate	OCN^-	pyrophosphate (diphosphate)	$P_2O_7^{4-}$
cyanide	CN^-	sulfate	SO_4^{2-}
dichromate	$Cr_2O_7^{2-}$	sulfite	SO_3^{2-}
dihydrogenphosphate	$H_2PO_4^-$	tetrafluoridoborate	BF_4^-
formate	HCO_2^-	thiocyanate	SCN^-
hydrogenphosphate	HPO_4^{2-}	triiodide	I_3^-

NOTES

1. Tiesinga, E.; Mohr, P.J.; Newell, D.B.; Taylor, B.N. CODATA Recommended Values of the Fundamental Physical Constants: 2018. *J. Phys. Chem. Ref. Data.*, **2021**, *50* (3), 033105-1–033105-61. DOI: 10.1063/5.0064853.
2. Prohaska, T., et al. Standard Atomic Weights of the Elements 2021 (IUPAC Technical Report). *Pure Appl. Chem.*, **2022**, *94* (5), 573–600. DOI: 10.1515/pac-2019-0603.
3. All reported configurations conform with those reported by NIST Atomic Spectra Database: Kramida, A.; Ralchenko, Y.; Reader, J.; NIST ASD Team. *NIST Atomic Spectra Database* (ver. 5.10), [Online]. Available: https://physics.nist.gov/asd. (Date Accessed: 4/28/2023). National Institute of Standards and Technology, Gaithersburg, MD. DOI: https://doi.org/10.18434/T4W30F. For meitnerium (Mt) and beyond the configurations are theoretical following expected trends.

Answers to In-text Problems

CHAPTER 1 ANSWERS

Problem 1.1. For each of the following, identify the claim, evidence, and reasoning in each statement.

a. The reaction produced heat (was exothermic) because the temperature of the water increased. The increase in temperature was the result of energy being released by the reaction, and releasing energy is exothermic.
Claim: The reaction produced heat (was exothermic).

Evidence: The temperature of the water increased.

Reasoning: The increase in temperature was the result of energy being released by the reaction, and releasing energy is exothermic.

b. Susan is happy. Susan has a beaming smile and whenever she is smiling like that it means that she is happy.
Claim: Susan is happy.

Evidence: Susan has a beaming smile.

Reasoning: Whenever she is smiling like that it means that she is happy.

c. The weather has been unnaturally warm and snowless for the past several winters. A long-term change in weather patterns is evidence of a change in climate. Together we can say that there is local evidence of global warming.
Claim: There is local evidence of global warming.

Evidence: The weather has been unnaturally warm and snowless for the past several winters.

Reasoning: A long-term change in weather patterns is evidence of a change in climate.

Problem 1.2. For each of the following, identify what piece (claim, evidence, or reasoning) or pieces are missing from each statement.

a. Anna is allergic to mangos.
This is a claim, but there is neither evidence nor reasoning.

b. As the reaction progressed, the temperature decreased from 25.0 °C to 18.2 °C.
This is evidence (observations), but there is neither a claim nor is there any reasoning.

c. Will is unhappy because he is frowning.
There is a claim (Will is unhappy) and evidence (he is frowning), but there is no reasoning to connect the two.

CHAPTER 2 ANSWERS

Problem 2.1. For each of the following items, identify whether they are matter (have mass and take up space).

a. Wood – matter (it has mass and takes up space)
b. Gold – matter (it has mass and takes up space)
c. Heat – not matter (heat is a form of energy, which does not have mass and occupies no space)
d. A cell phone – matter (it has mass and takes up space)
e. Light – not matter (light is a form of energy, which does not have mass and occupies no space)
f. Water – matter (it has mass and takes up space)
g. Air – matter (it has mass and takes up space). Air can be challenging to consider because we tend not to think of the mass or volume of air around us, but it is indeed matter.

Problem 2.2. Using Table 2.1, Figure 2.2, and Figure 2.4 answer the following questions.

a. What is the name for the element with the atomic symbol Na?
Sodium (the symbol is from the Latin word natrium)

b. What is the atomic symbol for the element sulfur?
S

c. Halogens is the name given to the elements in which group?
Group 17

d. What element is in period 4, group 8?
Iron, whose chemical symbol is Fe (from the Latin word ferrum)

e. What is the atomic number for carbon?
6

f. Gold is in what period and in what group?
Period 6, group 11

g. What type of element (metal, nonmetal, metalloid) is W?
Tungsten (W) is a metal (to the left of the bold, zigzag line)

h. What is the state of matter for Hg?
Mercury (Hg) is a liquid

i. What is the name for element Kr?
Krypton

j. What element has the atomic number 94?
Plutonium (chemical symbol Pu)

k. Identify the elements that are gases at standard temperature and pressure.
All the following are gases at STP: hydrogen (H), helium (He), nitrogen (N), oxygen (O), fluorine (F), neon (Ne), chlorine (Cl), argon (Ar), krypton (Kr), xenon (Xe), and radon (Rn).

l. Identify the element, other than mercury, that is a liquid at standard temperature and pressure.
Bromine (chemical symbol Br) is the only other element that is liquid at STP.

m. Identify the lanthanoid named for the mythological titan who stole fire from Mount Olympus.
Promethium (chemical symbol Pm)

n. What type of element is Samarium? Sodium? Silicon? Sulfur?
Samarium (chemical symbol Sm) is a metal and one of the lanthanoids.
Sodium (chemical symbol Na) is a metal and an alkali metal (group 1).
Silicon (chemical symbol Si) is a metalloid and a member of the carbon group (group 14).
Sulfur (chemical symbol S) is a nonmetal (group 16)

Problem 2.3. Identify each of the following as an element, compound, homogeneous mixture, or heterogeneous mixture.

a. Ocean water – a heterogeneous mixture that typically contains visual particulate matter dispersed through the salty water

b. Salt water – a homogeneous mixture of salt and water

c. Table salt (sodium chloride, NaCl) – a chemical compound that has a fixed ratio of one sodium particle for every one chloride particle

d. Water (H_2O) – a chemical compound that has a fixed ratio of two hydrogen particles for every one oxygen particle

e. Sodium metal (Na) – an element (atomic number 11)

f. Chlorine gas (Cl_2) – an element (atomic number 17). There are two chlorine atoms that constitute the fundamental molecular entity for chlorine.

g. Chocolate chip cookies – a heterogeneous mixture of many, delicious, things

h. Sugar cookies – this is up to interpretation. Store-bought cookies can be quite uniform looking and so that could be classified as a homogeneous mixture. Heterogeneous mixture is also acceptable if one imagines a cookie with differing levels of browning and/or the presence of raw sugar on the top of the cookie.

i. Laptop – a complex, heterogeneous mixture

j. Black ink – a homogeneous mixture, which looks uniform, but the separate (usually colorful) components can be seen when ink runs

k. A cell – a complex, heterogeneous mixture (though it requires a microscope to see)

l. M&M's candies – a heterogeneous mixture of candy coating and chocolate

m. Aluminium (Al) foil – an element (atomic number 13)

n. Gold ring – this is also open to interpretation because the quality of gold is not specified. If 24 karat, then it would be an element (atomic number 79). If less than 24 karat, or if it is white gold or rose gold, then it would be a homogenous mixture (an alloy).

o. Paint thinner (toluene, C_7H_8) – a compound that has a fixed ratio of seven carbon particles for every eight hydrogen particles

p. A brick – open to what one imagines as a brick. Red bricks can look quite uniform and so that would be a homogeneous mixture. If one imagines a brick with visual particulate matter, then heterogeneous mixture would be a better assignment.

q. Stainless steel, an alloy of iron (Fe), carbon (C), and chromium (Cr) – a homogeneous mixture (an alloy is typically a homogeneous mixture with no visible differences in the material)

r. A cupcake – if a plain yellow, white, or chocolate cake and no frosting then this would be a homogeneous mixture. However, if the cake is more elaborate (potentially including funfetti or chocolate chips) and/or if there is frosting, then it would be a heterogeneous mixture.

s. A block of wood – a heterogeneous mixture. There are visible differences (grains) to the appearance of most wood.

t. Borosilicate glass (containing SiO_2 and B_2O_3) – as phrased in this question, this would be a homogeneous mixture. Glasses are complex chemicals, however, and this would be better classified as a compound. The assignment of glass as a compound, however, is beyond the scope of this course and this document.

Problem 2.4. Smell is dependent upon molecules reaching olfactory receptors in our noses. Provide an explanation why hydrogen sulfide ($H_2S(g)$) is intensely smelly even in small amounts, but phenol ($C_6H_5OH(s)$) is only intensely smelly in large amounts (or when heated).

Hydrogen sulfide is a gas, which fills its container and moves through the space. As such, hydrogen sulfide is intensely smelly in small amounts because it is a gas. In contrast, phenol is a solid, whose particles do not move through space and do not fill a container. Phenol, therefore, is only intensely smelly in very large amounts because it depends on very small numbers of particles converting from solid to gas.

Problem 2.5. Identify each of the following as a chemical property or a physical property.

a. Blue color – physical property
b. Density – physical property
c. Flammability – chemical property
d. Solubility – physical property
e. State (solid, liquid, or gas) – physical property
f. Reacts with acid – chemical property
g. Sour taste – physical property

h. Boiling point – physical property
i. Odor – physical property
j. Reacts with water – chemical property

Problem 2.6. Identify each of the following as an example of physical change or evidence of chemical change.

a. Baking soda ($NaHCO_3$) bubbles with vinegar – evidence of chemical change
b. Table salt (NaCl) dissolves in water – a chemical change
c. Milk sours – evidence of chemical change
d. Grass grows – evidence of chemical change
e. Iron (Fe) rusts – evidence of chemical change
f. Sugar ($C_{12}H_{22}O_{11}$) caramelizes – evidence of chemical change
g. An apple is cut – a physical change
h. Wood rots – evidence of chemical change
i. Heat converts water (H_2O) to steam – a physical change
j. A tire is inflated – a physical change
k. Alcohol (CH_3CH_2OH) evaporates – a physical change
l. Food is digested – evidence of chemical change
m. Pancakes cook – evidence of chemical change
n. Ice melts – a physical change
o. Silver (Ag) tarnishes – evidence of chemical change
p. A paper towel absorbs water – a physical change
q. Two chemicals are combined, and gas bubbles form – evidence of chemical change
r. A solid is crushed into a powder – a physical change
s. Mixing salt and pepper – a physical change
t. A marshmallow is cut in half – a physical change
u. A marshmallow is toasted over a fire – evidence of chemical change

CHAPTER 3 ANSWERS

Problem 3.1. Rewrite each of the following numbers in scientific notation.

a. 0.0050 5.0×10^{-3}
b. 0.000 028 7 2.87×10^{-5}
c. 0.000 000 000 000 000 000 000 000 000 000 000 662 6 6.626×10^{-34}

Problem 3.2. Rewrite each number in scientific notation in general form.

a. 1.38×10^{-23} 0.000 000 000 000 000 000 000 0138
b. 2.54×10^{-2} 0.0254
c. 1.3×10^{-1} 0.13

Problem 3.3. Rewrite each of the following numbers, that are greater than one, in scientific notation.

a. 5 280 5.28×10^{3}
b. 10 973 731 1.0973731×10^{7}
c. 100 000 000 000 000 1×10^{14}

Problem 3.4. Rewrite each number in scientific notation in general form.

a. $9.648\,5 \times 10^4$ 96 485
b. 3.00×10^8 300 000 000
c. 5.5×10^1 55

Problem 3.5. Which of the following are exact numbers or should be treated as exact numbers? If they are not exact, how many significant figures does each have?

a. 7 computers Exact number
b. 12.5 gal gasoline Not exact, three significant figures
c. The atomic mass of any element Depends on the information source
d. 10 g sugar Not exact, one significant figure
e. 1 mi = 5280 ft Exact (defined) relationship

Problem 3.6. Identify the number of significant figures for each of the following

a. 6.7540×10^{-3} Five significant figures
b. 0.0204 Three significant figures
c. 124 people Exact number
d. 12 in/1 ft Exact relationship
e. 28.5 °C Three significant figures
f. 20 One significant figure
g. 20 paper clips Exact number
h. 900. Three significant figures
i. 5723.090 ft Seven significant figures
j. 0.000 357 g Three significant figures
k. 907.1 lb Four significant figures
l. 7.92×10^{-4} L Three significant figures
m. 3.141 59 in Six significant figures
n. 200.000 m Six significant figures
o. $0.000\,65 \times 10^3$ s Two significant figures
p. 1.0065×10^3 h Five significant figures
q. 6.022×10^{23} atoms Four significant figures

Problem 3.7. Rewrite each of the following numbers with three significant figures.

a. 100.000 100.
b. 8854.05 8850
c. 0.005 000 0.005 00
d. 5×10^{-3} 5.00×10^{-3}
e. 73 000 7.30×10^4

Problem 3.8. Complete each of the following calculations and express the answer with the correct number of significant figures.

a. $321.55 - \dfrac{6104.5}{2.3} =$ −2300

b. $(0.000\,45 \times 20\,000.0) + (2813 \times 12) =$ 34000

c. $863[1255-(3.45 \times 108)] =$ 762 000
d. $2.823 \times 10^5 - 1.220 \times 10^3 =$ 2.811×10^5

Problem 3.9. Rewrite each of the following numbers without either scientific notation or an SI prefix.

a. 1.30×10^6 g = 1 300 000 g
b. 4.4×10^{-6} g = 0.000 004 4 g
c. 1.1×10^{-4} L = 0.000 11 L
d. 1.9×10^2 J = 190 J
e. 7.41×10^{-10} s = 0.000 000 000 741 s

Problem 3.10. Rewrite each of the following numbers by using an appropriate SI prefix instead of scientific notation.

a. 1.30×10^6 g = 1.30 Mg or 1.30×10^3 kg
b. 4.4×10^{-6} g = 0.0044 mg or 4.4 μg
c. 1.1×10^{-4} L = 0.11 mL or 110 μL
d. 1.9×10^2 J = 0.19 kJ or 1.9 hJ or 19 da J
e. 7.41×10^{-10} s = 0.741 ns or 741 ps

Problem 3.11. Conduct the following mathematical operations. Round your final answer to the appropriate number of significant figures.

a. 1.2 mL + 2.7×10^{-4} L = 1.5 mL or 1.5×10^{-3} L or 0.001 5 L

b. $\dfrac{23.1000\,\text{g} - 22.0000\,\text{g}}{25.10\,\text{mL} - 25.00\,\text{mL}}$ = 11 g/mL

c. 500 032.1 cm + 3 cm = 500 035 cm
d. 500 032.1 cm + 3.00 cm = 500 035.1 cm
e. 7001 g + 6.001 kg = 13 002 g or 1.3002×10^4 g or 13.002 kg
f. 1459.3 Å + 9.77 Å + 4.32 Å = 1 473.4 Å
g. $4.1 \times 6.022 \times 10^{23}$ atom = 2.5×10^{24} atom
h. (1206.7 mm – 0.904 mm)89 mm = 110 000 mm^2 or 1.1×10^5 mm^2
i. 3.8×10^5 nm – 8.45×10^4 nm = 3.0×10^5 nm

j. $\dfrac{9.2 \times 10^{24}\,\text{atom}}{6.022 \times 10^{23}\,\text{atom}}$ = 15

k. $\dfrac{4.55\,\text{g}}{407\ 859\,\text{mL}}$ + 1.000 98 g/mL = 1.000 99 g/mL

Problem 3.12. Rewrite the following numbers in scientific notation and in E notation.

a. Altitude of Concord Academy Science Office: 52.98 m = 5.298×10^1 m or 5.298E1 m
b. Altitude of the summit of Mount Washington: 1917 m = 1.917×10^3 m or 1.917E3 m
c. Wavelength of green light: 0.000 000 558 m = 5.58×10^{-7} m or 5.58E–7 m
d. Number of galaxies in the universe: one hundred billion galaxies = 1×10^{11} or 1E11
e. Volume of an H atom: 0.000 000 000 000 000 000 000 000 000 621 L = 6.21×10^{-28} L or 6.21E–28 L

Problem 3.13. Rewrite each of the following (from Problem 3.12) using the indicated SI prefix.

a. The altitude of Concord Academy Science Office in km = 0.052 98 km
b. The altitude at the summit of Mt. Washington in km = 1.917 km
c. The wavelength of green light in nanometers = 558 nm
d. The volume of an H atom in quectoliters (1 qL = 1×10^{-30} L) = 621 qL

Problem 3.14. Carry out each of the following calculations using scientific notation.

a. Ratio of the mass of the Earth to the mass of the Moon:

$$\frac{5.974 \times 10^{27}\ \text{g}}{7.348 \times 10^{25}\ \text{g}} = 81.30$$

b. Difference between the mass of the Earth and the mass of the Moon:
5.974×10^{27} g − 7.348×10^{25} g = 5.901×10^{27} g or 5.901E27 g

c. Ratio of the mass of a proton and the mass of an electron:

$$\frac{1.672\ 621\ 9 \times 10^{-24}\ \text{g}}{9.109\ 389\ 7 \times 10^{-28}\ \text{g}} = 1836.1514$$

d. Difference between the mass of an ^{1}H atom and the mass of an electron:
$1.673\ 557\ 5 \times 10^{-24}$ g − $9.109\ 389\ 7 \times 10^{-28}$ g = $1.672\ 646\ 6 \times 10^{-24}$ g

e. Difference between the mass of an ^{1}H atom and the mass of a proton:
$1.673\ 557\ 5 \times 10^{-24}$ g − $1.672\ 621\ 9 \times 10^{-24}$ g = 9.356×10^{-28} g

CHAPTER 4 ANSWERS

Problem 4.1. Carry out each of the following unit conversion problems using Table 4.1 and any conversion factors provided in the problems. Each calculation provides the unrounded calculator value and the correctly rounded (to the correct number of significant figures) final answer.

a. Convert 675 cal to J.

675 ~~cal~~	4.184 J
	1 ~~cal~~

= 2824.2 J = 2820 J or 2.82×10^{3} J or 2.82 kJ

b. Convert 0.984 atm to mmHg.

0.984 ~~atm~~	101 325 ~~Pa~~	1 mmHg
	1 ~~atm~~	133.322 ~~Pa~~

= 747.842 067 mmHg = 748 mmHg

c. Convert 764 mmHg to kPa.

764 ~~mmHg~~	133.322 ~~Pa~~	1 kPa
	1 ~~mmHg~~	1×10^{3} ~~Pa~~

= 101.858 008 kPa = 102 kPa

d. How many milligrams are equal to 0.542 kg?

0.542 ~~kg~~	1×10^{3} ~~g~~	1 mg
	1 ~~kg~~	1×10^{-3} ~~g~~

= 542 000 mg

e. If you decide to switch from drinking soda to drinking milk at every meal, how many gallons of milk will you drink if you drink 2.5 L of soda every week?

2.5 ~~L~~	0.001 ~~m³~~	1 gal	= 0.660 430 13 gal = 0.66 gal
	1 ~~L~~	0.003 785 411 784 ~~m³~~	

f. You see a bumper sticker for 100.0 km, implying the owner of the car has run a 100 km race. 100.0 km corresponds to how many miles?

100.0 ~~km~~	1×10^3 ~~m~~	1 mi	= 62.137 119 2 mi = 62.14 mi
	1 ~~km~~	1 609.344 ~~m~~	

g. A red blood cell has a diameter of 8 μm.

i. What is this diameter in mm?

8 ~~μm~~	1×10^{-6} ~~m~~	1 mm	= 0.008 mm
	1 ~~μm~~	1×10^{-3} ~~m~~	

ii. Your body makes 17.500 million new red blood cells (RBCs) every second. How many new red blood cells do you make in a week?

1 ~~week~~	7 ~~d~~	86400 ~~s~~	17.500×10^6 new RBCs	= 1.0584×10^{13} new RBCs
	1 ~~week~~	1 ~~d~~	1 s	

h. What is the volume (in liters) of 10. kg of bismuth (density is 9.747 g/mL)?

10. ~~kg~~	1×10^3 ~~g~~	1 ~~mL~~	1×10^{-3} L	= 1.025 956 7 L = 1.0 L
	1 ~~kg~~	9.747 ~~g~~	1 ~~mL~~	

i. The estimated volume of the world's oceans is 1.5×10^{21} L. If one milliliter of seawater contains 4.0×10^{-12} g of gold and gold is valued at nearly \$2000 per ounce, how much is the gold in the ocean worth? (1 lb = 16 oz).

1.5×10^{21} ~~L~~	1 ~~mL~~	4.0×10^{-12} ~~g~~	1 ~~kg~~	1 ~~lb~~	16 ~~oz~~	\$2000	= $\$4.2 \times 10^{14}$
	1×10^{-3} ~~L~~	1 mL	1×10^3 ~~g~~	0.453 592 37 ~~kg~~	1 ~~lb~~	oz	(\$420 trillion)

j. An average home in Concord, MA uses 28 kWh of electricity every day. How much energy (kJ) does an average home in Concord use every day?

28 ~~kWh~~	3 600 000 ~~J~~	1 kJ	= 100 800 kJ = 1.0×10^5 kJ
	1 ~~kWh~~	1×10^3 ~~J~~	

k. During Operation Sandblast, the first completely submerged circumnavigation of the ocean by a nuclear-powered submarine, the crew completed a trip of 49 491 km. If the average speed was 33 km/h, how many days was the trip?

49 491 ~~km~~	1 ~~h~~	1 d	= 62.488 636 4 d = 62 d
	33 ~~km~~	24 ~~h~~	

l. Your heart beats 80 beats per minute and pumps 80 mL per beat. Over 24 hours, what volume (mL) of blood does the heart circulate?

24 ~~h~~	60 ~~min~~	80 ~~beat~~	80 mL	= 9 216 000 mL = 9×10^6 mL
	1 ~~h~~	1 ~~min~~	1 ~~beat~~	

Problem 4.2. The specific heat capacity of water is 4.184 J/(g °C). How much energy (J) does it take to warm up 155 kg of water (mass of water in a bathtub) from 25 °C to 38 °C? Note that in this problem, the change in temperature (final temperature minus initial temperature) is used to cancel the unit of temperature (°C).

The change in temperature is 38 °C – 25 °C = 13 °C

13 ~~°C~~	4.184 J	= 54.392 J/g (54 J/g, if rounded, but we can wait and round at the end)
	g ~~°C~~	

155 ~~kg~~	1×10^3 ~~g~~	54.392 J	= 8 430 760 J = 8 400 000 J or 8.4×10^6 J
	1 ~~kg~~	~~g~~	

Problem 4.3. Use Figure 4.2 and Table 4.3 to answer the following questions.

a. What is the mass (in kg) of 5.0×10^{25} carbon atoms?

5.0×10^{25} C ~~atom~~	1 ~~mol~~	12.011 ~~g~~	1 kg	= 0.997 260 05 kg C
	6.022×10^{23} ~~atom~~	1 ~~mol~~	1×10^3 ~~g~~	= 1.0 kg C

b. What volume (L at STP) does 1.23 g of helium gas occupy?

1.23 ~~g~~ He	1 ~~mol~~	22.7 L	= 6.975 715 78 L = 6.98 L
	4.0026 ~~g~~	1 ~~mol~~	

c. If a solution is 0.900 mol/L NaCl, how many grams of NaCl are in 500.0 mL?

500.0 ~~mL~~	1×10^{-3} ~~L~~	0.900 ~~mol~~	58.44 g	= 26.298 g = 26.3 g
	1 ~~mL~~	1 ~~L~~	1 ~~mol~~	

d. What is the mass (in kilograms) of 5.0×10^{25} uranium atoms?

5.0×10^{25} U ~~atom~~	1 ~~mol~~	238.03 g	1 kg	= 19.763 367 7 kg U
	6.022×10^{23} ~~atom~~	1 ~~mol~~	1×10^3 g	= 20. kg U

e. If a solution is 0.014 mol/L HI, what mass of HI is in 5.500 L?

Molar mass of HI = 1.0080 g/mol + 126.90 g/mol = 127.91 g/mol

5.500 ~~L~~	0.014 ~~mol~~	127.91 g	= 9.849 07 g = 9.8 g
	1 ~~L~~	1 ~~mol~~	

f. If burning carbon (coal) produces 418 kJ/mol, how much energy (in kJ) is produced by a coal power plant that burns 9.00×10^9 g of coal? (This is the amount of coal typically combusted in a single day)?

9.00×10^9 ~~g~~	1 ~~mol~~	418 kJ
	12.011 ~~g~~	1 ~~mol~~

$= 3.132\ 13 \times 10^{11}$ kJ $= 3.13 \times 10^{11}$ kJ

Problem 4.4. The gas constant is $\frac{0.082\ 06 \text{ L atm}}{\text{mol K}}$, what is the volume (L) of 1.00 mol of water vapor at 373 K (100. °C) and 0.987 atm?

1.00 ~~mol~~	0.082 06 L atm
	~~mol~~ K

= 0.082 06 L atm/K (0.0821 L atm/K, if rounded)

373 ~~K~~	0.082 06 L atm
	K

= 30.608 38 L atm (30.6 L atm, if rounded)

30.608 38 L atm	1
	0.987 atm

= 31.011 529 9 L = 31.0 L

Problem 4.5. You are tasked with determining the density (g/mL) of an unknown metal. You place a piece of metal onto a balance and find that it weighs 45.001 g. To determine the volume, you place the cube into a graduated cylinder of water. Before adding the cube, you find the volume of water to be 22.5 mL. After adding the piece of metal to the cylinder, you find the water level to now be 24.5 mL. What is the density of the unknown metal?

Mass = 45.001 g

Volume = 24.5 mL – 22.5 mL = 2.0 mL

$$\text{Density} = \frac{45.001\text{g}}{2.0\text{mL}} = \frac{22.5005\text{g}}{\text{mL}} = \frac{23\text{g}}{\text{mL}} \text{ or } 23 \text{ g/mL}$$

Problem 4.6. You come across a solution of sodium carbonate (Na_2CO_3) in a chemistry supply room. The label is missing the amount concentration of the solution, but there is an ingredients list. Contents: 1.52 g sodium carbonate mixed with enough water to make a 250 mL solution.

a. What amount (mol) of sodium carbonate is dissolved in the solution?

Molar mass = 2(22.990 g/mol) + 1(12.011 g/mol) + 3(15.999 g/mol) = 105.988 g/mol

1.52 ~~g~~ Na_2CO_3	1 mol
	105.988 ~~g~~

= 0.014 341 25 mol Na_2CO_3 (0.0143 mol, if rounded)

b. What is the volume of the solution in liters?

250 ~~mL~~	1×10^{-3} L
	1 ~~mL~~

= 0.25 L

c. Using your answers to (a) and (b), what is the amount concentration of the unknown solution?

$$\text{Amount concentration} = \frac{0.014\ 341\ 25\,\text{mol Na}_2\text{CO}_3}{0.25\,\text{L}} = \frac{0.057\ 536\ 498\,\text{mol Na}_2\text{CO}_3}{\text{L}}$$

$$= \frac{0.058\,\text{mol Na}_2\text{CO}_3}{\text{L}}$$

Problem 4.7. 9.75 g potassium is added to 500. g water. After all the potassium reacted (producing a pink flame on the water's surface), the water (which was at 25.00 °C before adding the potassium) was now 36.70 °C. If water has a specific heat capacity of 4.184 J/(g °C), calculate the following.

a. What is the difference in temperature (final temperature minus initial temperature)?
Difference in temperature = 36.70 °C – 25.00 °C = 11.70 °C

b. Based on this temperature change and the specific heat capacity of water, how much thermal energy (kJ) does the water take in?

11.70 ~~°C~~	4.184 J
	g ~~°C~~

= 48.9528 J/g (48.95 J/g, if rounded)

500. ~~g~~	48.9528 ~~J~~	1 kJ
	~~g~~	1×10^3 ~~J~~

= 24.476 4 kJ (24.5 kJ, if rounded)

c. Note that numerically the water has a positive thermal energy transfer value because it took in energy. That energy came from the metal. What is the value of thermal energy transfer from the reaction (note the sign should be negative because it is energy given off)?
The water absorbed (positive) 24.4764 kJ, which means that the reaction gave off (negative) –24.4764 kJ.

d. Calculate the amount (mol) of potassium given that we start with 9.75 g.

9.75 ~~g~~ K	1 mol
	39.098 ~~g~~

= 0.249 373 37 mol potassium (0.249 mol, if rounded)

e. Determine the enthalpy (kJ/mol potassium) using your answers from c. and d.

$$\text{Enthalpy} = \frac{-24.4764\,\text{kJ}}{0.249\ 373\ 37\,\text{mol potassium}} = -98.151\ 619\ 2\ \text{kJ/mol potassium}$$

$$= -98.2\ \text{kJ/mol potassium}$$

CHAPTER 5 ANSWERS

Problem 5.1. Calculate the kinetic energy (E_k) in each of the following examples.

a. A horse (421 kg) galloping at 12 m/s.

$$E_k = \frac{1}{2} mv^2 = \frac{1}{2} (421 \text{ kg})(12 \text{ m/s})^2 = 3.0 \times 10^4 \text{ J}$$

b. A baseball (145 g) flying through the air at 120 km/h.

120 ~~km~~	1000 m	1 ~~h~~	1 ~~min~~	
h	~~km~~	60 ~~min~~	60 s	= 33 m/s

145 g (1 kg/1000 g) = 0.145 kg

$$E_k = \frac{1}{2} mv^2 = \frac{1}{2} (0.145 \text{ kg})(33 \text{ m/s})^2 = 79 \text{ J}$$

c. An electron (9.11×10^{-31} kg) moving at 3.00×10^7 m/s.

$$E_k = \frac{1}{2} mv^2 = \frac{1}{2} (9.11 \times 10^{-31} \text{ kg})(3.00 \times 10^7 \text{ m/s})^2 = 4.10 \times 10^{-16} \text{ J}$$

Problem 5.2. For each situation define the system and its surroundings and give the direction of heat (thermal energy) transfer (endothermic or exothermic).

a. Ice cubes are placed into a glass of water and slowly melt.
System: ice cubes
Surroundings: water
Endothermic: thermal energy transfer from surroundings to system

-or-

System: ice water in glass
Surroundings: air around glass and surface it is on
Endothermic: thermal energy transfer from surroundings to system.

b. Hydrogen gas in a balloon explosively combusts with oxygen, producing a bright flame.
System: hydrogen gas balloon
Surroundings: air
Exothermic: thermal energy transfer from system to surroundings.

c. During exercise, metabolism in our cells provides energy for movement.
System: cells
Surroundings: body
Exothermic: thermal energy transfer from system to surroundings.

-or-

System: body
Surroundings: air
Exothermic: thermal energy transfer from system to surroundings.

d. Propane is burning in a Bunsen burner in the laboratory.
System: flame
Surroundings: air
Exothermic: thermal energy transfer from system to surroundings.

e. After you have a swim, water droplets on your skin evaporate.
System: body
Surroundings: water droplets/air
Exothermic: thermal energy transfer from system to surroundings.

-or-

System: water droplets
Surroundings: body/air
Endothermic: thermal energy transfer from surroundings to system.

f. Water, originally at 25 °C, is placed in a freezing compartment of a refrigerator.
System: water in freezer
Surroundings: freezer
Exothermic: thermal energy transfer from system to surroundings.

g. Two chemicals are mixed in a flask on a laboratory bench. A reaction occurs and feels hot to the touch.
System: chemical mixture
Surroundings: laboratory bench and someone's hand
Exothermic: thermal energy transfer from system to surroundings.

h. Thorium burns to produce a white flame.
System: burning thorium
Surroundings: air and surface the thorium is on
Exothermic: thermal energy transfer from system to surroundings.

i. An ice pack is placed on a sore ankle.
System: body
Surroundings: cold pack
Exothermic: thermal energy transfer from system to surroundings.

-or-

System: cold pack
Surroundings: body/air
Endothermic: thermal energy transfer from surroundings to system.

j. A student (with normal 37 °C body temperature) sits in a classroom.
System: student
Surroundings: classroom
Exothermic: thermal energy transfer from system to surroundings.

Problem 5.3. For each of the following, calculate the thermal energy transfer (J) in each of the following examples, identify whether each is endothermic or exothermic.

a. 0.125 kg of hydrogen fluoride is cooled from 19.5 °C to 0.0 °C. The specific heat capacity of hydrogen fluoride is 1.456 J/(g °C).
0.125 kg(1000 g/kg) = 125 g

$$q = (125\ \cancel{g})\left(\frac{1.456\,\text{J}}{\cancel{g}\,\cancel{^\circ C}}\right)(0.0 - 19.5\ \cancel{^\circ C})$$

$q = -3550$ J, exothermic

b. 535 g iron is heated from 25.2 °C (room temperature) to 900 °C (red hot). The specific heat capacity of iron is 0.451 J/(g °C).

$$q = (535\ \cancel{g})\left(\frac{0.451\,\text{J}}{\cancel{g}\,\cancel{^\circ C}}\right)(900 - 25.2\ \cancel{^\circ C})$$

$q = 200\,000$ J, endothermic

c. 200.00 g solid sulfur goes from 20.15 °C to 34.27 °C. The specific heat capacity of solid sulfur is 0.732 J/(g °C).

$$q = (200.00\ \cancel{g})\left(\frac{0.732\,\text{J}}{\cancel{g}\,\cancel{^\circ C}}\right)(34.27 - 20.15\ \cancel{^\circ C})$$

$q = 2067$ J, endothermic

d. 10.0 g helium gas goes from 15.20 °C to −35.32 °C. The specific heat capacity of helium gas is 5.19 J/(g °C).

$$q = (10.0\ \cancel{g})\left(\frac{5.19\,\text{J}}{\cancel{g}\,\cancel{^\circ C}}\right)(-35.32 - 15.20\ \cancel{^\circ C})$$

$q = -2620$ J, exothermic

e. 270 mg dihydrogen gas goes from 288.35 K to 373.38 K. The specific heat capacity of dihydrogen gas is 14.31 J/(g K).
270 mg(1 g/1000 mg) = 0.270 g

$$q = (0.270\ \cancel{g})\left(\frac{14.31\,\text{J}}{\cancel{g}\,\cancel{^\circ C}}\right)(373.38 - 288.35\ \cancel{^\circ C})$$

$q = 329$ J, endothermic

Problem 5.4. For each of the following, calculate the thermal energy transfer (J) in each of the following examples, identify whether each is endothermic or exothermic.

a. 0.4535 kg of water freezes (enthalpy of solidification is −6.01 kJ/mol).

$$n = \frac{0.4535\ \cancel{kg}}{} \left|\frac{1000\ \cancel{g}}{1\ \cancel{kg}}\right| \frac{1\ \text{mol}}{18.015\ \cancel{g}} = 25.17\ \text{mol}$$

$q = (25.17\ \cancel{mol})(-6.01\ \text{kJ}/\cancel{mol}) = -151$ kJ, exothermic

b. 150 g of dry ice (CO_2) sublimates (enthalpy of sublimation is 27.2 kJ/mol).

$n =$

150 ~~g~~	1 mol
	44.009 ~~g~~

$= 3.4$ mol

$q = (3.4\ \cancel{\text{mol}})(27.2\ \text{kJ}/\cancel{\text{mol}}) = 93$ kJ, endothermic

c. 52 L of dihydrogen (H_2) gas (at STP) combusts (enthalpy of combustion is −286 kJ/mol).

$n =$

52 ~~L~~	1 mol
	22.7 ~~L~~

$= 2.3$ mol

$q = (2.3\ \cancel{\text{mol}})(-286\ \text{kJ}/\cancel{\text{mol}}) = -660$ kJ, exothermic

d. 750 mg of ethanol (CH_3CH_2OH) evaporates (enthalpy of vaporization is 42.3 kJ/mol).

$n =$

750 ~~mg~~	1 ~~g~~	1 mol
	1000 ~~mg~~	45.061 ~~g~~

$= 0.017$ mol

$q = (0.017\ \cancel{\text{mol}})(42.3\ \text{kJ}/\cancel{\text{mol}}) = 0.70$ kJ, endothermic

e. 47.2 L of methane (CH_4) condenses into a liquid (enthalpy of condensation is −8.9 kJ/mol).

$n =$

47.2 ~~L~~	1 mol
	22.7 ~~L~~

$= 2.08$ mol

$q = (2.08\ \cancel{\text{mol}})(-8.9\ \text{kJ}/\cancel{\text{mol}}) = -19$ kJ, exothermic

Problem 5.5. For each of the following, calculate the work (J) in each of the following examples, identify whether each is endoworkic or exoworkic.

a. A piston expands from 0.37 L to 2.41 L with an external pressure of 1.05 atm.
$w = -(1.05\ \text{atm})(2.41\ \text{L} - 0.37\ \text{L})$
$w = -2.14\ \text{L atm}$

−2.14 ~~L atm~~	101.325 J
	~~L atm~~

$= -217$ J, exoworkic

b. A sample of 1.5 L liquid water evaporates and becomes 2600 L of steam with an external pressure of 0.985 atm.
$w = -(0.985\ \text{atm})(2600\ \text{L} - 1.5\ \text{L})$
$w = -2600\ \text{L atm}$

−2600 ~~L atm~~	101.325 J
	~~L atm~~

$= -260\,000$ J, exoworkic

c. 498 mL of solid water converts to 454 mL of liquid water with an external pressure of 1.1 atm.
$w = -(1.1\ \text{atm})((454\ \text{mL} - 498\ \text{mL})(1\ \text{L}/1000\ \text{mL}))$
$w = 0.048\ \text{L atm}$

$$\frac{0.048\ \cancel{\text{L atm}}}{} \left|\ \frac{101.325\ \text{J}}{\cancel{\text{L atm}}}\right. = 4.9\ \text{J, endoworkic}$$

d. 22.7 L of ammonia gas condenses to become 25 mL of liquid ammonia with an external pressure of 8.51 atm. 25 mL(1 L/1000 mL) = 0.025 L
$w = -(8.51\ \text{atm})(0.025\ \text{L} - 22.7\ \text{L})$
$w = 193\ \text{L atm}$

$$\frac{193\ \cancel{\text{L atm}}}{} \left|\ \frac{101.325\ \text{J}}{\cancel{\text{L atm}}}\right. = 19\ 500\ \text{J, endoworkic}$$

e. 14 mL of white tin becomes 17 mL of gray tin with an external pressure of 0.991 atm.
$w = -(0.991\ \text{atm})((17\ \text{mL} - 14\ \text{mL})(1\ \text{L}/1000\ \text{mL}))$
$w = -0.003\ \text{L atm}$

$$\frac{0.003\ \cancel{\text{L atm}}}{} \left|\ \frac{101.325\ \text{J}}{\cancel{\text{L atm}}}\right. = 0.3\ \text{J, endoworkic}$$

CHAPTER 6 ANSWERS

Problem 6.1. The mass-to-charge quotient of H^+ is 1.0×10^{-8} kg/C.

a. What is special about the element hydrogen?
The element hydrogen is the lightest element and so it is unique because there is no element that is smaller than hydrogen.

b. Given your answer to part (a) and the mass-to-charge ratio magnitudes, what are the two possible inferences one could make about cathode rays?
Because the value is the ratio of mass to charge, there are two possible inferences as to why the value for cathode rays is smaller than the value for H^+. The first inference is that if the mass is the same (same numerator), then the magnitude of the charge (the denominator) of the cathode ray must be larger. The second inference is that if the magnitude of charge is the same (same denominator), then the mass (the numerator) of the cathode ray must be smaller.

Problem 6.2. Complete the following table using the information provided and Figure 6.2.

Atomic symbol	Isotope designation	Atomic number (Z)	Mass number (A)	Charge number (z)	Number of protons	Number of neutrons	Number of electrons
^{10}B	boron-10	5	10	0	5	5	5
$^{23}Na^+$	sodium-23(1+)	11	23	1+	11	12	10
$^{48}Ti^{4+}$	titanium-48(4+)	22	48	4+	22	26	18

Atomic symbol	Isotope designation	Atomic number (Z)	Mass number (A)	Charge number (z)	Number of protons	Number of neutrons	Number of electrons
^{79}Br	bromine-79	35	79	0	35	44	35
$^{79}Br^{-}$	bromide-79(1–)	35	79	1–	35	44	36
^{83}Kr	krypton-83	36	83	0	36	47	36
$^{32}S^{2-}$	sulfide-32(2–)	16	32	2–	16	16	18
$^{56}Fe^{2+}$	iron-56(2+)	26	56	2+	26	30	24
$^{200}Hg^{2+}$	mercury-200(2+)	80	200	2+	80	120	78
$^{127}I^{-}$	iodide-127(1–)	53	127	1–	53	74	54
^{54}Cr	chromium-54	24	54	0	24	30	24

CHAPTER 7 ANSWERS

Problem 7.1. For each of the following, predict the α decay product isotopes or starting isotopes.

a. $^{212}_{84}Po \rightarrow\ ^{208}_{82}Pb +\ ^{4}_{2}He$

b. $^{239}_{94}Pu \rightarrow\ ^{235}_{92}U +\ ^{4}_{2}He$

c. $^{214}_{86}Rn \rightarrow\ ^{210}_{84}Po +\ ^{4}_{2}He$

d. $^{293}_{118}Og \rightarrow\ ^{289}_{116}Lv +\ ^{4}_{2}He$

e. $^{243}_{95}Am \rightarrow\ ^{239}_{93}Np +\ ^{4}_{2}He$

f. $^{252}_{101}Md \rightarrow\ ^{248}_{99}Es +\ ^{4}_{2}He$

g. $^{209}_{83}Bi \rightarrow\ ^{205}_{81}Tl +\ ^{4}_{2}He$

h. $^{8}_{4}Be \rightarrow\ ^{4}_{2}He +\ ^{4}_{2}He$

Beryllium-8 is unstable and, uniquely for nuclei with fewer than 82 protons, undergoes α decay. On Earth, this is an interesting anomaly, but this has important ramifications in how the fusion of stars works.

Problem 7.2. Predict the β^{-} decay product isotope or starting isotope.

a. $^{28}_{13}Al \rightarrow\ ^{28}_{14}Si +\ ^{0}_{-1}e$

b. $^{98}_{43}Tc \rightarrow\ ^{98}_{44}Ru +\ ^{0}_{-1}e$

c. $^{3}_{1}H \rightarrow\ ^{3}_{2}He +\ ^{0}_{-1}e$

d. $^{20}_{9}F \rightarrow ^{20}_{10}Ne + ^{0}_{-1}e$

e. $^{24}_{11}Na \rightarrow ^{24}_{12}Mg + ^{0}_{-1}e$

f. $^{206}_{81}Tl \rightarrow ^{206}_{82}Pb + ^{0}_{-1}e$

g. $^{133}_{54}Xe \rightarrow ^{133}_{55}Cs + ^{0}_{-1}e$

h. $^{233}_{90}Th \rightarrow ^{233}_{91}Pa + ^{0}_{-1}e$

Problem 7.3. Identify the decay products for each step of the uranium series. The uranium decay series starts with uranium-238 and undergoes the following decay steps to produce lead-206:

$\alpha, \beta^-, \beta^-, \alpha, \alpha, \alpha, \alpha, \alpha, \beta^-, \beta^-, \alpha, \beta^-, \beta^-, \alpha.$

$$^{238}_{92}U \xrightarrow{\alpha} ^{234}_{90}Th \xrightarrow{\beta^-} ^{234}_{91}Pa \xrightarrow{\beta^-} ^{234}_{92}U \xrightarrow{\alpha} ^{230}_{90}Th \xrightarrow{\alpha} ^{226}_{88}Ra \xrightarrow{\alpha} ^{222}_{86}Rn \xrightarrow{\alpha} ^{218}_{84}Po \xrightarrow{\alpha} ^{214}_{82}Pb \xrightarrow{\beta^-}$$

$$^{214}_{83}Bi \xrightarrow{\beta^-} ^{214}_{84}Po \xrightarrow{\alpha} ^{210}_{82}Pb \xrightarrow{\beta^-} ^{210}_{83}Bi \xrightarrow{\beta^-} ^{210}_{84}Po \xrightarrow{\alpha} ^{206}_{82}Pb$$

CHAPTER 8 ANSWERS

Problem 8.1. Explain what each number and letter mean in the term: $3p^3$.

3 is the shell number – gives an indication of how far away an electron is from the nucleus.

p is the subshell identifier – a specific region of space where an electron is likely to be found.

^{3} is the number of electrons in the specific subshell.

Problem 8.2. Let's consider phosphorus.

a. How many electrons does a neutral phosphorus atom have? Explain your answer.
   15 electrons. Phosphorus's atomic number is 15, which means it has 15 protons. A neutral atom will have the same number of electrons as there are protons, and so it has 15 electrons.

b. Consider the expanded electron configuration for phosphorus: $1s^2 2s^2 2p^6 3s^2 3p^3$.
   i. How many electrons are in each shell?
      Two electrons are in shell 1.
      Eight electrons are in shell 2.
      Five electrons are in shell 3.
   ii. For each shell, in what subshells are the electrons located? How many electrons in each subshell?
      Two electrons are in shell 1 and both in the s subshell.
      Eight electrons are in shell 2 and two are in the s subshell and six are in the p subshell.
      Five electrons are in shell 3 and two are in the s subshell and three are in the p subshell.
   iii. Consider the noble gas notation electron configuration for phosphorus: $[Ne]3s^2 3p^3$
      1. Comparing the expanded electron configuration and the noble gas notation, what electrons are core electrons for phosphorus?
         The core electrons for phosphorus are $1s^2 2s^2 2p^6$ (the neon core).

2. Comparing the expanded electron configuration and the noble gas notation, what electrons are valence electrons for phosphorus?
   The valence electrons are $3s^23p^3$ (those not part of the neon core).

Problem 8.3. How many valence electrons does each of the following have: N, P, As, Sb, Bi? Do you notice a trend?

They all have five valence electrons. They're all in the same group and have the same number of valence electrons.

Problem 8.4. Using the periodic table, identify the element that each configuration corresponds to and the group that that element is in.

a. $1s^22s^22p^6$ – Neon (Ne), group 18

b. $1s^22s^22p^4$ – Oxygen (O), group 16

c. $[Ar]3d^{10}4s^24p^5$ – Bromine (Br), group 17

d. $[Ar]3d^24s^2$ – Titanium (Ti), group 4

e. $[Ne]3s^2$ – Magnesium (Mg), group 2

f. $1s^22s^22p^63s^23p^63d^{10}4s^24p^64d^{10}5s^25p^6$ – Xenon (Xe), group 18

Problem 8.5. Identify the number of valence electrons in each element. Do you notice any trends?

H 1	He 2					
Li 1	Be 2	B 3	C 4	N 5	O 6	F 7
Na 1	Mg 2	Al 3	Si 4	P 5	S 6	Cl 7
K 1	Ca 2	Ga 3	Ge 4	As 5	Se 6	Br 7
Rb 1	Sr 2	In 3	Sn 4	Sb 5	Te 6	I 7
Cs 1	Ba 2	Tl 3	Pb 4	Bi 5	Po 6	At 7

All elements in the same group have the same number of valence electrons. The group number counted across (skipping the transition metals and subtracting 10 from groups 13–18) corresponds to how many valence electrons there are for each of the main group elements.

Problem 8.6. Elements in the same group on the periodic table show similar chemical properties. Provide an explanation for this similarity in chemical properties (see Problem 8.5).

Chemical properties are the tendency of an element to undergo chemical reactions and reactions occur when bonds are made and/or broken. Ionic bonds occur when electron(s) are donated and accepted, and covalent bonds are formed when electrons are shared between atoms. It makes sense that the number of valence electrons corresponds to if and how an element will make bonds and thus the element's chemical properties.

Problem 8.7. Why are the noble gases electronically stable? What rule do we use as a shorthand to explain their stability?

Noble gases are electronically stable because they have filled s and p subshells (filled valence shells). Since it takes eight electrons (two for s and six for p) to fill the outermost shell of an atom, the "octet rule" was developed as a way of explaining what makes atoms/ions electronically stable.

Problem 8.8. For the following electron configurations, is the configuration stable? If not, would the atom with this electronic configuration tend to lose electrons or gain electrons? Explain your answer.

a. $1s^22s^22p^6$ – This is a stable electron configuration because the valence s and p subshells are filled.

b. $1s^22s^22p^4$ – This is an unstable electron configuration because the valence s and p subshells are not completely filled. An atom with this configuration will tend to gain two electrons.

c. $[Ar]3d^{10}4s^24p^5$ – This is an unstable electron configuration because the valence s and p subshells are not completely filled. An atom with this configuration will tend to gain one electron.

d. $[Ar]3d^24s^2$ – This is an unstable electron configuration because the valence s and p subshells are not completely filled. An atom with this configuration will tend to lose four electrons.

e. $[Ne]3s^2$ – This is an unstable electron configuration because the valence s and p subshells are not completely filled. An atom with this configuration will tend to lose two electrons.

f. $1s^22s^22p^63s^23p^63d^{10}4s^24p^64d^{10}5s^25p^6$ – This is a stable electron configuration because the valence s and p subshells are filled.

Problem 8.9. The common charges the elements take when they make ionic compounds are as follows: 1+ for group 1, 2+ for group 2, 3+ for group 13, 3– for group 15, 2– for group 16, and 1– for group 17. Explain why these groups of atoms take on these charges (make sure to consider their electron configurations).

All group 1 elements have the valence configuration [noble gas]ns^1 and so those elements will tend to lose one electron, which leads to an ion with a 1+ charge.

All group 2 elements have the valence configuration [noble gas]ns^2 and so those elements will tend to lose two electrons, which leads to an ion with a 2+ charge.

All group 13 elements have the valence configuration [noble gas]ns^2np^1 and so those elements will tend to lose three electrons, which leads to an ion with a 3+ charge.

All group 15 elements have the valence configuration [noble gas]ns^2np^3 and so those elements will tend to gain three electrons, which leads to an ion with a 3– charge.

All group 16 elements have the valence configuration [noble gas]ns^2np^4 and so those elements will tend to gain two electrons, which leads to an ion with a 2– charge.

All group 17 elements have the valence configuration [noble gas]ns^2np^5 and so those elements will tend to gain one electron, which leads to an ion with a 1– charge.

Problem 8.10. Group 14 elements can readily take on charges between 4+ and 4–. In terms of their electron configurations, why would these elements show this ambivalent behavior?

All group 14 elements have the valence configuration [noble gas]ns^2np^2 and so those elements could lose up to four electrons or gain up to four electrons, which leads to ions with charges between 4+ or 4–.

CHAPTER 9 ANSWERS

Problem 9.1. Looking at equation 9.2, what will make the magnitude of the attractive force ($F_{electrostatic}$) between a valence electron and the nucleus stronger (in terms of Z_{eff} and r)?

The attractive force ($F_{electrostatic}$) is directly related to the effective nuclear charge (Z_{eff}) and inversely related to the distance between particles (r). Therefore, the $F_{electrostatic}$ will be stronger (a larger value) when Z_{eff} is larger and when r is smaller. Given that the distance term is squared (r^2), distance changes will have a larger impact than will changes in the effective nuclear charge (Z_{eff}).

Problem 9.2. Looking at equation 9.2, what will make the magnitude of the attractive force ($F_{electrostatic}$) between a valence electron and the nucleus weaker (in terms of Z_{eff} and r)?

The attractive force ($F_{electrostatic}$) is directly related to the effective nuclear charge (Z_{eff}) and inversely related to the distance between particles (r). Therefore, the $F_{electrostatic}$ will be weaker (a smaller value) when Z_{eff} is smaller and when r is bigger. Given that the distance term is squared (r^2), distance changes will have a larger impact than will changes in the effective nuclear charge (Z_{eff}).

Problem 9.3. Looking at a period, sodium to chlorine, what is changing in the atoms to cause Z_{eff} to increase?

Moving across period 3, the shell number (3) does not change, but the number of protons in the nucleus is increasing. More protons, for a fixed shell number, mean a greater positive charge which is reflected in a greater value for Z_{eff}.

Problem 9.4. Looking at a group, why does Z_{eff} for silicon (2.32) show only a 27% increase compared to carbon (1.82) while Z shows a 133% increase?

Silicon, atomic number 14, has 14 protons while carbon, atomic number six, only has six protons (which means a 133% increase in Z for silicon over carbon). Our first instinct, then, is that silicon's nucleus should exert 133% more pull on its valence electrons than carbon does, but the actual or effective charge (Z_{eff}) for valence electrons in silicon is 2.32 while in carbon it is 1.82. The effective charge for both carbon and silicon is less than the actual nuclear charge because of shielding from the core electrons. So, while the silicon nucleus has 133% more protons than carbon does, there is also more shielding due to more core electrons: the carbon core is $1s^2$ and the silicon core is $1s^22s^22p^6$. Altogether, the increase in Z_{eff} is less than the increase in Z due to shielding.

Problem 9.5. What happens to the magnitude of the attractive force ($F_{electrostatic}$) between the nucleus and a valence electron as shell number increases (Table 9.1)?

Looking at Table 9.1, we can see that as the shell number increases, the distance between an electron and the nucleus also increases. That is a larger shell number corresponds to a greater distance (r). Since r^2 is in the denominator, the attractive force between nucleus and electrons decreases as shell number increases. That is an electron in shell 3 (50.27 pm) experiences a much stronger attraction to the nucleus than does an electron in shell 4 (196.85 pm).

Problem 9.6. Thinking about valence electron shell numbers, what would you predict for the trend in atomic size for an atom in period 1 compared to an atom in period 2 compared to an atom in period 3? Explain.

For an atom in period 1, the valence electron(s) are in shell 1. For an atom in period 2, their valence electron(s) are in shell 2. For an atom in period 3, the valence electron(s) are in shell 3. Since higher shell numbers correspond to greater distances from the nucleus (Table 9.1), atoms in period

1 should be smaller (electrons are closer to the nucleus) than atoms in period 2. Atoms in period 2 should be smaller than atoms in period 3.

Problem 9.7. Consider the elements in group 18, Figure 9.2. Use the electron configuration of each element and equation 9.2 to provide an explanation for the trend r_{cov}(He) < r_{cov}(Ne) < r_{cov}(Ar) < r_{cov}(Kr) < r_{cov}(Xe).

Considering the electron configurations (here just of the first three elements):

Helium: $1s^2$
Neon: [He]$2s^22p^6$
Argon: [Ne]$3s^23p^6$

As you move down group 81, the valence electron is in a successively higher shell number. Since higher shell numbers correspond to greater distances from the nucleus, atoms in period 1 should be smaller (electrons are closer to the nucleus) than atoms in period 2. Atoms in period 2 should be smaller than atoms in period 3.

Problem 9.8. Consider the elements in period 3, Na to Ar, Figure 9.2. Using the electron configuration of each element and equation 9.2, provide an explanation for the general decrease in covalent radius from sodium to argon.

For elements sodium (11) through argon (18), the highest shell number is 3. The number of shells does not change. What does change is the increasing number of protons, which means that the effective nuclear charge will increase. As Z_{eff} increases the attractive force between the nucleus and the valence electrons ($F_{electrostatic}$) will increase. This means the electrons are pulled in closer to the nucleus and the radius decreases.

Problem 9.9. Provide an explanation for why cations are smaller than their corresponding neutral atoms. Think about how the electron configuration changes, for example, when sodium becomes sodium(1+).

Compare the electron configuration of sodium ($1s^22s^22s^63s^1$) to a sodium cation ($1s^22s^22s^6$). The highest shell number decreases from 3 to 2. This means that the sodium cation radius will be substantially smaller than the neutral sodium atom because radius is proportional to the highest shell number of the valence electrons.

Problem 9.10. Provide an explanation for why anions are similar in size to their corresponding neutral atoms. Think about how the electron configuration changes, for example, when chlorine becomes chloride(1–).

If we look at the electron configuration of chlorine ($1s^22s^22s^63s^23p^5$) and chloride(1–) ($1s^22s^22s^63s^23p^6$), we can see that there is not a change in the amount of shielding (the core is $1s^22s^22s^6$ before and after) nor is there a change in the valence shell number. Together, there is not a significant reason for the atomic and anionic size to be different. If we consider Figure 9.3, we can see that the anion radius and the neutral atom radius are more or less the same (sometimes the anion radius is a little smaller, sometimes the same size, and sometimes a little bigger).

Problem 9.11. Order the following ions in order of radius size (from smallest to largest): F^-, Na^+, O^{2-}, Mg^{2+}, N^{3-}, Al^{3+}. What is the determining factor (Z_{eff} or shell number) in this order?

They all have the same electron configuration and so the shell number is not the dominant factor. The only significant change is Z_{eff}. More protons will mean higher $F_{electrostatic}$ and so the order is (from smallest to largest): Al^{3+}, Mg^{2+}, Na^+, F^-, O^{2-}, N^{3-} (see Figure 9.3 for actual values).

Problem 9.12. Notice that the E_i(Li) < E_i(Be). Provide an explanation for this.

Because it has one more proton, beryllium has a larger effective nuclear charge (1.66) than lithium (1.26). This means that there is a greater attractive force between the nucleus and the valence electrons. To remove an electron from beryllium, then, more energy would have to be added (greater E_i) to overcome the greater attraction between the beryllium nucleus and its valence electrons than between the lithium nucleus and its valence electron.

Problem 9.13. Provide an explanation for the trend in ionization: E_i(He) > E_i(Ne) > E_i(Ar) > E_i(Kr) > E_i(Xe).

We will closely examine helium and neon, but the following trend applies for the series. For helium, the valence electrons are in shell 1 (configuration $1s^2$); while for neon, the valence electrons are in shell 2 (configuration $[He]2s^22p^6$). The greater shell number corresponds to a larger distance between the nucleus and the electrons (Table 9.1). Because the attractive force between the nucleus and the valence electrons is ($F_{electrostatic}$) is inversely related to the square of the distance (r^2), the larger the distance between the valence electrons and the nucleus, the lower the value for $F_{electrostatic}$. Because neon's valence electrons are in a higher shell number than helium's valence electrons, neon's electrons experience a weaker attraction to the nucleus and less energy would have to be added (lesser E_i) to overcome the lower attraction between the neon nucleus and its valence electrons than between the helium nucleus and its valence electrons.

Problem 9.14. Moving from sodium to argon we cover the entire third period of elements. Can you make a generalization for the overall trend for E_i within a period?

In general, E_i increases across a period. The increase in ionization energy is because of the increase in the attractive force between the nucleus and the valence electrons ($F_{electrostatic}$). Across a period, as the number of protons in the nucleus increases, the effective nuclear charge (Z_{eff}) increases, which leads to an increase in $F_{electrostatic}$.

Problem 9.15. Does the generalized trend you proposed in Problem 9.14 hold up for the other periods in Figure 9.4?

It does. Each period shows, with some irregularities, an increase in E_i across the period.

Problem 9.16. If we consider E_{ea}, is there really a trend (that is, could you draw a linear line) for elements 3–10 (period 2)?

No there is not as regular of a trend for electron affinity as there is for ionization energy. The irregularities are more irregular and so there is not a period-wide trend. There is, however, a linear increase in E_{ea} within the s, p, d, and f blocks.

Problem 9.17. We can, however, analyze E_{ea} values for different groups.

a. Explain why group 17 elements have the largest electron affinity values.
   The elements with the biggest E_{ea} values are the halogens (group 17). They are all only one electron away from a noble gas configuration and so they have a very high affinity for electrons as they become significantly more stable upon the addition of one electron.
b. While not as high as group 17, explain why groups 1 and 11 both have significant electron affinity values.
   The group 1 and 11 elements have a significant E_{ea} because they have a partially filled s and d subshell, respectively. If one electron is added that completes the subshell. Filled subshells are local energy minima (compared to the global energy minimum of a filled shell), and so group 1 and group 11 elements show appreciable electron affinity values.

c. Explain why groups 2, 12, and 18 have an electron affinity of zero.
Groups 2 and 12 have filled subshells and group 18 has a filled shell. These are all energy minima and adding an electron would be destabilizing as each element would be shifted away from an energy minimum by the addition of one electron.

Problem 9.18. When potassium chloride (KCl) forms from potassium and chlorine, the final compound consists of potassium cations (K^+) and chloride anions (Cl^-). Potassium chloride will never form a potasside anion (K^-) and chlorine cation (Cl^+). Provide a quantitative explanation (using Figure 9.4) why potassium is always the cation and chloride always the anion in potassium chloride in terms of:

a. The ionization energy values (E_i) of potassium and chlorine.
The ionization energy (E_i) for potassium (the energy to convert potassium to potassium(1+)) is about 400 kJ/mol and the ionization energy for chlorine (the energy to convert chlorine to chlorine(1+)) is about 1100 kJ/mol. That means that it requires almost three times as much energy to make a chlorine(1+) cation as it does to make a potassium(1+) cation. Energetically, then, it is much lower energy to produce K^+ than it is to produce Cl^+.

b. The electron affinity values (E_{ea}) of potassium and chlorine.
The electron affinity (E_{ea}) for potassium (the energy given off when potassium is converted to potasside(1–)) is about –50 kJ/mol and the electron affinity of chlorine (the energy given off to convert chlorine to chloride(1–)) is about –300 kJ/mol. That means that converting chlorine to an anion releases six times as much energy as converting potassium to an ion. Energetically, then, it is much lower energy to produce Cl^- than it is to produce K^-.

c. The overall energy of forming K^+ and Cl^- versus K^- and Cl^+ (a and b).
Forming K^+ and Cl^- requires a net energy input of about 100 kJ/mol (+400 kJ/mol to ionize potassium and –300 kJ/mol when chlorine gains an electron). In contrast forming K^- and Cl^+ requires a net energy input of about 1100 kJ/mol (+1100 kJ/mol to ionize chlorine and –50 kJ/mol when potassium gains an electron). Energetically, then, forming K^+ and Cl^- is much lower energy and therefore much more likely than forming K^- and Cl^+.

Problem 9.19. Consider the electrostatic potential maps of sodium hydride, dihydrogen, and hydrogen chloride.

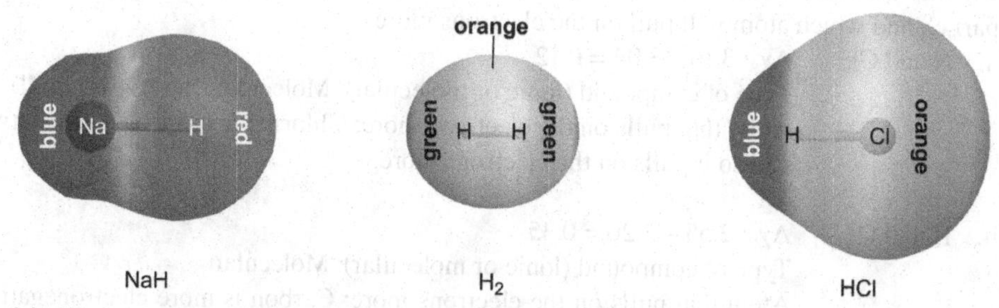

Red indicates areas of high electron density, blue indicates areas of low electron density, and green indicates a middling amount of electron density.

Using the Pauling electronegativity values in Figure 9.5, can you explain the differences in color for H in NaH, H_2, and HCl.

The difference in the images is due to the electronegativity of the atom H is combined with. If we look at the middle image, H is combined with H and so there is no electronegativity difference, and they equally share the electrons (which is why the surface is green with a yellow band of electrons exactly in the middle).

If we look at NaH, hydrogen ($\chi_P = 2.20$) is more electronegative than sodium ($\chi_P = 0.93$) and so the H atom attracts the electrons towards itself (giving it a red color and larger size) and away from the Na atom (leaving it electron poor, smaller, and blue in the image).

If we look at HCl, hydrogen ($\chi_P = 2.20$) is less electronegative than chlorine ($\chi_P = 3.16$) and so the Cl atom attracts the electrons towards itself (giving it an orange color and a larger size) and away from the H atom (leaving it electron poor, smaller, and blue in the image).

Problem 9.20. Look at the electronegativity values in Figure 9.5.

a. LiCl and CaF_2 are example ionic compounds. What is the difference in electronegativity ($\Delta\chi_P$) for Li and Cl? For Ca and F?
The electronegativity difference ($\Delta\chi_P$) between lithium ($\Delta\chi_P$(Li) = 0.98) and chlorine ($\Delta\chi_P$(Cl) = 3.16) is 2.18.
The electronegativity difference ($\Delta\chi_P$) between calcium ($\Delta\chi_P$(Ca) = 1.00) and fluorine ($\Delta\chi_P$(F) = 3.98) is 2.98.

b. Now consider NH_3 and CCl_4, example molecular compounds. What is the difference in electronegativity ($\Delta\chi_P$) for N and H? For C and Cl?
The electronegativity difference ($\Delta\chi_P$) between nitrogen ($\Delta\chi_P$(N) = 3.04) and hydrogen ($\Delta\chi_P$(H) = 2.20) is 0.84.
The electronegativity difference ($\Delta\chi_P$) between carbon ($\Delta\chi_P$(C) = 2.55) and chlorine ($\Delta\chi_P$(Cl) = 3.16) is 0.61.

c. Provide a generalization about the difference in electronegativity ($\Delta\chi_P$) for ionic compounds (compounds composed of a metal and nonmetal) compared to the difference in electronegativity for molecular compounds (compounds composed of nonmetals only)?
In ionic compounds is there is a significant difference in electronegativity (as we shall see in Chapter 11, $\Delta\chi_P > 1.70$ is the cutoff), which leads to the electrons to be attracted so strongly to the more electronegative atom that the more electronegative atom takes the electrons away from the less electronegative atom. In contrast, in molecular compounds, the difference in electronegativity is usually small to moderate ($\Delta\chi_P \leq 1.70$), and so electrons may be shared unequally but they are still shared between the two atoms that bond.

d. For each pair of elements, determine the electronegativity difference ($\Delta\chi_P$), indicate whether the compound formed is most likely ionic or molecular (based on your answer to part c), and which atom will pull on the electrons more.

i. N and Cl
$\Delta\chi_P$: 3.16 – 3.04 = 0.12
Type of compound (ionic or molecular): Molecular (the $\Delta\chi_P$ is small)
Atom that pulls on the electrons more: Chlorine is more electronegative and so it pulls on the electrons more.

ii. H and C
$\Delta\chi_P$: 2.55 – 2.20 = 0.35
Type of compound (ionic or molecular): Molecular
Atom that pulls on the electrons more: Carbon is more electronegative and so it pulls on the electrons more.

iii. H and Si
$\Delta\chi_P$: 2.20 – 1.90 = 0.30
Type of compound (ionic or molecular): Molecular
Atom that pulls on the electrons more: Hydrogen is more electronegative and so it pulls on the electrons more.

iv. Br and Na $\Delta\chi_P$: 2.96 – 0.93 = 2.03
Type of compound (ionic or molecular): Ionic
Atom that pulls on the electrons more: Bromine is more electronegative and so it pulls on the electrons more.

v. Al and P $\Delta\chi_P$: 2.19 – 1.61 = 0.58
Type of compound (ionic or molecular): Molecular
Atom that pulls on the electrons more: Phosphorus is more electronegative and so it pulls on the electrons more.

CHAPTER 10 ANSWERS

Problem 10.1. Nitrogen trichloride decomposes to produce 0.47 L nitrogen gas, $N_2(g)$, and 1.41 L chlorine gas, $Cl_2(g)$ at STP. Using this information, determine the empirical formula of the compound. Mass spectrometry gives a molar mass value of 120.36 g/mol.

0.47 L N_2	1 mol	2 mol N	= 0.041 mol N
	22.7 L	1 mol N_2	

1.41 L Cl_2	1 mol	2 mol Cl	= 0.12 mol Cl
	22.7 L	1 mol Cl_2	

The amount (mol) ratio 0.041 mol N:0.12 mol Cl simplifies to 1 mol N:3 mol Cl, which gives an empirical formula of NCl_3. The empirical formula mass is 120.35 g/mol (1(14.007 g/mol) + 3(35.45 g/mol)), which is in close agreement with the mass spectrometry value.

Problem 10.2. Determine the empirical formula for each compound.

a. Cisplatin, the common name for a platinum compound used in medicine, has the following percent composition values: 65.02% Pt, 9.34% N, 2.02% H, 23.63% Cl. Determine the empirical formula for cisplatin. The mass spectrometry molar mass is 300.1 g/mol.

65.02 g Pt	1 mol	= 0.3332 mol Pt
	195.08 g	

9.34 g N	1 mol	= 0.667 mol N
	14.007 g	

2.02 g H	1 mol	= 2.00 mol H
	1.0080 g	

23.63 g Cl	mol	= 0.6666 mol Cl
	35.45 g	

The amount (mol) ratio 0.333 2 mol Pt:0.667 mol N:2.00 mol H:0.6666 mol Cl simplifies to 1 mol Pt:2 mol N:6 mol H:2 mol Cl, which gives an empirical formula of $PtN_2H_6Cl_2$. The empirical formula mass is 300.04 g/mol, which is in close agreement with the mass spectrometry value.

b. Calculate the empirical formula for calcium given the percent composition values: 29.44% Ca, 23.55% S, and 47.01% O. The experimental molar mass is 136.13 g/mol.

$$\frac{29.44\ \text{g Ca}}{} \cdot \frac{1\ \text{mol}}{40.078\ \text{g}} = 0.7346\ \text{mol Ca}$$

$$\frac{23.55\ \text{g S}}{} \cdot \frac{1\ \text{mol}}{32.06\ \text{g}} = 0.7346\ \text{mol S}$$

$$\frac{47.01\ \text{g O}}{} \cdot \frac{1\ \text{mol}}{15.999\ \text{g}} = 2.938\ \text{mol O}$$

The amount (mol) ratio 0.7346 mol Ca:0.7346 mol S:2.938 mol O simplifies to 1 mol Ca:1 mol S:4 mol O, which gives an empirical formula of $CaSO_4$. The empirical formula mass is 136.13 g/mol, which is in close agreement with the experimental molar mass value.

Problem 10.3. Determine the empirical formula and molecular formula for each compound.

a. Ribose, with a molar mass of 150.13 g/mol, has a chemical composition of 40.00 %C, 6.72 %H, and 53.28 %O.

$$\frac{40.00\ \text{g C}}{} \cdot \frac{1\ \text{mol}}{12.011\ \text{g}} = 3.330\ \text{mol C}$$

$$\frac{6.72\ \text{g H}}{} \cdot \frac{1\ \text{mol}}{1.0080\ \text{g}} = 6.67\ \text{mol H}$$

$$\frac{53.28\ \text{g O}}{} \cdot \frac{1\ \text{mol}}{15.999\ \text{g}} = 3.330\ \text{mol O}$$

Based on the amount (mol) ratios 3.330 mol C:6.667 mol H:3.330 mol O, the empirical formula for glucose is CH_2O. Now when we compare our empirical formula molar mass (30.026 g/mol) with the molar mass of the compound (150.13 g/mol), the empirical formula molar mass is one-fifth of the compound molar mass $\left(\frac{30.026\ \text{g/mol}}{150.13\ \text{g/mol}} = \frac{1}{5}\right)$. This tells us that the empirical formula is one-fifth of the molecular formula and that the molecular formula of ribose is found by multiplying each subscript in the empirical formula by five: $C_5H_{10}O_5$.

b. Ethylene glycol (antifreeze), with a molar mass of 62.07 g/mol, has a chemical composition of 38.70 %C, 9.74 %H, and 51.55 %O.

$$\frac{38.70\ \text{g C}}{} \cdot \frac{1\ \text{mol}}{12.011\ \text{g}} = 3.222\ \text{mol C}$$

$$\frac{9.74\ \cancel{g}\ H \;\Big|\; 1\ \cancel{mol}}{\;\Big|\; 1.0080\ \cancel{g}} = 9.66\ mol\ H$$

$$\frac{51.55\ \cancel{g}\ O \;\Big|\; 1\ \cancel{mol}}{\;\Big|\; 15.999\ \cancel{g}} = 3.222\ mol\ O$$

Based on the amount (mol) ratios 3.222 mol C:7.68 mol H:3.222 mol O, the empirical formula for glucose is CH_3O. Now when we compare our empirical formula molar mass (31.034 g/mol) with the molar mass of the compound (62.07 g/mol), the empirical formula molar mass is one half of the compound molar mass $\left(\frac{31.034\ g/mol}{62.07\ g/mol} = \frac{1}{2}\right)$. That tells us that the empirical formula is one half of the molecular formula, and that the molecular formula of ethylene glycol is found by multiplying each subscript in the empirical formula by two: $C_2H_6O_2$.

Problem 10.4. Using the rules of IUPAC nomenclature, write a chemical formula for each compound.

calcium chloride

$CaCl_2$

hydrogen selenide

H_2Se

magnesium sulfide

MgS

lithium phosphide

Li_3P

rubidium selenide

aluminium oxide

Al_2O_3

caesium nitride

Cs_3N

hydrogen chloride

HCl

potassium fluoride

KF

barium iodide Rb_2Se BaI_2

Problem 10.5. Name the following compounds according to the rules for IUPAC nomenclature.

$SrCl_2$

strontium chloride

Na_2S

sodium sulfide

LiI

lithium iodide

H_3P

hydrogen phosphide

Mg_3P_2

Magnesium phosphide

BaO

barium oxide

Rb_3N

rubidium nitride

MgF_2

magnesium fluoride

HF

hydrogen fluoride

$CaSe$

calcium selenide

Problem 10.6. Using the rules of IUPAC nomenclature, write a chemical formula for each compound.

cobalt(3+) oxide

Co_2O_3

copper(II) bromide

$CuBr_2$

chromium(III) sulfide

Cr_2S_3

sulfur(II) fluoride

SF_2

potassium nitride

K_3N

sodium selenide

Na_2Se

tin(2+) oxide

SnO

iron(III) chloride

$FeCl_3$

lithium phosphide

Li_3P

phosphorus(V) chloride

PCl_5

osmium(4+) nitride

Os_3N_4

iron(II) phosphide

Fe_3P_2

silver(II) fluoride

AgF_2

gold(III) chloride

$AuCl_3$

Problem 10.7. Name the following compounds according to the rules for IUPAC nomenclature.

CrO_3

chromium(6+) oxide or chromium(VI) oxide

PbO_2

lead(4+) oxide or lead(IV) oxide

$ScCl_3$

scandium(3+) chloride or scandium(III) chloride or scandium chloride

PCl_3

phosphorus(III) chloride

$PbCl_2$

lead(2+) chloride or lead(II) chloride

RuO_3

ruthenium(6+) oxide or ruthenium(VI) oxide

GaN

gallium nitride

Ni_3P_2

nickel(2+) phosphide or nickel(II) phosphide

Cs_2O

caesium oxide

MgI_2

magnesium iodide

SnF_4

tin(4+) fluoride or tin(IV) fluoride

CuSe

copper(2+) selenide or copper(II) selenide

LiBr

lithium bromide

CoN

cobalt(3+) nitride or cobalt(III) nitride

Problem 10.8. Using the rules of IUPAC nomenclature, write a chemical formula for each compound.

tin(IV) oxide	chlorine dioxide
SnO_2	ClO_2
phosphorus triiodide	calcium arsenide(3–)
PI_3	Ca_3As_2
boron trifluoride	dioxygen difluoride
BF_3	O_2F_2
carbon dioxide	sulfur hexafluoride
CO_2	SF_6
iron trichloride	osmium tetroxide
$FeCl_3$	OsO_4
osmium(VIII) oxide	bromine pentafluoride
OsO_4	BrF_5

Problem 10.9. Name the following compounds according to the rules for IUPAC nomenclature.

N_2F_2	OF_2
dinitrogen difluoride	oxygen(II) fluoride or oxygen difluoride
SBr_2	BCl_3
sulfur(II) bromide or sulfur dibromide	boron chloride or boron trichloride
XeF_4	$BiCl_3$
xenon(IV) fluoride or xenon tetrafluoride	bismuth(3+) chloride or bismuth(III) chloride or bismuth trichloride
P_4O_3	CS_2
tetraphosphorus trioxide	carbon(IV) sulfide or carbon disulfide
Na_2O	ClF_3
sodium oxide or disodium monoxide	chlorine(III) fluoride or chlorine trifluoride
Co_2O_3	HBr
cobalt(3+) oxide or cobalt(III) oxide or dicobalt trioxide	hydrogen bromide or hydrogen monobromide

Problem 10.10. Using the rules of IUPAC nomenclature, write a chemical formula for each compound.

iron triacetate

$Fe(CH_3CO_2)_3$

strontium sulfate

$SrSO_4$

hydrogen telluride(2–)

H_2Te

potassium permanganate

$KMnO_4$

copper(I) sulfide

Cu_2S

potassium cyanide

KCN

aluminium hydroxide

$Al(OH)_3$

lead(II) sulfate

$PbSO_4$

cobalt(2+) nitrate

$Co(NO_3)_2$

magnesium carbonate

$MgCO_3$

ammonium carbonate

$(NH_3)_2CO_3$

dimercury(2+) chloride

Hg_2Cl_2

Problem 10.11. Name the following compounds according to the rules for IUPAC nomenclature.

$Cu(OH)_2$

copper(2+) hydroxide or copper(II) hydroxide or copper dihydroxide

$Zn_3(PO_4)_2$

zinc phosphate or zinc(2+) phosphate or zinc(II) phosphate or trizinc bis(phosphate)

H_2S

hydrogen sulfide or dihydrogen monosulfide

NH_4F

ammonium fluoride or ammonium monofluoride

$NaHCO_3$

sodium hydrogencarbonate or sodium bicarbonate

$LiBrO_4$

lithium perbromate or lithium mono(perbromate)

$Sn(CH_3CO_2)_4$

tin(4+) acetate or tin(IV) acetate or tin tetraacetate

$Pb(NO_3)_2$

lead(2+) nitrate or lead(II) nitrate or lead dinitrate

Ni_2O_3

nickel(3+) oxide or nickel(III) oxide or dinickel trioxide

$MgSO_4$

magnesium sulfate or magnesium monosulfate

Li_2O_2

lithium dioxide(2–) or lithium peroxide

$AgNO_3$

silver nitrate or silver mononitrate

Problem 10.12. Using the rules of IUPAC nomenclature, write a chemical formula for each compound.

acetic acid

CH_3CO_2H

sodium hypobromite

$NaBrO$

silver fluoride

AgF

perbromic acid

$HBrO_4$

hypobromous acid

$HBrO$

dichromic acid

$H_2Cr_2O_7$

phosphoric acid

H_3PO_4

sodium acetate

$NaCH_3CO_2$

bromous acid

$HBrO_2$

bromic acid

$HBrO_3$

tin(II) bromide

$SnBr_2$

tin tetrabromide

$SnBr_4$

Problem 10.13. Name the following compounds according to the rules for IUPAC nomenclature.

$HMnO_4$

permanganic acid

HNO_2

nitrous acid

H_2CO_3

carbonic acid

HBr

Hydrogen bromide

Na_2CO_3

sodium carbonate

$LiMnO_4$

lithium permanganate

H_2CrO_4

chromic acid

$BaCl_2$

barium chloride or barium dichloride

$Fe(NO_3)_3$

iron(3+) nitrate or iron(III) nitrate
or iron trinitrate

$Ca(CN)_2$

calcium cyanide or calcium dicyanide

HNO_3

nitric acid

CS_2

carbon(IV) sulfide or carbon disulfide

CHAPTER 11 ANSWERS

Problem 11.1. Draw Lewis structures for each of the following chemical formulae: CH_2O, nitrogen trifluoride, hydrogen iodide, SH_3^+, NF_4^+, $AlCl_4^-$, hydroxide, and OH.

CH_2O

The carbon atom has four valence electrons:	$1C(4\ e^-) =$	$4\ e^-$
Each hydrogen atom has one valence electron:	$2H(1\ e^-) =$	$2\ e^-$
The oxygen atom has six valence electrons:	$+1(O)6\ e^- =$	$6\ e^-$
The total number of valence electrons is:		$12\ e^-$ total

The carbon atom needs eight electrons:	$1C(8\ e^-) =$	$8\ e^-$
Each hydrogen atom needs two electrons:	$2H(2\ e^-) =$	$4\ e^-$
The oxygen atom needs eight electrons:	$+1(O)8\ e^- =$	$8\ e^-$
The total number of electrons required is:		$20\ e^-$ needed

$20\ e^-$ needed
$-\ 12\ e^-$ total
$8\ e^-$ shared (or four bonds)

$12\ e^-$ total
$-\ 8\ e^-$ shared
$4\ e^-$ unshared

Lewis structure

```
  ..
  O:
  ||
H-C-H
```

NF_3

The nitrogen atom has five valence electrons:	$1N(5\ e^-) =$	$5\ e^-$
Each fluorine atom has seven valence electrons:	$+3F(7\ e^-) =$	$21\ e^-$
The total number of valence electrons is:		$26\ e^-$ total
The nitrogen atom needs eight electrons:	$1N(8\ e^-) =$	$8\ e^-$
Each fluorine atom needs eight electrons:	$+3F(8\ e^-) =$	$24\ e^-$
The total number of electrons required is:		$32\ e^-$ needed

$32\ e^-$ needed
$-\ 26\ e^-$ total
$6\ e^-$ shared (or three bonds)

$26\ e^-$ total
$-\ 6\ e^-$ shared
$20\ e^-$ unshared

Lewis structure

```
 ··  ··  ··
:F—N—F:
 ··  |  ··
    :F:
     ··
```

HI The iodine atom has seven valence electrons: $1I(7\ e^-) = 7\ e^-$

The hydrogen atom has one valence electron: $+1H(1\ e^-) = 1\ e^-$

The total number of valence electrons is: $8\ e^-$ total

The iodine atom needs eight electrons: $1I(8\ e^-) = 8\ e^-$

The hydrogen atom needs two electrons: $+1H(2\ e^-) = 2\ e^-$

The total number of electrons required is: $10\ e^-$ needed

$10\ e^-$ needed
$-\ 8\ e^-$ total
$2\ e^-$ shared (or one bond)

$8\ e^-$ total
$-\ 2\ e^-$ shared
$6\ e^-$ unshared

Lewis structure

```
   ··
H—I:
   ··
```

SH_3^+ The sulfur atom has six valence electrons: $1S(6\ e^-) = 6\ e^-$

Each hydrogen atom has one valence electron: $3H(1\ e^-) = 3\ e^-$

The positive charged removes one valence electron: $+1(+) = -1\ e^-$

The total number of valence electrons is: $8\ e^-$ total

The sulfur atom needs eight electrons: $1S(8\ e^-) = 8\ e^-$

Each hydrogen atom needs two electrons: $+3H(2\ e^-) = 6\ e^-$

The total number of electrons required is: $14\ e^-$ needed

$14\ e^-$ needed
$-\ 8\ e^-$ total
$6\ e^-$ shared (or three bonds)

$8\ e^-$ total
$-\ 6\ e^-$ shared
$2\ e^-$ unshared

Lewis structure

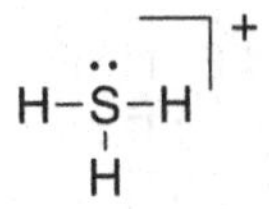

$$\left[\begin{array}{c} \text{H}-\ddot{\text{S}}-\text{H} \\ | \\ \text{H} \end{array}\right]^+$$

NF_4^+	The nitrogen atom has five valence electrons:	$1N(5\ e^-) = 5\ e^-$
	Each fluorine atom has seven valence electrons:	$+4F(7\ e^-) = 28\ e^-$
	The positive charged removes one valence electron:	$1(+) = \ -1\ e^-$
	The total number of valence electrons is:	$32\ e^-$ total
	The nitrogen atom needs eight electrons:	$1N(8\ e^-) = 8\ e^-$
	Each fluorine atom needs seven electrons:	$+4F(8\ e^-) = 32\ e^-$
	The total number of electrons required is:	$40\ e^-$ needed

$40\ e^-$ needed
$-\ 32\ e^-$ total
$8\ e^-$ shared (or four bonds)

$32\ e^-$ total
$-\ 8\ e^-$ shared
$24\ e^-$ unshared

Lewis structure

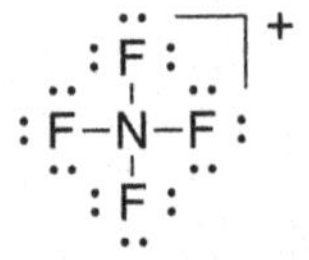

$$\left[\begin{array}{c} :\ddot{\text{F}}: \\ | \\ :\ddot{\text{F}}-\text{N}-\ddot{\text{F}}: \\ | \\ :\ddot{\text{F}}: \end{array}\right]^+$$

$AlCl_4^-$	The aluminium atom has three valence electrons:	$1Al(3\ e^-) = 3\ e^-$
	Each chlorine atom has seven valence electrons:	$4Cl(7\ e^-) = 28\ e^-$
	The negative charged adds one valence electron:	$+1(-) = 1\ e^-$
	The total number of valence electrons is:	$32\ e^-$ total
	The aluminium atom needs eight electrons:	$1S(8\ e^-) = 8\ e^-$
	Each chlorine atom needs eight electrons:	$+4Cl(8\ e^-) = 32\ e^-$
	The total number of electrons required is:	$40\ e^-$ needed

$40\ e^-$ needed
$-\ 32\ e^-$ total
$8\ e^-$ shared (or four bonds)

$32\ e^-$ total
$-\ 8\ e^-$ shared
$24\ e^-$ unshared

Lewis structure

```
      ..
     :Cl: ⌉ −
  ..  |  ..
 :Cl–Al–Cl:
  ..  |  ..
     :Cl:
      ..
```

OH^-

The oxygen atom has six valence electrons:	$1(O)6\ e^- = 6\ e^-$
The hydrogen atom has one valence electron:	$1H(1\ e^-) = 1\ e^-$
The negative charged adds one valence electron:	$+1(-) = 1\ e^-$
The total number of valence electrons is:	$8\ e^-$ total
The oxygen atom needs eight electrons:	$1(O)8\ e^- = 8\ e^-$
The hydrogen atom needs two electrons:	$+1H(2\ e^-) = 2\ e^-$
The total number of electrons required is:	$10\ e^-$ needed

$10\ e^-$ needed
$-\ 8\ e^-$ total
$2\ e^-$ shared (or one bond)

$8\ e^-$ total
$-\ 2\ e^-$ shared
$6\ e^-$ unshared

Lewis structure

```
      ..
     :Cl: ⌉ −
  ..  |  ..
 :Cl–Al–Cl:
  ..  |  ..
     :Cl:
      ..
```

OH

The oxygen atom has six valence electrons:	$1(O)6\ e^- = 6\ e^-$
The hydrogen atom has one valence electron:	$+1H(1\ e^-) = 1\ e^-$
The total number of valence electrons is:	$7\ e^-$ total
The oxygen atom needs eight electrons:	$1(O)8\ e^- = 8\ e^-$
The hydrogen atom needs two electrons:	$+1H(2\ e^-) = 2\ e^-$
The total number of electrons required is:	$10\ e^-$ needed

$10\ e^-$ needed
$-\ 7\ e^-$ total
$3\ e^-$ shared (or 1.5 bonds)

$7\ e^-$ total
$-\ 3\ e^-$ shared
$4\ e^-$ unshared

Lewis structure

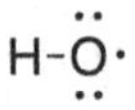

H-Ö·

Problem 11.2. For each of the following molecules, calculate the formal charge on each atom and indicate any formal charges by appropriately drawing in the formal charge(s).

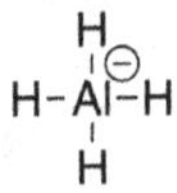

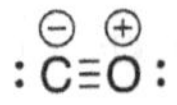

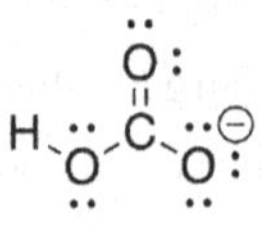

$H-\overset{\ominus}{Al}H_3$ $\quad$ $:\overset{\ominus}{C}\equiv\overset{\oplus}{O}:$ $\quad$ $H-O-C(=O)-O^{\ominus}$

Problem 11.3. Draw a Lewis structure for each of the following. Include all nonzero formal charges: sulfate, HPO_3^{2-}, hydrogen cyanide, HCO_2^-.

SO_4^{2-}	The sulfur atom has six valence electrons:	1S(6 e⁻) = 6 e⁻
	Each oxygen atom has six valence electrons:	4(O)6 e⁻ = 24 e⁻
	Each negative charged adds one valence electron:	+2(-) = 2 e⁻
	The total number of valence electrons is:	32 e⁻ total
	The sulfur atom needs eight electrons:	1S(8 e⁻) = 8 e⁻
	Each oxygen atom needs eight electrons:	+4(O)8 e⁻ = 32 e⁻
	The total number of valence electrons is:	40 e⁻ needed

40 e⁻ needed
– 32 e⁻ total
8 e⁻ shared (or four bonds)

32 e⁻ total
– 8 e⁻ shared
24 e⁻ unshared

Lewis structure

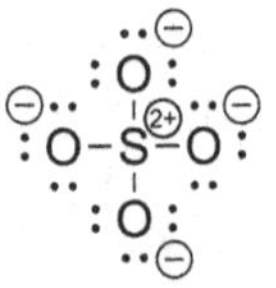

SO_4^{2-}: S^{2+} bonded to four $:\ddot{O}:^{\ominus}$

HPO_3^{2-}	The hydrogen atom has one valence electron:	1H(1 e⁻) = 1 e⁻
	The phosphorus atom has five valence electrons:	1P(5 e⁻) = 5 e⁻
	Each oxygen atom has six valence electrons:	3(O)6 e⁻ = 18 e⁻
	Each negative charged adds one valence electron:	+2(-) = 2 e⁻

The total number of valence electrons is: 26 e^- total
The hydrogen atom needs two electrons: 1H(2 e^-) = 2 e^-
The phosphorus atom needs eight electrons: 1P(8 e^-) = 8 e^-
Each oxygen atom needs eight electrons: +3(O)8 e^- = 24 e^-
The total number of valence electrons is: 34 e^- needed

34 e^- needed
− 26 e^- total
8 e^- shared (or four bonds)

26 e^- total
− 8 e^- shared
18 e^- unshared

Lewis structure

```
      ..⊖
     :O:
   ⊕  |   ..⊖
  H—P—O:
      |   ..
     :O:
      ..⊖
```

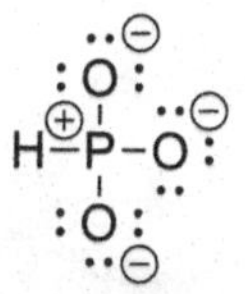

HCN The hydrogen atom has one valence electron: 1H(1 e^-) = 1 e^-
The carbon atom has four valence electrons: 1C(4 e^-) = 4 e^-
The nitrogen atom has five valence electrons: +1N(5 e^-) = 5 e^-
The total number of valence electrons is: 10 e^- total
The hydrogen atom needs two electrons: 1H(2 e^-) = 2 e^-
The carbon atom needs eight electrons: 1C(8 e^-) = 8 e^-
The nitrogen atom needs eight electrons: +1N(8 e^-) = 8 e^-
The total number of valence electrons is: 18 e^- needed

18 e^- needed
− 10 e^- total
8 e^- shared (or four bonds)

10 e^- total
− 8 e^- shared
2 e^- unshared

Lewis structure

H–C≡N

HCO_2^-

The hydrogen atom has one valence electron:	$1H(1\ e^-) =$	$1\ e^-$
The carbon atom has four valence electrons:	$1C(4\ e^-) =$	$4\ e^-$
The oxygen atom has six valence electrons:	$2(O)6\ e^- =$	$12\ e^-$
The negative charged adds one valence electron:	$+1(-) =$	$1\ e^-$
The total number of valence electrons is:		$18\ e^-$ total

The hydrogen atom needs two electrons:	$1H(2\ e^-) =$	$2\ e^-$
The carbon atom needs eight electrons:	$1C(8\ e^-) =$	$8\ e^-$
Each oxygen atom needs eight electrons:	$+2(O)(8\ e^-) =$	$16\ e^-$
The total number of valence electrons is:		$26\ e^-$ needed

$26\ e^-$ needed
$-\ 18\ e^-$ total
$8\ e^-$ shared (or four bonds)

$18\ e^-$ total
$-\ 8\ e^-$ shared
$10\ e^-$ unshared

Lewis structure

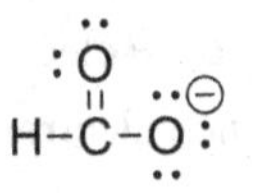

Problem 11.4. For each of the following chemical formulae, draw all contributing structures.

S_2O

Each sulfur atom has six valence electrons:	$2S(6\ e^-) =$	$12\ e^-$
The oxygen atom has six valence electrons:	$+1(O)6\ e^- =$	$6\ e^-$
The total number of valence electrons is:		$18\ e^-$ total

Each sulfur atom needs eight electrons:	$2S(8\ e^-) =$	$16\ e^-$
The oxygen atom needs eight electrons:	$+1(O)8\ e^- =$	$8\ e^-$
The total number of electrons required is:		$24\ e^-$ needed

$24\ e^-$ needed
$-\ 18\ e^-$ total
$6\ e^-$ shared (or three bonds)

$18\ e^-$ total
$-\ 6\ e^-$ shared
$12\ e^-$ unshared

Lewis structure

:S=S–O: ⟷ :S–S=O:

BO_3^{3-} The boron atom has three valence electrons: 1B(3 e^-) = 3 e^-
Each oxygen atom has six valence electrons: 3(O)6 e^- = 18 e^-
Each negative charged adds one valence electron: +3(-) = 3 e^-

The total number of valence electrons is: 24 e^- total
The boron atom needs eight electrons: 1B(8 e^-) = 8 e^-
Each oxygen atom needs eight electrons: +3(O)8 e^- = 24 e^-
The total number of electrons required is: 32 e^- needed

32 e^- needed
– 24 e^- total
8 e^- shared (or four bonds)

24 e^- total
– 8 e^- shared
16 e^- unshared

Lewis structure

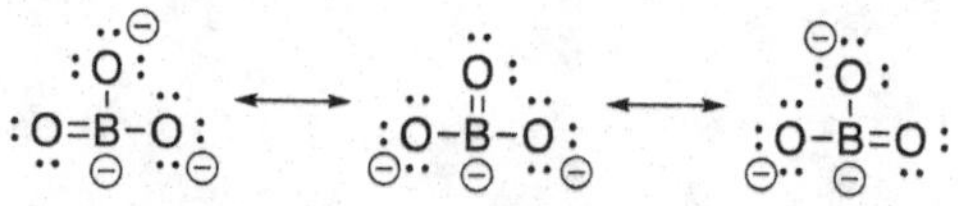

:O=B(–O:)–O: ⟷ :O–B(=O:)–O: ⟷ :O–B(–O:)=O:

CSN^- The carbon atom has four valence electrons 1C(4 e^-) = 4 e^-
The nitrogen atom has five valence electrons: 1N(5 e^-) = 5 e^-
The sulfur atom has six valence electrons: 1S(6 e^-) = 6 e^-
The negative charged adds one valence electron: +1(-) = 1 e^-
The total number of valence electrons is: 16 e^- total

The carbon atom needs eight electrons: 1C(8 e^-) = 8 e^-
Each nitrogen atom needs eight electrons: 1N(8 e^-) = 8 e^-
The sulfur atom needs eight electrons: +1S(8 e^-)= 8 e^-
The total number of electrons required is: 24 e^- needed

24 e^- needed
– 16 e^- total
8 e^- shared (or four bonds)

16 e^- total
– 8 e^- shared
8 e^- unshared

Lewis structure

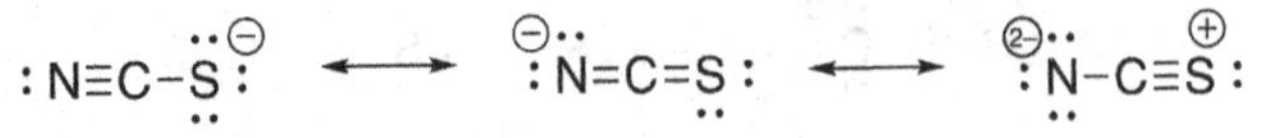

$:N{\equiv}C{-}\ddot{S}:^{\ominus} \longleftrightarrow {}^{\ominus}\ddot{N}{=}C{=}S: \longleftrightarrow {}^{2\ominus}\ddot{N}{-}C{\equiv}S:^{\oplus}$

COF_2

The carbon atom has four valence electrons:	$1C(4\ e^-) = 4\ e^-$
Each fluorine atom has seven valence electrons:	$2F(7\ e^-) = 14\ e^-$
The oxygen atom has six valence electrons:	$+1(O)6\ e^- = 6\ e^-$
The total number of valence electrons is:	$24\ e^-$ total
The carbon atom needs eight electrons:	$1C(8\ e^-) = 8\ e^-$
Each fluorine atom needs eight electrons:	$2F(8\ e^-) = 16\ e^-$
The oxygen atom needs eight electrons:	$+1(O)8\ e^- = 8\ e^-$
The total number of electrons required is:	$32\ e^-$ needed

$32\ e^-$ needed
$-\ 24\ e^-$ total
$8\ e^-$ shared (or four bonds)

$24\ e^-$ total
$-\ 8\ e^-$ shared
$16\ e^-$ unshared

Lewis structure

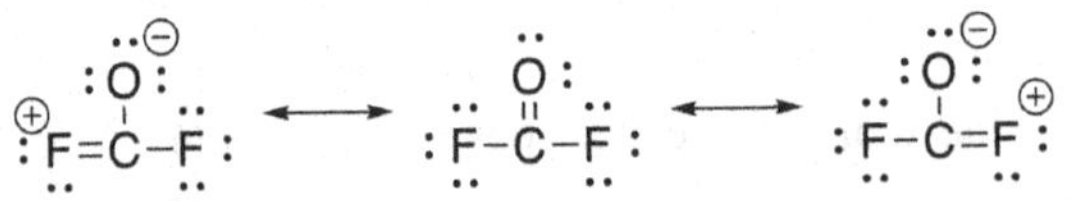

$F^{\oplus}{=}C(O^{\ominus}){-}F \longleftrightarrow F{-}C({=}O){-}F \longleftrightarrow F{-}C(O^{\ominus}){=}F^{\oplus}$

Problem 11.5. Draw the Lewis structures for each of the following chemical formulae: chlorine trifluoride, SO_2Cl_2, xenon trioxide, SeO_4^{2-}, ClO_3^-, and Br_3^-. Check the Lewis structures you have drawn in this problem against the results found online, where hypervalent structures are almost exclusively presented.

ClF_3

The chlorine atom has seven valence electrons:	$1Cl(7\ e^-) = 7\ e^-$
Each fluorine atom has seven valence electrons:	$+3F(7\ e^-) = 21\ e^-$
The total number of valence electrons is:	$28\ e^-$ total
The chlorine atom needs eight electrons:	$1Cl(8\ e^-) = 8\ e^-$
Each fluorine atom needs eight electrons:	$+3F(8\ e^-) = 24\ e^-$
The total number of electrons required is:	$32\ e^-$ needed

32 e^- needed
− 28 e^- total
4 e^- shared (or two bonds)

28 e^- total
− 4 e^- shared
24 e^- unshared

Lewis structure:

Hypervalent structure:

SO_2Cl_2

The sulfur atom has six valence electrons:	1S(6 e^-) = 6 e^-
Each oxygen atom has five valence electrons:	2(O)6 e^- = 12 e^-
Each chlorine atom has seven valence electrons:	+2Cl(7 e^-) = 14 e^-
The total number of valence electrons is:	32 e^- total

The sulfur atom needs eight electrons:	1S(8 e^-) = 8 e^-
Each oxygen atom has five valence electrons:	2(O)8 e^- = 16 e^-
Each chlorine atom needs eight electrons:	+2Cl(8 e^-) = 16 e^-
The total number of electrons required is:	40 e^- needed

40 e^- needed
− 32 e^- total
8 e^- shared (or four bonds)

32 e^- total
− 8 e^- shared
24 e^- unshared

Lewis structure:

Hypervalent structure:

XeO_3

The xenon atom has eight valence electrons:	$1Xe(8\ e^-) = 8\ e^-$
Each oxygen atom has six valence electrons:	$+3(O)6\ e^- = 18\ e^-$
The total number of valence electrons is:	$26\ e^-$ total

The xenon atom needs eight electrons:	$1Xe(8\ e^-) = 8\ e^-$
Each oxygen atom needs eight electrons:	$+3(O)8\ e^- = 24\ e^-$
The total number of electrons required is:	$32\ e^-$ needed

$32\ e^-$ needed
$-\ 26\ e^-$ total
$6\ e^-$ shared (or three bonds)

$26\ e^-$ total
$-\ 6\ e^-$ shared
$20\ e^-$ unshared

Lewis structure:

Hypervalent structure:

SeO_4^{2-}

The selenium atom has six valence electrons:	$1Se(6\ e^-) = 6\ e^-$
Each oxygen atom has six valence electrons:	$4(O)6\ e^- = 24\ e^-$
Each negative charged adds one valence electron:	$+2(-) = 2\ e^-$
The total number of valence electrons is:	$32\ e^-$ total

The selenium atom needs eight electrons:	$1Se(8\ e^-) = 8\ e^-$
Each oxygen atom needs eight electrons:	$+4(O)8\ e^- = 32\ e^-$
The total number of electrons required is:	$40\ e^-$ needed

$40\ e^-$ needed
$-\ 32\ e^-$ total
$8\ e^-$ shared (or four bonds)

$32\ e^-$ total
$-\ 8\ e^-$ shared
$24\ e^-$ unshared

Lewis structure:

Hypervalent structure:

ClO_3^-		
	The chlorine atom has seven valence electrons:	$1Cl(7\ e^-) = 7\ e^-$
	Each oxygen atom has six valence electrons:	$3(O)6\ e^- = 18\ e^-$
	Each negative charge adds one valence electron:	$+1(-) = 1\ e^-$
	The total number of valence electrons is:	$26\ e^-$ total
	The chlorine atom needs eight electrons:	$1Cl(8\ e^-) = 8\ e^-$
	Each oxygen atom needs eight electrons:	$+3(O)8\ e^- = 24\ e^-$
	The total number of electrons required is:	$32\ e^-$ needed

$32\ e^-$ needed
$-\ 26\ e^-$ total
$6\ e^-$ shared (or three bonds)

$26\ e^-$ total
$-\ 6\ e^-$ shared
$20\ e^-$ unshared

Lewis structure: Hypervalent structure:

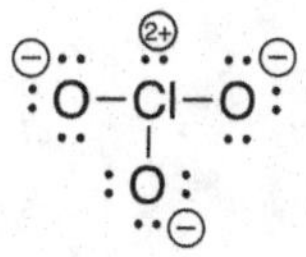

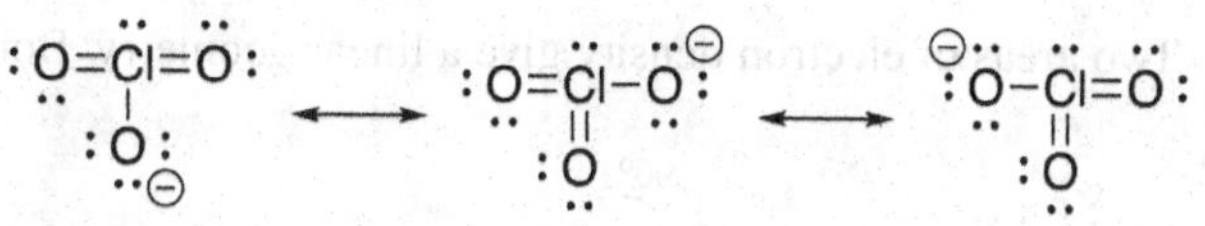

Br_3^-		
	Each bromine atom has seven valence electrons:	$3Br(7\ e^-) = 21\ e^-$
	The negative charged adds one valence electron:	$+1(-) = 1\ e^-$
	The total number of valence electrons is:	$22\ e^-$ total
	Each bromine atom needs eight electrons:	$3Br(8\ e^-) = 24\ e^-$
	The total number of electrons required is:	$24\ e^-$ needed

$24\ e^-$ needed
$-\ 22\ e^-$ total
$2\ e^-$ shared (or one bond)

$22\ e^-$ total
$-\ 2\ e^-$ shared
$20\ e^-$ unshared

Lewis structure: Hypervalent structure:

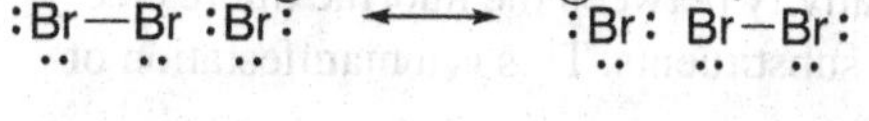

:Br—Br :Br: ⟷ :Br: Br—Br: :Br—Br—Br:

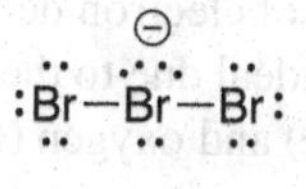

CHAPTER 12 ANSWERS

Problem 12.1. For each of the following Lewis structures, count the number of areas of electron density and redraw the Lewis structure (using Table 12.2) to appropriately show the correct three-dimensional shape. Provide the electron geometry shape name and identify the bond angle(s).

Molecular models are provided for clarity.

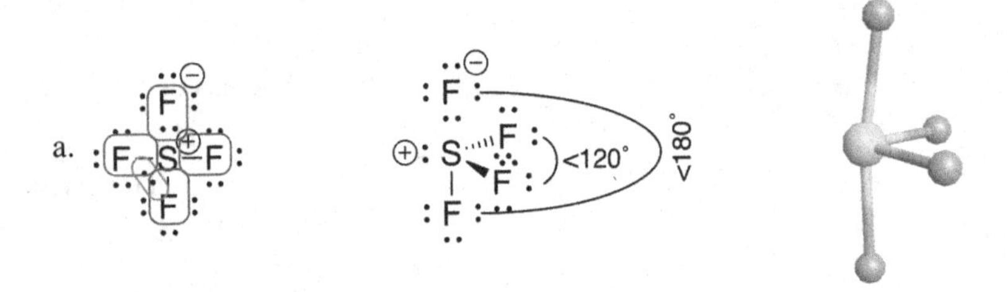

Five areas of electron density give a trigonal bipyramidal geometry. Bond angles <120° for equatorial and <180° for axial (less than ideal because of the lone pair).

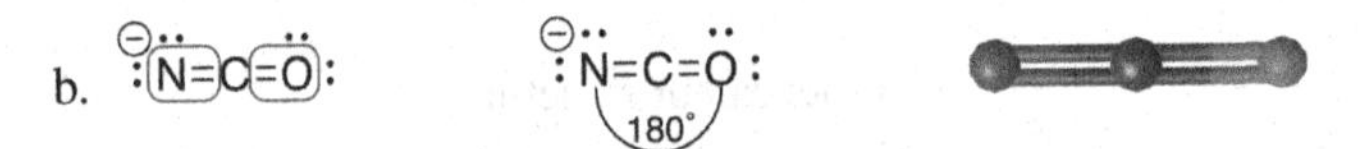

Two areas of electron density give a linear geometry. Bond angles 180°

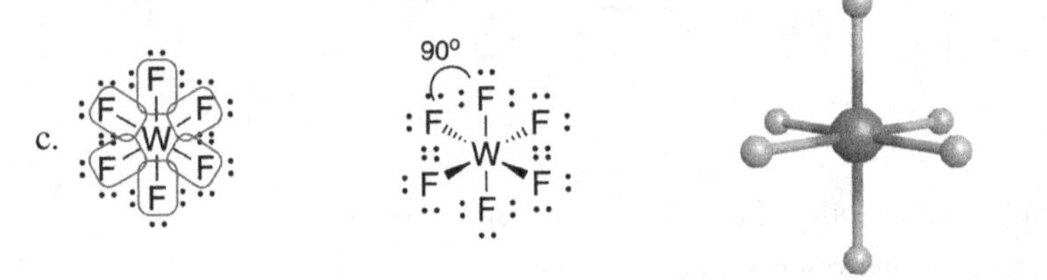

Six areas of electron density give an octahedral geometry. Bond angles 90°.

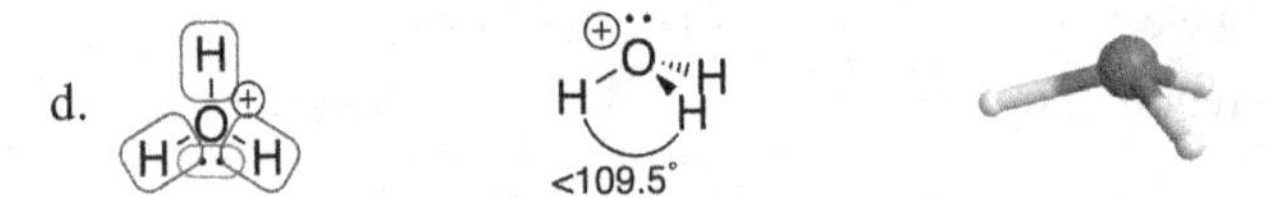

Four areas of electron density give a tetrahedral geometry. Bond angles <109.5° (less than ideal because of the lone pair).

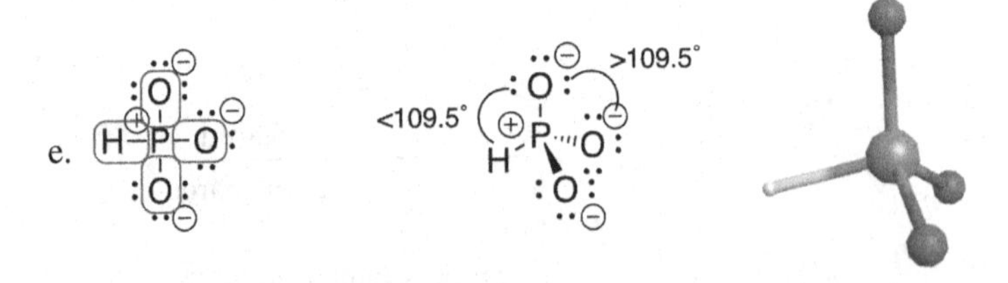

Four areas of electron density give a tetrahedral geometry. Bond angles are distorted from the 109.5° ideal due to the differences in electronegativity between the fluorine (more electronegative) and oxygen (less electronegative) atom substituents. This is a manifestation of Bent's rule.

f. 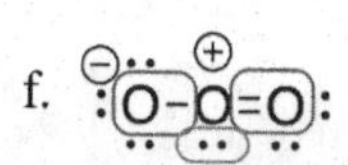 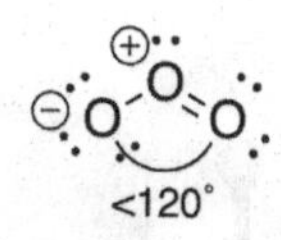

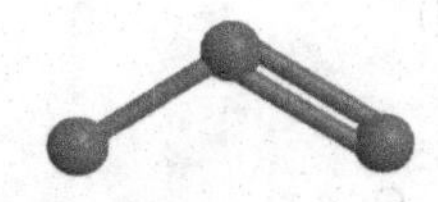

Three areas of electron density give a trigonal planar geometry. Bond angles <120° (less than ideal because of lone pair.

g. 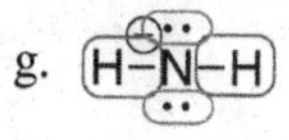 

Four areas of electron density give a tetrahedral geometry. Bond angle <109.5° (less than ideal because of lone pair.

Problem 12.2. For each of the following chemical formulae, draw a Lewis structure that shows the correct three-dimensional shape of the molecule.

Molecular models are provided for clarity.

a. nitrogen dioxide – NO_2 (trigonal planar)

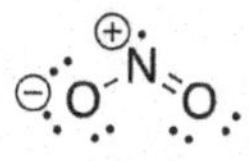
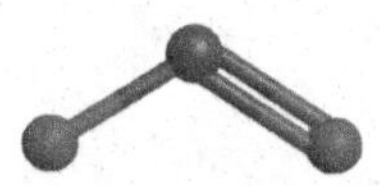

b. iodine pentafluoride – IF_5 (octahedral)

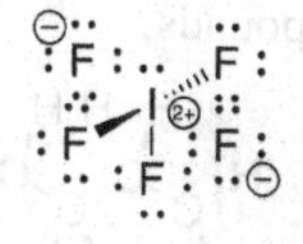
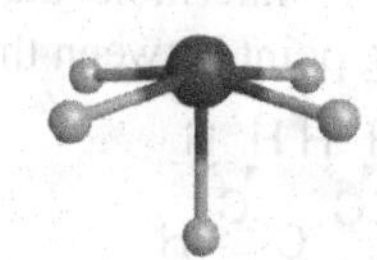

c. boron tribromide – BBr_3 (trigonal planar)

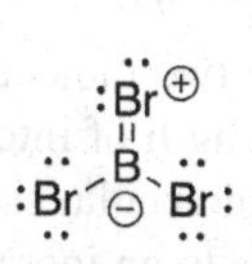
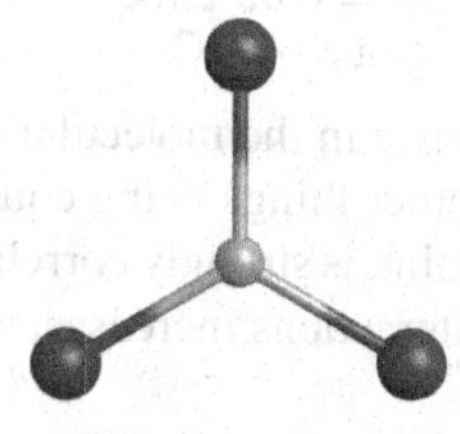

d. silicon tetraiodide – SiI_4 (tetrahedral)

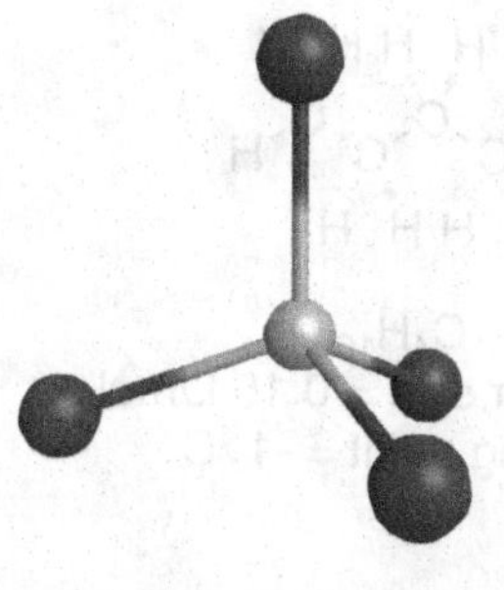

e. sulfite – SO_3^{2-} (tetrahedral)

f. beryllium chloride – $BeCl_2$ (linear)

g. bromine trifluoride – BrF_3 (trigonal bipyramidal)

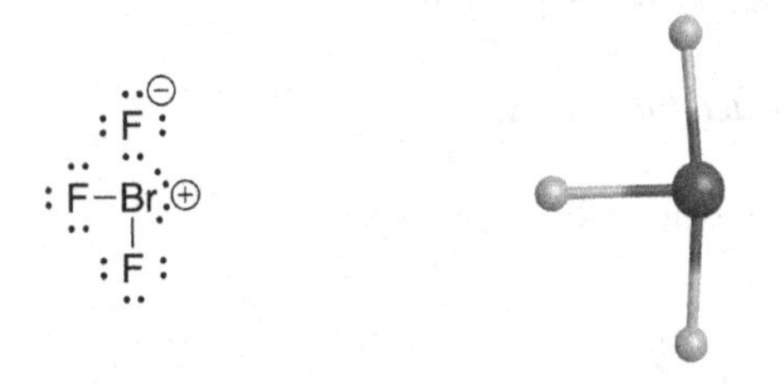

CHAPTER 13 ANSWERS

Problem 13.1. Considering the intermolecular forces involved, can you provide an explanation for the difference in the boiling point between these two compounds?

C_3H_8
Molecular size = 0.09 L/mol
Boiling point = -42 °C

C_4H_{10}
Molecular size = 0.15 L/mol
Boiling point = -1 °C

There is a significant difference in the molecular size of the two molecules. The molecule C_4H_{10} is 67% larger than C_3H_8. All other things being equal, the strength of intermolecular interactions, as represented by the boiling point, is strongly correlated with molecular size: as the size increases, the strength of the dispersion interactions increases, which leads to an increase in boiling point.

Problem 13.2. Considering the intermolecular forces involved, can you provide an explanation for the difference in the boiling point for the two compounds below?

C_4H_{10}
Molecular size = 0.15 L/mol
Boiling point = -1 °C

C_4H_8O
Molecular size = 0.13 L/mol
Boiling point = 79 °C

The molecule C_4H_8O is slightly smaller (0.13 L/mol) than C_4H_{10} (0.15 L/mol). A larger size means stronger dispersion interactions and stronger intermolecular forces and so we would expect C_4H_{10} to have a higher boiling point, but C_4H_8O has a higher boiling point. This means that there are stronger intermolecular forces between C_4H_8O molecules. C_4H_{10} has no polar bonds ($\chi_C = 2.55$ and $\chi_H = 2.20$), while C_4H_8O does have a polar C=O bond ($\chi_O = 3.44$ and $\chi_C = 2.55$). This means that C_4H_8O interacts with other C_4H_8O molecules through both dispersion and dipole–dipole interactions, a more strongly attracting set of interactions, while C_4H_{10} interacts through only dispersion, a weaker interaction. The strength of the intermolecular forces between particles is directly related to the boiling point of a chemical species; therefore, the stronger intermolecular interactions (due to dipole-dipole interactions) between C_4H_8O molecules lead to a higher boiling point.

Problem 13.3. Considering the intermolecular interactions involved, can you provide an explanation for the difference in the boiling point for the two isomers below? Isomers are compounds that have the same formula but a different arrangement of atoms.

Methyl acetate
$C_3H_6O_2$
Molecular size = 0.12 L/mol
Boiling point = 57 °C

Propionic acid
$C_3H_6O_2$
Molecular size = 0.13 L/mol
Boiling point = 140 °C

There is no significant difference in molecular size. Methyl acetate does not have the ability to form a significant hydrogen bond (no N–H, O–H, nor F–H), while propionic acid can form a significant hydrogen bond (it contains an O–H). This means that propionic acid interacts with other propionic acid molecules through dispersion, dipole–dipole, and hydrogen bond interactions, a more strongly attracting set of interactions, while methyl acetate interacts with other methyl acetate molecules only with dispersion and dipole–dipole interactions, a weaker set of interactions. The strength of the intermolecular forces between particles is directly related to the boiling point of a chemical species; therefore, the stronger intermolecular interactions (due to hydrogen bonding) lead to a higher boiling point for propionic acid.

Problem 13.4. For the following questions, we will consider two different vitamins – vitamin C and vitamin E – and their physical properties.

a. Consider the Lewis structure for Vitamin C.

Vitamin C

When two vitamin C molecules interact, what intermolecular forces hold two vitamin C molecules together?

Vitamin C interacts with vitamin C through dispersion (the universal IMF), dipole–dipole interactions (there are a number of polar bonds C=O, C–O, O–H), and hydrogen bonds (there are four O–H groups).

b. Consider the Lewis structure for Vitamin E.

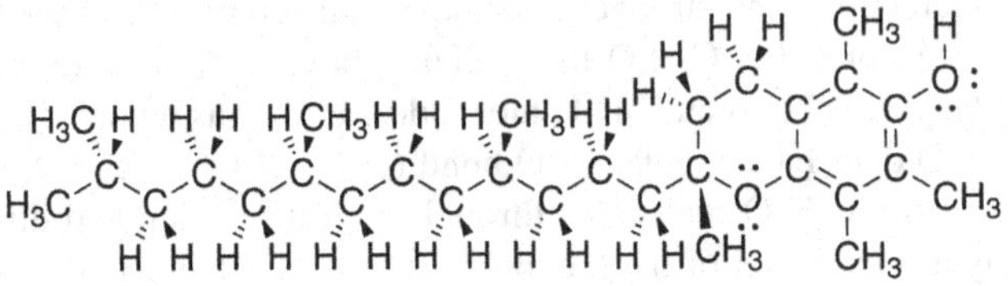

Vitamin E

When two vitamin E molecules interact, what intermolecular forces hold two vitamin E molecules together?
Vitamin E interacts with vitamin E through dispersion (the universal IMF), dipole-dipole interactions (there are a number of polar bonds C=O, C–O, O–H), and hydrogen bonds (there is one O–H group).

c. Vitamin C is a water-soluble vitamin and vitamin E is a fat-soluble vitamin, which means that Vitamin C will dissolve in water while vitamin E will not. Provide an explanation for the difference in water solubility of vitamin C and vitamin E.
Vitamin C is small, has many polar bonds, and many hydrogen bonding groups and so vitamin C can interact with water (which also can hydrogen bond). Vitamin E is a big molecule which is mostly nonpolar (C–C and C–H bonds) and has only a small portion that is polar (C–O and O–H bonds) and only one hydrogen bonding group. Given that most of the molecule is nonpolar it will not dissolve in water. Put another way, water can interact with other water molecules more strongly than water can interact with vitamin E.

d. At normal temperature and pressure, vitamin C is a solid and vitamin E is a liquid. Provide an explanation for this difference in their state of matter.
Vitamin C can interact with vitamin C through multiple hydrogen bonds and through dipole-dipole interactions (it is highly polar). This constellation of strong interactions leads to vitamin C being a solid.

9. In contrast, vitamin E is mostly nonpolar (with only a small section that is polar and can hydrogen bond), which means that it interacts with other vitamin E molecules mostly through dispersion, which is a weaker interaction and causes vitamin E to be a liquid.

Problem 13.5. Whether or not two compounds will mix together depends on whether they have similar intermolecular forces. This is often summarized as "like dissolves like." For each of the following the first molecule will be the liquid, solvent, and the second will be a solid, which may or may not dissolve in the liquid, a potential solute. State whether the solid will, or will not, dissolve in the liquid and explain your reasoning.

a. Liquid: H–Ö–H Solid:

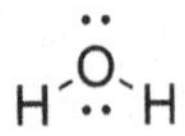

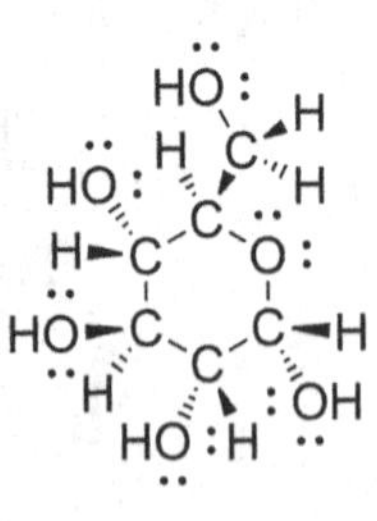

The liquid, water, interacts through hydrogen bonds (O–H), dipole-dipole, and dispersion interactions. The solid, glucose, interacts through hydrogen bonds (O–H), dipole-dipole, and dispersion interactions. Since both solid and liquid have similar intermolecular forces ("they are like"), the solid will dissolve in the liquid ("like dissolves like").

b. Liquid: H–Ö–H Solid:

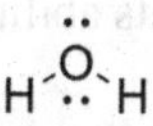

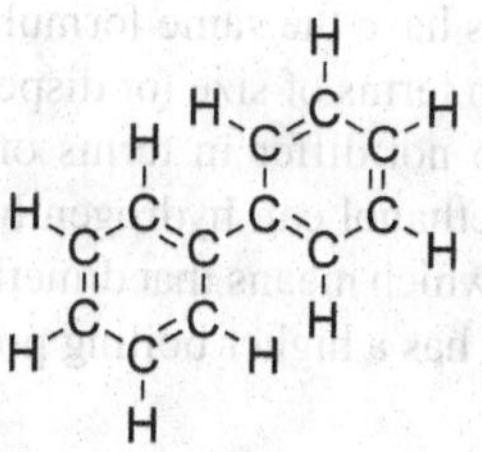

The liquid, water, interacts through hydrogen bonds (O–H), dipole-dipole, and dispersion interactions. The solid, biphenyl, interacts through dispersion interactions only. The solid and liquid do not have similar intermolecular forces, and so the solid will not dissolve in the liquid.

c. Liquid: 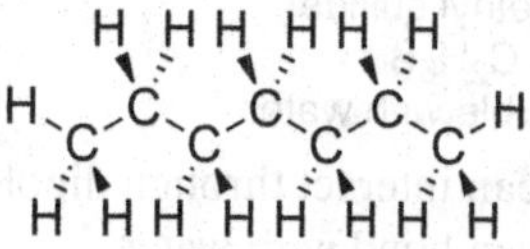 Solid:

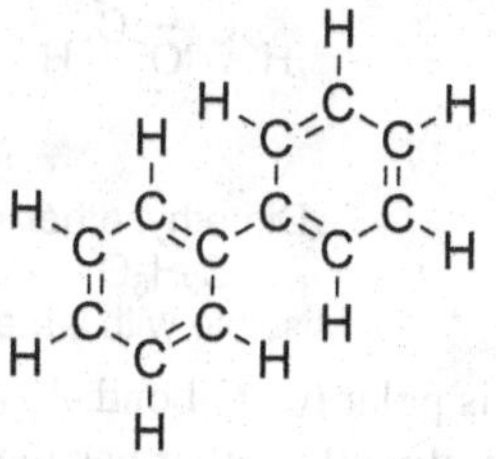

The liquid, heptane, interacts through dispersion interactions only. The solid, biphenyl, interacts through dispersion interactions only. Since both solid and liquid have similar intermolecular forces ("they are like"), the solid will dissolve in the liquid ("like dissolves like").

d. Liquid: 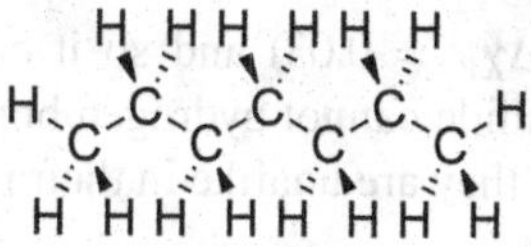 Solid:

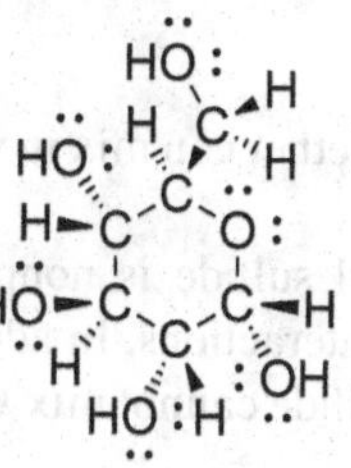

The liquid, heptane, interacts through dispersion interactions only. The solid, glucose, interacts through hydrogen bonds (O–H), dipole-dipole, and dispersion interactions. The solid and liquid do not have similar intermolecular forces, and so the solid will not dissolve in the liquid.

Problem 13.6. Ethanol and dimethyl ether are isomers.

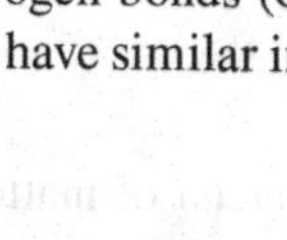

Ethanol
Formula C_2H_6O
Boiling point 78 °C

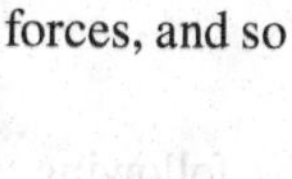

Dimethyl ether
Formula C_2H_6O
Boiling point -11 °C

Provide an explanation for the difference in the boiling points of ethanol and dimethyl ether.

Both compounds have the same formula and so there is no difference in size, which means that they do not differ in terms of size (or dispersion interactions). Both molecules are also polar, which means that they do not differ in terms of dipole-dipole interactions. Ethanol has an O–H bond, which means that ethanol can hydrogen bond with other ethanol molecules; while dimethyl ether has no O–H bond, which means that dimethyl ether cannot hydrogen bond with other dimethyl ether molecules. Ethanol has a higher boiling point than dimethyl ether because of its ability to hydrogen bond.

Problem 13.7. As we will see in the future, molecules mix more readily if they share non-covalent interactions. Why is it that dimethyl ether is miscible (mixes with) water, but dimethyl sulfide is not miscible with water?

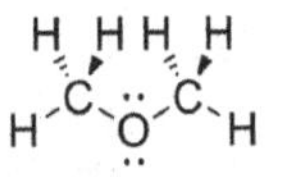

Dimethyl ether
C_2H_6O
Miscible with water

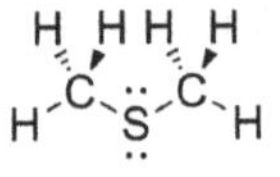

Dimethyl sulfide
C_2H_6S
Immiscible with water

Dimethyl ether is polar (C–O bond $\Delta\chi_P = 0.89$) and so can interact through dipole-dipole interactions. In addition, dimethyl ether has the ability to hydrogen bond with water:

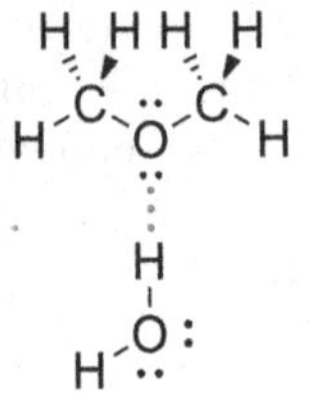

Altogether, dimethyl ether can mix in water because of the similar intermolecular forces between the two molecules.

In contrast, dimethyl sulfide is nonpolar (C–S bond $\Delta\chi_P = 0.03$) and so it cannot interact through dipole-dipole interactions. In addition, dimethyl sulfide cannot hydrogen bond with water. Altogether, dimethyl sulfide cannot mix with water because they are unalike in their intermolecular interactions.

CHAPTER 14 ANSWERS

Problem 14.1. For each of the following, identify the state(s) of matter (solid, liquid, or gas) and whether the material is a singular phase or more than one phase.

a. Seltzer water
   The seltzer water contains both liquid (single phase consisting of a homogeneous mixture of water, dissolved carbon dioxide, carbonic acid, and hydrogen carbonate) and gas states (small bubbles that are a single phase consisting of a homogeneous mixture of water vapor and carbon dioxide).
b. A typical school desk
   The desk is a solid with several different phases (with different chemical compositions): the metal legs, the wood surface, the clear enamel coating on the wood, and any plastic or rubber components.

Problem 14.2. Consider Figure 14.2. Rank a solid (ice) versus a liquid (water) versus a gas (steam) from highest to lowest energy. Explain your answer.

From highest to lowest energy the phases are, in order: gas, liquid, and solid.

A gas has the highest energy because the particles are moving the fastest, which is evidenced by the wide spacing between particles.

A solid has the lowest energy because the particles only vibrate (and do not rotate nor translate) and they are closely packed.

A liquid is intermediate between the gas and solid phases because the particles are spread out more than a solid and do vibrate, rotate, and translate, but they translate at a much slower speed than a gas.

Problem 14.3. Use Table 14.1 to answer the following questions about different types of solids.

a. Consider the two ionic compounds NaF and MgO. For both compounds, the cations (Na^+ and Mg^{2+}) and anions (F^- and O^{2-}) have the same number of electrons and both have a similar atomic arrangement. Why, then, are the melting points of NaF (992 °C) and MgO (2825 °C) so different?
   MgO has a higher melting point because the ions are 2+/2– ions, which means there is a stronger ionic bond between them than between the 1+/1– ions in NaF.
b. Titanium(IV) bromide is a brown, crystalline solid with a melting point of 39 °C and a boiling point of 230 °C. It does not conduct electricity as a solid or liquid. Chromium(III) bromide is a lustrous black, crystalline solid with a melting point of 1130 °C and it is soluble in water and does not conduct electricity, but molten chromium(III) bromide does conduct electricity.
   i. Is titanium(IV) bromide an ionic, molecular, metallic, network, or amorphous solid? Explain your selection. Molecular – low melting and boiling point, poor conductor.
   ii. Is chromium(III) chloride an ionic, molecular, metallic, network, or amorphous solid? Explain your selection. Ionic – high melting point, soluble in water, good conductor when molten.

Problem 14.4. Water is unique because it is the only substance where its density as a solid is less than its density as a liquid. Think about life on Earth and provide an example of how this unique property is important for life.

The fact that ice is less dense than water is deeply important to aquatic life. If ice were denser than water, it would sink (rather than float), which could be potentially dangerous to any life it sank into.

Problem 14.5. The density of solid ice and liquid water differ by less than 10%, but there is a dramatic change in density for steam. Provide an explanation as to why the density of steam is so much less than the density of solid ice and liquid water.

Gases are a dispersed phase, which means that the particles are widely dispersed from one another. The dispersal of particles means that there are significantly fewer particles per volume, which means a much-reduced mass per volume (or a reduced density).

Problem 14.6. A researcher has three containers of gas at 27.0 °C are connected by a closed valve:

	Container 1	Container 2	Container 3
Gas	O_2	N_2	Ar
Volume	0.003 00 m^3	200. mL	5.00 L
Pressure	146 kPa	0.908 atm	630 mmHg
Gas constant	$\frac{8.314\,Pa\,m^3}{mol\,K}$	$\frac{0.082\,06\,L\,atm}{mol\,K}$	$\frac{62.36\,L\,mmHg}{mol\,K}$

Which container has the most amount (mol) of gas? $T = 27.0 + 273.15 = 300.2$ K
Container 1
$n = PV/RT = (146\,000 \text{ Pa} \times 0.00300 \text{ m}^3)/(8.314 \text{ (Pa m}^3\text{/mol K)} \times 300.2 \text{ K})$
$n = 0.175$ mol

Container 2
$n = PV/RT = (0.908 \text{ atm} \times 0.200 \text{ L})/(0.082\,06 \text{ (L atm/mol K)} \times 300.2 \text{ K})$
$n = 0.007\,37$ mol

Container 3
$n = PV/RT = (630 \text{ mmHg} \times 5.00 \text{ L})/(62.36 \text{ (L mmHg/mol K)} \times 300.2 \text{ K})$
$n = 0.168$ mol

Container 1 has the greatest amount (mol) of gas.

Problem 14.7. If you breathe out 240 mL of CO_2 per minute, what is the mass (in kg) of CO_2 you exhale in a 24.0-hour day? Assume normal conditions (1.00 atm, 25.0 °C).

24.0 ~~h~~	60 ~~min~~	240 ~~mL~~	1×10^{-3} L	= 346 L CO_2
	1 ~~h~~	1 ~~min~~	1 ~~mL~~	

$T = 25.0 + 273.15 = 298.2$ K
$n = PV/RT = (1.00 \text{ atm} \times 346 \text{ L})/(0.082\,06 \text{ (L atm/mol K)} \times 298.2 \text{ K})$
$n = 14.1$ mol

14.1 ~~mol~~	44.0 ~~g~~	1 kg	= 0.621 kg CO_2
	1 ~~mol~~	1×10^3 ~~g~~	

Problem 14.8. If you have ice at −1 °C and 1 atm and increase the pressure, what happens to the solid water? What does this tell you about the density of liquid water compared to solid water?

Since the line slopes backward, if you pressurize solid ice near the melting point, the solid will convert into a liquid. This occurs because liquid water is denser than solid water (unique compared to every other substance).

Problem 14.9. Consider the pressure-temperature diagram for sulfur (Figure 14.4) and answer the following questions.

a. Identify any triple point(s) (by listing their pressure and temperature) for sulfur. What phases are present at the triple point(s)?
90 °C, 1×10^{-3} kPa – rhombic, monoclinic, vapor
140 °C, 1×10^{5} kPa – rhombic, monoclinic, liquid
120 °C, 1×10^{-2} kPa – monoclinic, liquid, vapor

b. If a sample of sulfur gas at 130 °C and 10^{-4} atm is slowly pressurized up to 10^{4} atm, list the sequence of phases that would be visible as the pressure increased.
Vapor → liquid → monoclinic → rhombic

c. Which, if either, is denser, monoclinic or rhombic sulfur? Explain your answer.
Rhombic is denser. As you increase the pressure, eventually (at higher pressures) monoclinic changes to rhombic. This suggests that rhombic, which occurs at higher pressures, is denser.

Problem 14.10. If we compare hydrogen sulfide (boils at −60.28 °C at 1 atm) and hydrogen selenide (boils at −41.25 °C at 1 atm), we can see that hydrogen selenide must be given more thermal energy to vaporize.

a. What can we say about the strength of the intermolecular forces between hydrogen sulfide molecules versus the strength of the intermolecular forces between hydrogen selenide?
Because it requires a higher temperature to vaporize, we can conclude that hydrogen selenide has stronger intermolecular forces than hydrogen sulfide.

b. Can you provide an explanation for the differences in intermolecular forces?
Given the larger size of the selenium atom compared to the sulfur atom, hydrogen selenide has a larger surface area and therefore stronger van der Waals interactions.

Problem 14.11. Water requires 6.01 kJ/mol to melt and 40.65 kJ/mol to vaporize. Provide an explanation why more heat is required to vaporize water than is required to melt water. Consider the particles (Figure 14.2) and how they change upon vaporization versus melting.

Gas particles are widely separated so that there are no particle–particle interactions. Liquid and solid particles are close together and so there are particle-particle interactions. Given this difference, melting a solid only requires enough energy so that particles can move past one another, but particle-particle interactions still occur. To vaporize a substance, the particle-particle interactions (intermolecular forces) must be entirely disrupted, which requires a substantial amount of energy.

Problem 14.12. For each of the following aqueous compounds, identify what species will be present in solution.

a. HNO_3(aq), a strong acid: H^+(aq) and NO_3^-(aq)

b. $C_6H_{12}O_6$(aq): $C_6H_{12}O_6$(aq)

c. CH_3OCH_3(aq): CH_3OCH_3(aq)

d. Na_2SO_4(aq): 2 Na^+(aq) and SO_4^{2-}(aq)

e. H_2SO_4(aq), a strong acid and a weak acid: There are two H^+ ions in H_2SO_4. The first of the two fully dissociates to give H^+(aq) and HSO_4^-(aq). The H^+ in HSO_4^-(aq) does not fully dissociate and so there will be H^+(aq) and SO_4^{2-}(aq). In this way, sulfuric acid, which is diprotic (two H^+'s) is both strong and weak and in solution there will be H^+(aq), HSO_4^-(aq), and SO_4^{2-}(aq).

f. HBr, a strong acid: H^+(aq) and Br^-(aq)

g. $CaCl_2$(aq): Ca^{2+}(aq) and 2 Cl^-(aq)

Problem 14.13. Sodium bromide is soluble in water, methanol (CH_3OH), and ethanol (CH_3CH_2OH). The solubility of sodium bromide in these three solvents varies: water (94 g/mL at 25 °C), methanol (17 g/mL at 25 °C), ethanol (2 g/mL at 25 °C). Provide an explanation for the differences in solubility of sodium bromide in these three solvents.

To dissolve, sodium bromide must be separated into Na^+(aq) and Br^-(aq). This requires a hydrogen-bonding solvent that can stabilize the sodium ion (here through the coordination of an electron-rich atom) and the bromide ion (through hydrogen bonding). Water (H_2O) has two hydrogen-bonding O–H groups and an electron-rich oxygen atom. Methanol and ethanol have only one hydrogen-bonding O–H group, which account, in part, for the reduced solubility of sodium bromide in both compared to water. In addition, methanol, and ethanol both have nonpolar groups (CH_3 and CH_2CH_3), which do not help to stabilize the ions, and the larger nonpolar group in ethanol (CH_2CH_3) diminishes the solubility of sodium bromide in the solvent.

CHAPTER 15 ANSWERS

Problem 15.1. Provide an explanation of why the equation, as written, violates the law of conservation of mass. As written, the decomposition of sodium bicarbonate has more sodium, carbon, hydrogen, and oxygen atoms on the right than on the left. This would mean that atoms (matter) are being created, and since matter has volume and mass this would mean the spontaneous increase in mass, which violates the law of conservation of mass.

Problem 15.2. The reaction of thermite – iron(3+) oxide powder with aluminium powder – to produce solid aluminium oxide and solid iron metal is incredibly exothermic and dramatic to watch.

a. Write a chemical equation that shows all the reactants, products, and the appropriate phase notation.

$$Fe_2O_3(s) + Al(s) \rightarrow Fe(s) + Al_2O_3(s)$$

b. Initially 10.0 g iron(3+) oxide is mixed with 3.45 g aluminium to produce 6.40 g of aluminium oxide and 6.95 g iron. Determine the amount (mol) of each reactant and product.

10.0 g	mol	
	159.687 g	= 0.0626 mol Fe_2O_3(s)

3.45 g	mol	
	26.982 g	= 0.128 mol Al(s)

$$6.40\ \text{g} \times \frac{\text{mol}}{101.961\ \text{g}} = 0.0628\ \text{mol}\ Al_2O_3(s)$$

$$6.95\ \text{g} \times \frac{\text{mol}}{55.845\ \text{g}} = 0.124\ \text{mol}\ Fe(s)$$

c. Use the amounts determined in 14.2.b to find the appropriate coefficients to balance the reaction.

$$0.0626\ \text{mol}\ Fe_2O_3(s) + 0.128\ \text{mol}\ Al(s) \rightarrow 0.124\ \text{mol}\ Fe(s) + 0.0628\ \text{mol}\ Al_2O_3(s)$$

Simplifies to:

$$Fe_2O_3(s) + 2\ Al(s) \rightarrow 2\ Fe(s) + Al_2O_3(s)$$

Problem 15.3. For each problem, write a balanced chemical equation (making sure to include phase notation).

a. Liquid octane (C_8H_{18}) combines with oxygen gas to burn and yield carbon dioxide gas and water vapor.

$$2\ C_8H_{18}(l) + 25\ O_2(g) \rightarrow 16\ CO_2(g) + 18\ H_2O(g)$$

b. Dihydrogen reacts with octasulfur to produce hydrogen sulfide gas.

$$8\ H_2(g) + S_8(s) \rightarrow 8\ H_2S(g)$$

c. Solid aluminium chloride decomposes to produce aluminium metal and chlorine gas.

$$2\ AlCl_3(s) \rightarrow 2\ Al(s) + 3\ Cl_2(g)$$

d. Solid magnesium burns in the presence of oxygen gas to produce magnesium oxide.

$$2\ Mg(s) + O_2(g) \rightarrow 2\ MgO(s)$$

e. Dichlorine gas reacts with an aqueous solution of sodium bromide to produce liquid bromine and an aqueous solution of sodium chloride.

$$Cl_2(g) + 2\ NaBr(aq) \rightarrow Br_2(l) + 2\ NaCl(aq)$$

f. When lithium is added to solid sodium chloride, solid lithium chloride and sodium are produced.

$$Li(s) + NaCl(s) \rightarrow Na(s) + LiCl(s)$$

g. Aqueous rubidium hydroxide solution is combined with beryllium fluoride solid to produce aqueous rubidium fluoride and solid beryllium hydroxide.

$$2\ RbOH(aq) + BeF_2(g) \rightarrow 2\ RbF(aq) + Be(OH)_2(s)$$

h. A solution of aqueous silver nitrate combines with solid copper to produce elemental silver and copper(2+) nitrate solution.

$$2\ AgNO_3(aq) + Cu(s) \rightarrow Cu(NO_3)_2(aq) + 2\ Ag(s)$$

i. Lithium metal reacts with solid magnesium bromide to produce solid lithium bromide and pure magnesium metal.

$$2\ Li(s) + MgBr_2(s) \rightarrow 2\ LiBr(s) + Mg(s)$$

j. Chlorine gas reacts with aqueous sodium bromide solution to produce liquid bromine and aqueous sodium chloride.

$$Cl_2(g) + 2\ NaBr(aq) \rightarrow Br_2(l) + 2\ NaCl(aq)$$

k. Lead dihydroxide solid reacts with aqueous hydrogen chloride to produce water and solid lead dichloride.

$$Pb(OH)_2(s) + 2\ HCl(aq) \rightarrow 2H_2O(l) + PbCl_2(s)$$

l. Aqueous sodium phosphate and aqueous calcium chloride react to produce aqueous sodium chloride and solid calcium phosphate.

$$2\ Na_3PO_4(aq) + 3\ CaCl_2(aq) \rightarrow 6\ NaCl(aq) + Ca_3(PO_4)_2(s)$$

Problem 15.4. Consider the reaction of ClO_2 with water.

$$6\ ClO_2(g) + 3\ H_2O(l) \rightarrow 5\ HClO_3(aq) + HCl(aq)$$

a. How many milliliters of water (density = 1.00 g/mL) are needed to react with 2.00 g of ClO_2?

2.00 ~~g ClO_2~~	1 ~~mol~~	3 ~~mol~~ H_2O	18.015 ~~g~~	1 mL	= 0.267 mL H_2O
	67.448 g	6 mol ~~ClO_2~~	1 ~~mol~~	1.00 ~~g~~	

b. How many grams of $HClO_3$ are formed from 2.00 g of ClO_2?

2.00 ~~g ClO_2~~	1 ~~mol~~	5 ~~mol~~ $HClO_3$	84.46 ~~g~~	= 2.09 g $HClO_3$
	67.448 ~~g~~	6 ~~mol ClO_2~~	1 ~~mol~~	

Problem 15.5. Answer the following questions about the production of iron.

$$2\ Fe_2O_3(s) + 3\ C(s) \rightarrow 4\ Fe(s) + 3\ CO_2(g)$$

a. Balance the above equation.
b. How many grams of iron(III) oxide do you need to produce 500.0 g Fe(s)?

500.0 g Fe	1 ~~mol~~	2 ~~mol~~ Fe_2O_3	159.687 g	= 714.9 g Fe_2O_3
	55.845 ~~g~~	4 ~~mol Fe~~	1 ~~mol~~	

c. How many liters of CO_2 are produced (at STP) when 3.45 kg iron(III) oxide is reacted with excess carbon?

3.45 kg Fe_2O_3	1×10^3 g	1 mol	3 mol CO_2	22.7 L	
	1 kg	159.687 g	2 mol Fe_2O_3	1 mol	= 736 L CO_2

Problem 15.6. Answer the following questions about the production of ammonia (NH_3).

a. Balance the following combination reaction

$$N_2(g) + 3\ H_2(g) \rightarrow 2\ NH_3(g)$$

b. How many grams of nitrogen gas are needed to produce 10.0 g NH_3?

10.0 g NH_3	1 mol	1 mol N_2	28.014 g	
	17.031 g	2 mol NH_3	1 mol	= 8.22 g N_2

c. How many liters of hydrogen gas (at STP) are needed to produce 135 g NH_3?

135 g NH_3	1 mol	3 mol H_2	22.7 L	
	17.031 g	2 mol NH_3	1 mol	= 270. L H_2

Problem 15.7. Answer the following questions about the reaction of aluminium and copper(II) chloride.

$$2\ Al(s) + 3\ CuCl_2(aq) \rightarrow 3\ Cu(s) + 2\ AlCl_3(aq)$$

a. If 4.5 mol Al reacts with excess copper(II) chloride, what amount of copper (mol) will be produced?

4.5 mol Al	3 mol Cu	
	2 mol Al	= 6.8 mol Cu

b. If 69.5 g copper(II) chloride are reacted with excess aluminium, how many grams of aluminium chloride would be produced?

69.5 g $CuCl_2$	1 mol	2 mol $AlCl_3$	133.332 g	
	134.446 g	3 mol $CuCl_2$	1 mol	= 46.0 g $AlCl_3$

Problem 15.8. Answer the following question about aluminium chloride.

$$2\ AlCl_3(s) \rightarrow 2\ Al(s) + 3\ Cl_2(g)$$

a. Balance the equation above

b. If 43.0 g aluminium chloride is decomposed according to the equation above, how many chlorine atoms will be produced?

43.0 ~~g $AlCl_3$~~	1 ~~mol~~	3 ~~mol Cl_2~~	2 ~~mol~~ Cl	6.022×10^{23} atoms	$= 5.83\times10^{23}$ Cl atoms
	133.332 ~~g~~	2 mol ~~$AlCl_3$~~	1 ~~mol Cl_2~~	1 ~~mol~~	

Problem 15.9. Using the following balanced chemical equation, what volume (in L) of 0.790 mol/L lead(II) nitrate is required to react with 1.25 L of 0.550 mol/L potassium chloride solution?

$$2\ KCl(aq) + Pb(NO_3)_2(aq) \rightarrow PbCl_2(s) + 2\ KNO_3(aq)$$

1.25 ~~L KCl solution~~	0.550 ~~mol~~	1 ~~mol~~ $Pb(NO_3)_2$	L	$= 0.435$ L $Pb(NO_3)_2$
	L	2 ~~mol KCl~~	0.790 ~~mol~~	

Problem 15.10. 5.0 mL of 0.100 mol/L calcium chloride is combined with 7.5 mL of 0.060 mol/L lithium carbonate. Determine the mass of solid calcium carbonate produced (in grams) and identify the limiting reagent and the reagent present in excess.

$$CaCl_2(aq) + Li_2CO_3(aq) \rightarrow CaCO_3(s)\downarrow + 2\ LiCl(aq)$$

5.0 ~~mL $CaCl_2$~~	1×10^{-3} ~~L~~	0.100 ~~mol~~	1 ~~mol~~ $CaCO_3$	100.086 g	$= 0.050$ g $CaCO_3$
	1 ~~mL~~	~~L~~	1 ~~mol $CaCl_2$~~	~~mol~~	

7.5 ~~mL Li_2CO_3~~	1×10^{-3} ~~L~~	0.060 ~~mol~~	1 ~~mol~~ $CaCO_3$	100.086 g	$= 0.045$ g $CaCO_3$
	1 ~~mL~~	~~L~~	1 ~~mol Li_2CO_3~~	~~mol~~	

Lithium carbonate is the limiting reagent and calcium chloride is present in excess, and 0.045 g calcium carbonate will be produced.

Problem 15.11. For each of the following, use the information provided, the balanced reaction, and the enthalpy of reaction to determine the energy produced (or consumed).

a. $2\ S(s) + C(s, graphite) \rightarrow CS_2(l)$, $\Delta_r H^\circ = 89.0$ kJ/mol

1.0 kg of graphite is converted into carbon disulfide.

1.0 kg C(s, graphite)	1×10^3 g	1 mol	1 mol rxn	89 kJ	= 7400 kJ (energy consumed)
	1 kg	12.011 g	1 mol C(s, graphite)	mol rxn	

b. $3\ Ag(s) + AuCl_3(aq) \rightarrow 3\ AgCl(s) + Au(s)$, $\Delta_r H^\circ = -288.6$ kJ/mol

450 g of silver is converted into silver chloride.

450 g Ag(s)	1 mol	1 mol rxn	−288.6 kJ	$= -4.0\times10^2$ kJ (energy produced)
	107.87 g	3 mol Ag(s)	mol rxn	

CHAPTER 16 ANSWERS

Problem 16.1. Using the common charges of elements in ionic compounds (Appendix 4), identify the charge of each element before and after the chemical reaction. Then identify which element is reduced and which element is oxidized.

a. $MgI_2(aq) + Br_2(l) \rightarrow I_2(s) + MgBr_2(aq)$

Reactants	Products
Mg = 2+	Mg = 2+
I = 1–	I = 0
Br = 0	Br = 1–

Iodine is oxidized and bromine is reduced.

b. $TiCl_4(l) \rightarrow Ti(s) + 2\ Cl_2(g)$

Reactants	Products
Ti = 4+	Ti = 0
Cl = 1–	Cl = 0

Chlorine is oxidized and titanium is reduced.

c. $Fe_2O_3(s) + 3\ Mg(s) \rightarrow 3\ MgO(s) + 2\ Fe(s)$

Reactants	Products
Fe = 3+	Fe = 0
O = 2–	O = 2–
Mg = 0	Mg = 2+

Magnesium is oxidized and iron is reduced.

d. $2\ Cu(s) + S(s) \rightarrow Cu_2S(s)$

Reactants	Products
Cu = 0	Cu = 1+
S = 0	S = 2–

Copper is oxidized and sulfur is reduced.

Problem 16.2. Answer the following questions about the synthesis of sodium chloride from solid sodium and dichlorine gas.

$$2\ Na(s) + Cl_2(g) \rightarrow 2\ LiCl(s)$$

a. What element is oxidized and what element is reduced?

Reactants	Products
Na = 0	Na = +1
Cl = 0	Cl = –1

Sodium is oxidized and chlorine is reduced.

b. This reaction is very favorable. Favorable reactions tend to release energy and this reaction releases 411 kJ/mol. Consider the electron configuration of sodium and the electron configuration of chlorine. And explain why the electron transfer you identified in 16.2.a is so favorable?

The electron configuration of sodium is $[Ne]3s^1$ and the electron configuration of chlorine is $[Ne]3s^23p^5$. The transfer of an electron from sodium to chlorine will result in two ions (sodium(1+) and chloride(1–)) that each have a noble gas configuration, which is highly favorable. The transfer of an electron from sodium to chlorine releases a lot of energy because it creates two ions with noble gas configurations and equal charge. The sodium(1+) cation and chloride(1–) anion are then attracted to each other, creating an ionic bond, which also releases energy. We could also explain this in terms of electronegativity, which highlights the strong attractive power of chlorine (3.16) and the significantly weaker attractive power of sodium (0.93). This reaction is favorable because an electron goes from the less electronegative sodium to the more electronegative chlorine.

c. The reaction of sodium and chlorine is very exothermic. Would the reaction of caesium and dichlorine be more or less energetic than the reaction of sodium and dichlorine? Consider periodic trends at play for sodium versus caesium and how that might impact the reaction. The reaction of caesium and dichlorine would likely be more exothermic than the reaction of sodium and dichlorine. Caesium has a lower ionization energy, which is reflected in its lower electronegativity, which means that less energy is required to remove an electron from caesium than is required to remove an electron from sodium. This means that the energy debit (ionization energy) is lower in the case of caesium and the energy credit (electron affinity) is the same, which should produce a more exothermic reaction overall.

d. Would the reaction of sodium and diiodine be more or less energetic than the reaction of sodium and dichlorine? Consider periodic trends at play for iodine versus chlorine and how that might impact the reaction.
The reaction of sodium and diiodine would likely be less exothermic than the reaction of sodium and dichlorine. Iodine has a lower electron affinity than chlorine, which is reflected in its lower electronegativity, which means that less energy is released when iodine gains an electron compared to when chlorine gains an electron. This means that the energy credit (electron affinity) for iodine is less, and the energy debit (ionization energy) is the same, which should produce a less exothermic reaction overall.

Problem 16.3. Assign the oxidation number for each atom or ion.

a. $Mn(s)$ — $Mn = 0$
b. $Ca^{2+}(aq)$ — $Ca = +2$
c. $S_8(s)$ — $S = 0$
d. $N_2(g)$ — $N = 0$
e. $I^-(aq)$ — $I = -1$

Problem 16.4. Assign the oxidation number for each element in the following binary compounds.

a. $C_2H_6(g)$ — $C = -3, H = +1$
b. $C_2H_4(g)$ — $C = -2, H = +1$
c. $C_2H_2(g)$ — $C = -1, H = +1$
d. $MnCl_2(s)$ — $Mn = +2, Cl = -1$
e. $IF_7(g)$ — $I = +7, F = -1$

Problem 16.5. Assign the oxidation number for each element in the following polyatomic ions.

a. $NO_2^-(aq)$ — $N = +3, O = -2$
b. $NO_3^-(aq)$ — $N = +5, O = -2$
c. $Hg_2^{2+}(aq)$ — $Hg = +1$
d. $OH^-(aq)$ — $O = -2, H = +1$
e. $PO_4^{3-}(aq)$ — $P = +5, O = -2$

Problem 16.6. Assign the oxidation number for each element in the following compounds.

a. $Fe(NO_3)_3(s)$	Fe = +3, N = +5, O = −2
b. $H_3PO_4(l)$	H = +1, P = +5, O = −2
c. $KMnO_4(s)$	K = +1, Mn = +7, O = −2
d. KOH(s)	K = +1, O = −2, H = +1
e. $Al_2(SO_4)_3(s)$	Al = +3, S = +6, O = −2

Problem 16.7. What is the oxidation number of carbon in each of the following compounds?

CH_4 (−4); CH_3OH (−2); $H_2C{=}O$ (0); $HCOOH$ (+2); $O{=}C{=}O$ (+4)

Problem 16.8. For each of the following, identify which element is oxidized and which element is reduced.

a. $Hg_2O(s) \rightarrow HgO(s) + Hg(l)$

Reactants	Products
Hg = +1	Hg = +2 (HgO), 0 (Hg)
O = −2	O = −2

Mercury is oxidized and mercury is reduced. A reaction where one element is both oxidized and reduced is called a disproportionation reaction.

b. $CH_3Cl(g) + 3\,Cl_2(g) \rightarrow CCl_4(g) + 3\,HCl(g)$

Reactants	Products
C = −2	C = +4
H = +1	H = +1
Cl = −1 (CH_3Cl), 0 (Cl_2)	Cl = −1

Carbon is oxidized and chlorine is reduced.

c. $2\,KMnO_4(aq) + 5\,HCO_2H(aq) + 6\,HCl(aq) \rightarrow 2\,MnCl_2(aq) + 8\,H_2O(l) + 5\,CO_2(g) + 2\,KCl(aq)$

Reactants	Products
K = +1	K = +1
Mn = +7	Mn = +2
O = −2	O = −2
H = +1	H = +1
C = +2	C = +4
Cl = −1	Cl = −1

Carbon is oxidized and manganese is reduced.

Problem 16.9. For each of the following, predict whether you think the reaction enthalpy ($\Delta_r H^\circ$) would be positive (endothermic) or negative (exothermic).

a. $NH_2Cl(g) + Cl_2(g) \rightarrow NHCl_2(g) + HCl(g)$

Nitrogen ($\chi_P = 3.04$) is oxidized and chlorine ($\chi_P = 3.16$) is reduced. Because electrons are going in the favorable direction (from a less electronegative element to a more electronegative element), the reaction should be exothermic ($\Delta_r H^\circ < 0$).

b. $2\ NaCl(s) + Br_2(l) \rightarrow 2\ NaBr(s) + Cl_2(g)$
Chlorine ($\chi_P = 3.16$) is oxidized and bromine ($\chi_P = 2.96$) is reduced. Because electrons are going in the unfavorable direction (from a more electronegative element to a less electronegative element), the reaction should be endothermic ($\Delta_r H^\circ > 0$).

c. $3\ Ag(s) + AuCl_3(aq) \rightarrow 3\ AgCl(s) + Au(s)$
Silver ($\chi_P = 1.93$) is oxidized and gold ($\chi_P = 2.54$) is reduced. Because electrons are going in the favorable direction (from a less electronegative element to a more electronegative element), the reaction should be exothermic ($\Delta_r H^\circ < 0$).

d. $2\ HCl(aq) + Pt(s) \rightarrow PtCl_2(s) + H_2(g)$
Platinum ($\chi_P = 2.28$) is oxidized and hydrogen ($\chi_P = 2.20$) is reduced. Because electrons are going in the unfavorable direction (from a more electronegative element to a less electronegative element), the reaction should be endothermic ($\Delta_r H^\circ > 0$).

e. $CH_4(g) + 2\ O_2(g) \rightarrow CO_2(g) + 2\ H_2O(g)$
Carbon ($\chi_P = 2.55$) is oxidized and oxygen ($\chi_P = 3.44$) is reduced. Because electrons are going in the favorable direction (from a less electronegative element to a more electronegative element), the reaction should be exothermic ($\Delta_r H^\circ < 0$).

Problem 16.10. For each of the following reactions, identify the acid and the base. For any reactions involving hydron transfer, also identify the conjugate acid and conjugate base.

a.
base (electron-pair donor) — acid (electron-pair acceptor)

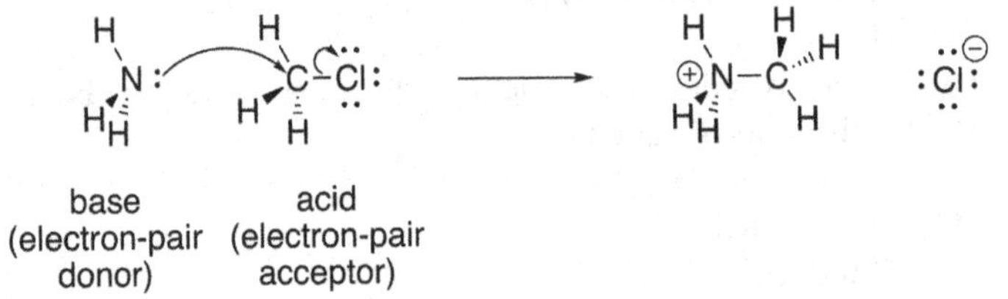

b.
base (hydron acceptor) — acid (hydron donor) — conjugate acid — conjugate base

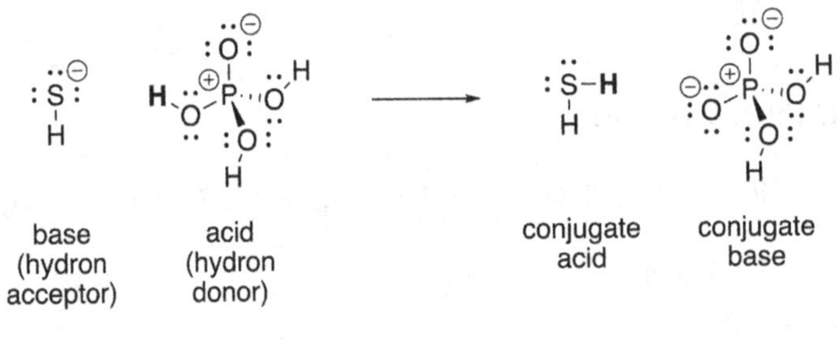

c.
acid (hydron donor) — base (hydron acceptor) — conjugate base — conjugate acid

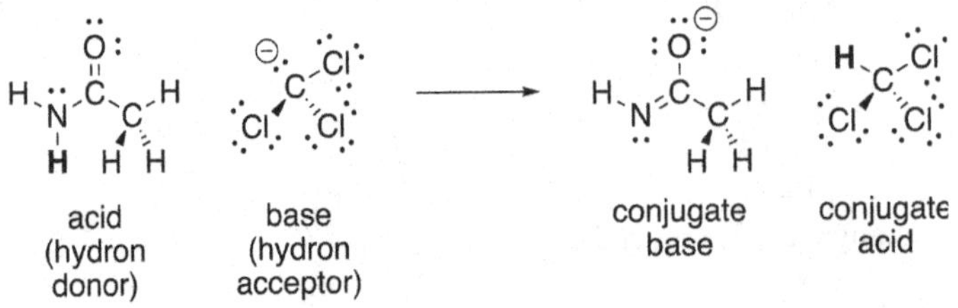

d.
base (electron-pair donor) — acid (electron-pair acceptor)

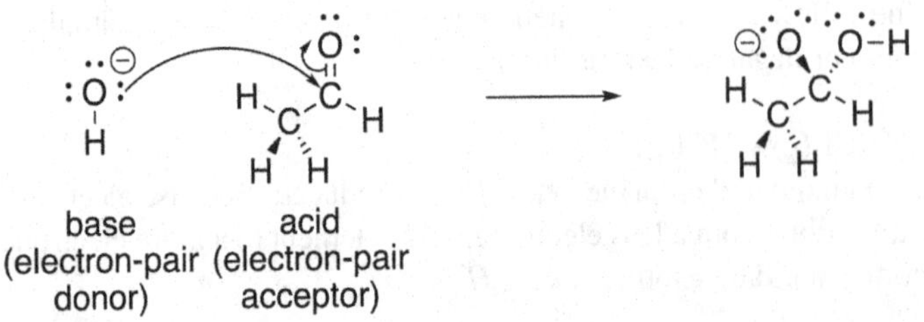

e.

acid (hydron donor) | base (hydron acceptor) | → | conjugate base | conjugate acid

Problem 16.11 Given the following pH values, determine the amount concentration of hydrons, $H^+(aq)$, and hydroxide, $OH^-(aq)$. Identify each as acidic or basic.

a. $pH = 1.00 = -\log[H^+(aq)]$
$[H^+(aq)] = 10^{-1.00} = 1.0 \times 10^{-1}$
$[1.0 \times 10^{-1}][OH^-(aq)] = 1.0 \times 10^{-14}$
$[OH^-(aq)] = 1.0 \times 10^{-13}$
Given that the concentration of hydrons is greater than the concentration of hydroxide, this is an acidic solution.

b. $pH = 2.00 = -\log[H^+(aq)]$
$[H^+(aq)] = 10^{-2.00} = 1.0 \times 10^{-2}$
$[1.0 \times 10^{-2}][OH^-(aq)] = 1.0 \times 10^{-14}$
$[OH^-(aq)] = 1.0 \times 10^{-12}$
Given that the concentration of hydrons is greater than the concentration of hydroxide, this is an acidic solution.

c. $pH = 8.18 = -\log[H^+(aq)]$
$[H^+(aq)] = 10^{-8.18} = 6.6 \times 10^{-9}$
$[6.6 \times 10^{-9}][OH^-(aq)] = 1.0 \times 10^{-14}$
$[OH^-(aq)] = 1.5 \times 10^{-6}$
Given that the concentration of hydroxide is greater than the concentration of hydrons, this is a basic solution.

Problem 16.12 Given the following hydroxide amount concentrations, determine the amount concentration of hydrons and the pH of the solution. Identify each as acidic or basic.

a. $[OH^-(aq)] = 1.8 \times 10^{-5}$
$[H^+(aq)][1.8 \times 10^{-5}] = 1.0 \times 10^{-14}$
$[H^+(aq)] = 5.6 \times 10^{-10}$
$pH = -\log[H^+(aq)] = -\log(5.6 \times 10^{-10}) = 9.26$
Given that the concentration of hydroxide is greater than the concentration of hydrons, this is a basic solution.

b. $[OH^-(aq)] = 4.1 \times 10^{-2}$
$[H^+(aq)][4.1 \times 10^{-2}] = 1.0 \times 10^{-14}$
$[H^+(aq)] = 2.4 \times 10^{-13}$
$pH = -\log[H^+(aq)] = -\log(2.4 \times 10^{-13}) = 12.61$
Given that the concentration of hydroxide is greater than the concentration of hydrons, this is a basic solution.

c. $[OH^-(aq)] = 9.8{\times}10^{-8}$
$[H^+(aq)][9.8 \times 10^{-8}] = 1.0 \times 10^{-14}$
$[H^+(aq)] = 1.0 \times 10^{-7}$
$pH = -\log[H^+(aq)] = -\log(1.0 \times 10^{-7}) = 6.99$
Because the concentration of hydrons is greater than the concentration of hydroxide, this technically an acidic solution. But the concentration difference is very slight and this is close to a neutral solution.

Problem 16.13. Given the following hydron amount concentrations, determine the amount concentration of hydroxide and the pH of the solution. Identify each as acidic or basic.

a. $[H^+(aq)] = 1.0$
$[1.0][OH^-(aq)] = 1.0 \times 10^{-14}$
$[OH^-(aq)] = 1.0 \times 10^{-14}$
$pH = -\log[H^+(aq)] = -\log(1.0) = 0.00$
Given that the concentration of hydrons is greater than the concentration of hydroxide, this is an acidic solution.

b. $[H^+(aq)] = 5.0 \times 10^{-7}$
$[5.0 \times 10^{-7}][OH^-(aq)] = 1.0 \times 10^{-14}$
$[OH^-(aq)] = 2.0 \times 10^{-8}$
$pH = -\log[H^+(aq)] = -\log(5.0 \times 10^{-7}) = 6.30$
Given that the concentration of hydrons is greater than the concentration of hydroxide, this is an acidic solution.

c. $[H^+(aq)] = 5.0 \times 10^{-6}$
$[5.0 \times 10^{-6}][OH^-(aq)] = 1.0 \times 10^{-14}$
$[OH^-(aq)] = 2.0 \times 10^{-9}$
$pH = -\log[H^+(aq)] = -\log(5.0 \times 10^{-6}) = 5.30$
Given that the concentration of hydrons is greater than the concentration of hydroxide, this is an acidic solution.

Problem 16.14. For each of the following precipitation reactions, write a net-ionic equation and identify the Lewis acid and identify the Lewis base.

a. $K_2CO_3(aq) + SrCl_2(aq) \rightarrow 2\ KCl(aq) + SrCO_3(s)\downarrow$
Net-ionic equation: $CO_3^{2-}(aq) + Sr^{2+}(aq) \rightarrow SrCO_3(s)\downarrow$
Lewis acid: $Sr^{2+}(aq)$
Lewis base: $CO_3^{2-}(aq)$

b. $CrCl_3(aq) + 3\ LiOH(aq) \rightarrow 3\ LiCl(aq) + Cr(OH)_3(s)\downarrow$
Net-ionic equation: $Cr^{3+}(aq) + 3\ OH^-(aq) \rightarrow Cr(OH)_3(s)\downarrow$
Lewis acid: $Cr^{3+}(aq)$
Lewis base: $OH^-(aq)$

c. $Co(CH_3CO_2)_2(aq) + 2\ HF(aq) \rightarrow 2\ CH_3CO_2H(aq) + CoF_2(s)\downarrow$
Net-ionic equation: $Co^{2+}(aq) + 2\ F^-(aq) \rightarrow CoF_2(s)\downarrow$
Lewis acid: $Co^{2+}(aq)$
Lewis base: $F^-(aq)$

Problem 16.15. For each of the following, predict whether you think the reaction enthalpy ($\Delta_r H°$) would be positive (endothermic) or negative (exothermic).

a. $HS^-(aq) + H_2O(l) \rightarrow H_2S(aq) + OH^-(aq)$
Sulfur and oxygen are in the same group and so we use atomic size to understand the reactivity. Sulfur (r_{cov} = 105 pm) is larger than oxygen (r_{cov} = 66 pm). Because sulfur is a larger atom it shares its electrons less (is a weaker base), and oxygen, a smaller atom, shares its electrons more (is a stronger base). Because this reaction is going in an unfavorable direction – from a weaker base (HS^-) to a stronger base (OH^-) – the reaction should be endothermic ($\Delta_r H° > 0$).

b. $CH_3^-(g) + SiH_4(g) \rightarrow CH_4(g) + SiH_3^-(g)$
Carbon and silicon are in the same group and so we use atomic size to understand the reactivity. Carbon (r_{cov} = 76 pm) is smaller than silicon (r_{cov} = 111 pm). Because carbon is a smaller atom it shares its electrons more (is a stronger base), and silicon, a larger atom, shares its electrons less (is a weaker base). Because this reaction is going in a favorable direction – from a stronger base (CH_3^-) to a weaker base (SiH_3^-) – the reaction should be exothermic ($\Delta_r H° < 0$).

c. $Br^-(aq) + H_2Se(aq) \rightarrow HBr(aq) + HSe^-(aq)$
Bromine and selenium are in the same period and so we use electronegativity to understand the reactivity. Bromine ($\chi_P = 2.96$) is more electronegative than selenium ($\chi_P = 2.55$). Because bromine is more electronegative it shares its electrons less (is a weaker base), and selenium, less electronegative, shares its electrons more (is a stronger base). Because this reaction is going in an unfavorable direction – from a weaker base (Br^-) to a stronger base (HSe^-) – the reaction should be endothermic ($\Delta_r H° > 0$).

d. $HS^-(aq) + HCl(aq) \rightarrow H_2S(aq) + Cl^-(aq)$
Sulfur and chlorine are in the same period and so we use electronegativity to understand the reactivity. Sulfur ($\chi_P = 2.58$) is less electronegative than chlorine ($\chi_P = 3.16$). Because sulfur is less electronegative it shares its electrons more (is a stronger base), and chlorine, more electronegative, shares its electrons less (is a weaker base). Because this reaction is going in a favorable direction – from a stronger base (HS^-) to a weaker base (Cl^-) – the reaction should be exothermic ($\Delta_r H° < 0$).

Index

Printed in the United States
by Baker & Taylor Publisher Services

Printed in the United States
by Baker & Taylor Publisher Services